Brill's Companion to Aineias Tacticus

Brill's Companions in Classical Studies

The titles published in this series are listed at *brill.com/bccs*

Brill's Companion to Aineias Tacticus

Edited by

Maria Pretzler
Nick Barley

BRILL

LEIDEN | BOSTON

Cover illustration: City walls of Messene, photo by Maria Pretzler

The Library of Congress Cataloging-in-Publication Data is available online at http://catalog.loc.gov
LC record available at http://lccn.loc.gov/2017037587

Typeface for the Latin, Greek, and Cyrillic scripts: "Brill". See and download: brill.com/brill-typeface.

ISSN 1872-3357
ISBN 978-90-04-22351-6 (hardback)
ISBN 978-90-04-35285-8 (e-book)

This book is printed on acid-free paper and produced in a sustainable manner.

Contents

List of Figures and Photos

Figures

Photos

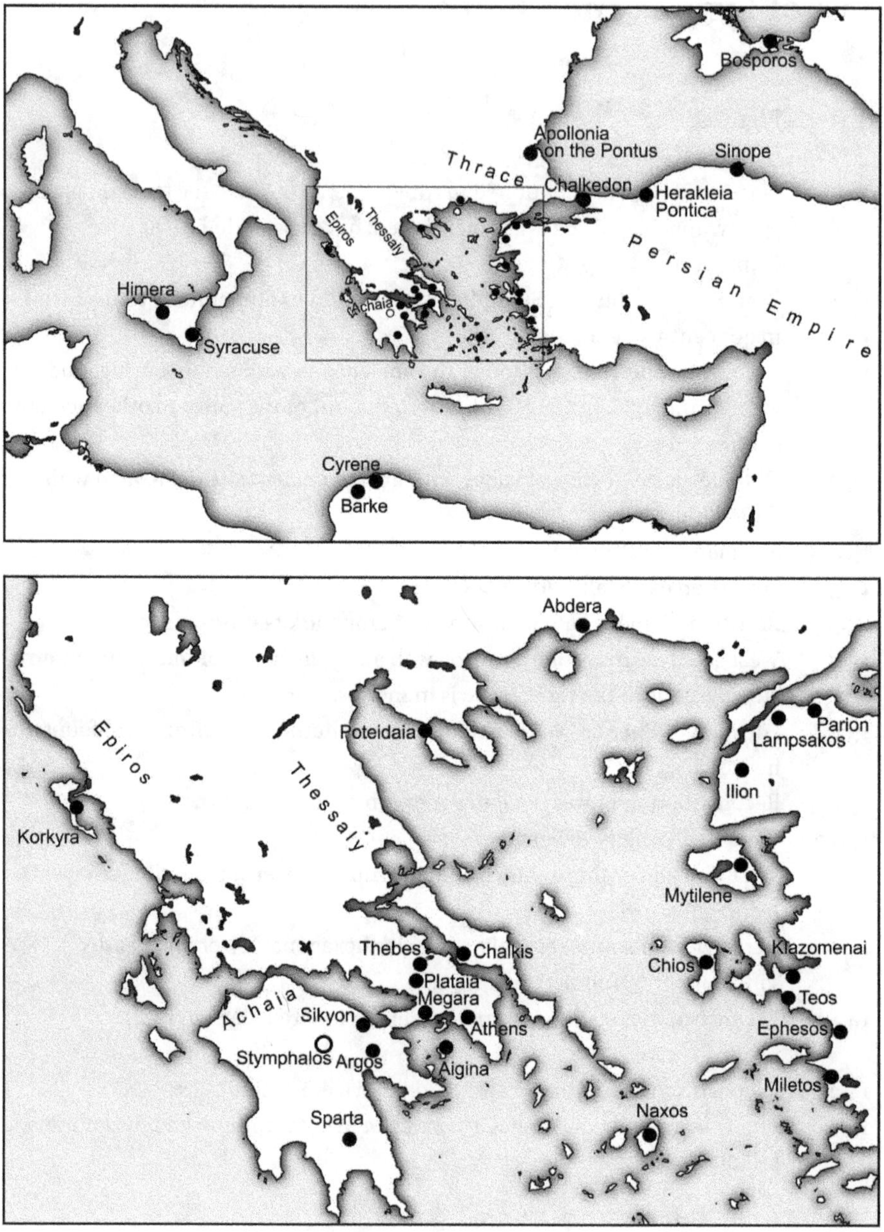

FIGURE 0.1 *Aineias' World. Places and regions mentioned in Aineias'* Poliorketika*—see table on pp. 89–93. Stymphalos, which is* not *mentioned, has been added to the map.*

CHAPTER 1

Introduction

Maria Pretzler and Nick Barley

This volume examines a unique example of classical Greek Literature: the *Poliorketika* or *How to Survive under Siege* by Aineias 'the Tactician'. Aineias' sole surviving work shows us an ancient Greece far from the well-documented affairs of Athens or the legendary society of Sparta. We are taken to places with small populations and modest resources which only experience large scale warfare at the whim of greater powers. Many of these smaller *poleis* would not have had much in the line of a grand military tradition or accumulated military experience, and in an emergency the right person for the job may simply be whoever got there first. Aineias' advice is directed to a Greece where survival of family and *polis* was balanced on a spear-point, where innovation and flexibility of doctrine and personnel were paramount, and where raiders on the very edge of territory represented not just an armed incursion, but an existential threat.

The *Poliorketika* defies genre conventions, even within the field of military writing, since it is a manual about sieges which barely talks about sieges as we would usually understand the term, and which focuses on the one universal weakness of any *polis*, its people, as well as presenting us with a dazzling array of vignettes illustrating the everyday life of a fourth-century *polis*. Aineias' perspective of the Greek world does not match the viewpoints we are used to seeing in classical literature, something emphasised by the fact that although it is a classical Greek text, it is not written in the usual classical (Attic) Greek literary language. In short, Aineias' *Poliorketika* is an exceptional work which rewards repeated reading. It offers a perspective on the Greek world that, armed as we are with our knowledge of the other great writers of the time, may be considered unusual, but which, on closer inspection, reveals the lives of citizens of 'everyday' states in all their chaotic glory. Further revisiting of the text raises many questions and reveals a frustrating discrepancy between what Aineias is telling us and what we would dearly like to know, but intensive study of the text reveals an author with a keen, inquisitive mind, a curiosity for innovation, an unsentimental pragmatism, astute psychological insight and a great gift for observation.

Here is a classical author, a close contemporary of Xenophon, Plato and Isocrates, who seems to be quite determined to remind us of 'the other Greece',

 | DOI: 10.1163/9789004352858_002

of life in the many *poleis* beyond the focus of our canonically Attic authors.[1] Aineias' perspective can offer insights into Greek lives and aspects of Greek culture we can only speculate about when we read the historical accounts by Thucydides or Xenophon: we rarely even get to wonder what it was like to live behind the city walls of a small town, hoping to avoid conquest and massacre, using what was at hand, and relying on what the small community could muster in terms of manpower, resources and ingenuity. Here is a different voice, standing on its own in terms of genre, viewpoint and approach. Aineias supplies a human scale to the grand narrative of classical Greek history: he homes in on the all-too-human failures and everyday problems found in any community, and regards his fellow Greeks with an eye both soberly pragmatic and attuned to what one might be able to ask of 'normal' people faced with real existential threats. Finally, and despite the serious subject matter, the text is simply great fun to read, rewarding newcomers and returning visitors alike with an entertaining insight into the psyche of a Classical Greek city and (some of) its population; while we would not dare say it to his face, the editors find Aineias' text charmingly engaging, and feel that his work is all the more important because of this.

We know very little about the author of this unusual military work. The manuscript tradition identifies the work as "Aelian's tactical treatise on how to survive under siege", with a scribal note at the end of the work suggesting that the name of the author is actually Aineias.[2] A further hint is contained in the work itself, where the author uses the name 'Aineias' in his explanation of a complex method to encrypt messages.[3] The date of the work has to be determined from the historical episodes mentioned to illustrate Aineias' advice: those stories which can be dated cluster in the 370s and 360s, with the latest episode probably in 360 BC.[4] A composition of the work in the 350s BC therefore seems very likely. This was a very turbulent time, after the final collapse of the dominance of Athens and Sparta, and the failure of Thebes to assert lasting influence over a areas well beyond its accustomed sphere of influence in and around Boiotia; in Macedonia, Philip II had started to build his empire, but the impact of Macedonian influence on the southern mainland of Greece was not yet widely recognised. Old alliances had become obsolete, and for many,

1 E.g. Whitehead (1990) 2–4, Hansen (2007).

2 Whitehead (1990) 5–6.

3 Aineias 33.18, the sequence of letters is in turn based on a conjecture, Haase (1835), cf. Whitehead (1990) 7–8.

4 Whitehead (1990) 8–10, Hunter and Handford (1927) xxxiv–xxxvii; latest episode probably 24.3–14, Ilion captured by Charidemos.

conflict was less predictable than during most of the classical period. For the generic city Aineias is thinking of in his work, these were uncertain times, when a readiness to defend the community against attack was perhaps more necessary than ever. Aineias' pragmatism and his refusal to focus on Athens, Sparta or even Mainland Greece fits the historical context of his work particularly well.

The *Poliorketika* is classified as a military manual, hence the author's name Aineias *Tacticus*, the Tactician, to distinguish him from namesakes, mythical or historical, and to define his genre. However, readers looking for grand strategic schemes or impressive war machinery will be disappointed, and the genre definition does not quite do the real value of the work justice. Aineias' treatise is very different from the other military books which have come down to us from antiquity, be they collections of stratagems (e.g. Polyainos' *Strategemata*), works providing advice on warfare with large well-organised armies (e.g. Vegetius' *De Re Militaria*) or various *Poliorketika* with a focus on large-scale siege technology. While Aineias was by no means immune to enthusiasm for complex technologies, particularly with respect to communications,[5] the main focus of his extant work is on exploiting the modest resources of a small or middle-sized community. Aineias uses what is available, rather than relying on a complex infrastructure or machinery which, in his own time, only a few large cities such as Syracuse and Athens were capable of maintaining until, soon after he was writing, kingdoms such as Macedonia could develop them to their full advantage. Thus, the greatest assets Aineias' commander can draw on are the morale of his fellow-citizens, the advantages of local knowledge, and everyday resources the community can provide even in a crisis; while the greatest liabilities are also found in the very real limitations to morale, expertise and resources in any small Greek *polis*, which might be a close-knit community of just a few thousand, even, in some cases, merely a few hundred people.

In Aineias' work we also catch a glimpse of intellectual activity beyond Athens in the mid-fourth century, a period which, from our limited perspective, is completely dominated by the literature, traditions and philosophy of this one city. This is not a coincidence, since the educated elites of later periods, particularly during the high Roman Empire, favoured texts in pure Attic Greek, which meant that apart from some rare exceptions, notably Herodotos' *Histories*, works which were not written in the preferred literary dialect never had a good chance of survival: most were probably lost before the end of antiquity, and the few that survived had few prospects to be copied and distributed

5 Aineias 31; Polyb. 10.44.

sufficiently to make it through the Middle Ages and into the age of the printing press. Aineias' work therefore appears almost unique, so that we see him as a pioneer in many respects, particularly in terms of genre and language. The *Poliorketika* is a practical handbook, a genre which is considered much more typical for the Hellenistic period. Aineias' language deviates from the standard classical Attic literary Greek, using an idiom which looks like an early forerunner of the *koine* Greek favoured in Hellenistic literature. In fact, it is difficult to determine how unique or ahead of its time this text actually was: because of the later selection of classical texts, any comparative material for Aineias' work has been lost and left his treatise as a unique reminder that Athens was not the only place producing literary works in the later classical period. At the same time, we can appreciate that Aineias was aware of books which are still available to us, not least the historical accounts of Herodotos and Thucydides, and we can observe how he responds to some of the major intellectual trends we also detect in the Athenian literature of the same period, especially the earliest development of didactic writing which is also represented in a number of Xenophon's works some of which (*On Horsemanship, Poroi*) were written around the same time, perhaps even in the same decade.[6]

Aineias Tacticus' *Poliorketika* is a fascinating mine of information for anybody wanting to get a sense of classical Greece beyond the dominant voice of Athens. We are not the first to notice Aineias' usefulness as a source for Greek lives, cultures and intellectual development in the many cities beyond Athens and Sparta: the special quality of the text as an alternative voice for classical Greece has been appreciated for a long time, although it is perhaps surprising that even after becoming widely available in translation in the early twentieth century, the *Poliorketika* has not had a more profound impact on how Greek culture in the fourth century has been described by historical scholars. This has begun to change in more recent times, not least following Whitehead's 1990 translation and commentary, which has had a particular impact on scholarship in the English-speaking world and beyond. The changing focus of undergraduate teaching in universities has played an important part in this development, too, since ancient history courses are increasingly moving beyond the narrow canon of classical Athenian literature. Anecdotal evidence would suggest that more aspiring scholars of the classical world are now getting to know Aineias during their undergraduate studies, and are encouraged to read his work as an alternative introduction to life in ancient Greece.

6 Gray (1985) 162–172, Marchant (1968) xxviii–xli, Delebecque (1950) 12. The *Cynegeticus* was perhaps written as early as in the 390s.

Scholarly work on Aineias is still rather limited, although there has been a steady trickle of publications for many years. There is a surprisingly wide range of translations of Ainieias' work into different languages, some with extensive commentaries.[7] Apart from translations and commentaries, many publications dealing with Aineias focus on specific passages rather than discussing the text as a whole. Classical scholarship, both in Greek literature and in ancient history, is increasingly open to the detailed study of texts outside the traditional literary canon. In this context, it seemed particularly striking that this one Greek text from the classical period has still received comparatively little attention—in fact, a comparison with the volume of publications discussing any other ancient text of comparative importance is quite staggering. Quantity is, of course, not the sole measure of a lively scholarly discussion, and it was clear that a number of people had recently been doing some fascinating work on Aineias or on relevant aspects of ancient Greek literature and culture. At the same time, no attempt had been made to compare notes and to bring together different approaches to Aineias' work, which seemed to be a particularly good subject for discussion across the usual boundaries of scholarly interests. Recent collections of articles on literary works, partculary Robin Lane Fox's edited volume *The Long March* on Xenophon's *Anabasis*[8] served as a template for this project.

A two-day conference on Aineias Tacticus took place in September 2010 at Gregynog in mid-Wales which allowed a wide-ranging discussion and an investigation of Aineias' text from many different perspectives. We hoped to encourage discussion across disciplines by focusing on this one text which invites questioning from many angles, and with different methods, and the resulting conversation was as lively as it was enlightening. This volume brings together most of the papers which were presented at this conference. The aim of this collection is to showcase, compare and contrast different approaches to Aineias' work and to provide crucial background reading and analysis for anybody with interest in this important ancient text. This book therefore contains a series of scholarly arguments related to different disciplines of classical studies, showcasing the complexity and depth of philological, literary and historical analysis which can be applied to Aineias' text. While some chapters pursue detailed arguments based on the original Greek text, the aim was to make at least the

---

7 Casaubon (1609), Orellius (1818), Köchly and Rüstow (1853), Hercher (1870), Hug (1874), Schoene (1911), Oldfather (1923) Hunter and Handsford (1927), Belyaev (1965), Dain and Bon (1967), Marek (1977), Whitehead (1990), Bettalli (1990), Vela (1991), Vela and Martín (1991), Battistini (1994), Burliga (2007), Brodersen (2017); see list below, pp. 12–13.

8 Lane Fox (2004).

conclusions accessible for a wide readership, bringing together discussions of Aineias' linguistic and literary qualities with studies of the historical context and informative content of his work. The approach to Greek in the text is therefore mixed: in most of the chapters, translations in the text are combined with the Greek text in the footnotes, but we have left the original Greek in the main text where the discussion specifically deals with the language rather than the content, and we have decided not to provide translations for every detail where the argument is about very specific aspects of the Greek language which cannot be captured in a translation.

David Whitehead was invited to give the keynote address at the conference, to reflect upon developments since the publication of his 1990 edition and commentary of Aineias Tacticus which made the work much more widely read. Much of the detailed introduction to our author and his work, complete with a definitive discussion of the scholarly tradition surrounding these questions, is therefore left to him. In his chapter David Whitehead returns to a discussion of scholarly conventions for the name of our author, re-iterating and expanding on his reasons for using the Greek version of the name, Aineias, instead of opting for the more commonly used Latin spelling Aeneas. In this volume we follow his advice and adopt the same style throughout. We have been less strict in imposing a uniform title on the work, because, as Whitehead also shows in his chapter, the original title is much less certain. This chapter also begins telling the story of Aineias' readers in antiquity, which is taken up again in the final chapter, with Rance's discussion of the reception of the *Poliorketika* through Byzantine period. Whitehead picks up the thread again at the end of the Middle Ages, with an overview of scholarship from the Renaissance up to the present day, and issues a challenge, not just for this volume, but also for scholars who are working on Aineias at this point in time, or will do so in the future.

Robin Lane Fox revisits the question of what we can find out about Aineias from his own work. Who was he? Where was he from, or where was he writing? These questions cannot be answered conclusively, as Lane Fox himself notes, but a re-opening of the question nevertheless yields new insights. In this chapter, we get to know a man with knowledge and perhaps experience of many parts of the Greek world who pays as much attention to cities by the Black Sea, in Asia Minor and in the West as he does to Mainland Greece. Linguistic details are pursued as indications of Aineias' geographical but also cultural and social context, and we begin to see an intellectual world beyond the more familiar ground of Attica which Aineias ignores almost completely.

During the conference, we also returned to the question of the identification of our author as the Aineias of Stymphalos in Arcadia who was active around the mid-fourth century, and there was no consensus. Whitehead remains convinced that Aineias was probably identical with Aineias of Stymphalos.[9] Arcadia was a region famous for its mercenaries, and Lane Fox shows how the historical examples in the *Poliorketika* might map onto an Arcadian commander's and mercenary's career in this period, with experience of action in Asia Minor and Sicily, describing Aineias as a 'global Greek, from Arcadia to Pontus to Syracuse'. At the same time, Lane Fox's chapter shows that the book is by no means particularly focused on the Peloponnese, or even Mainland Greece. In fact, the editors do not quite agree with each other but subscribe to varying degrees of agnosticism on the issue, and we have therefore left the authors to their own devices when it comes to this question. Some, most notably Shipley, Vela and Bettalli, take account of Aineias' Peloponnesian background in their arguments, while others do not; in any case, readers should not look for consistency in this matter.

The search for Aineias and his context continues in *Graham Shipley*'s chapter, which investigates the author's intellectual context. Basic assumptions can be gleaned from the advice and examples in the *Poliorketika*. Shipley discusses the social and political context of the work, emphasising the almost exclusive influence of the Greek *polis* on Aineias' thinking, but also posing the question whether some of his more authoritarian prescriptions would have been possible in the context of most Greek cities and their political structures. Shipley also investigates the literature and intellectual discourse of the period within which we need to place Aineias' work.

The *Poliorketika* was written in a period of significant developments in historiography. *Maria Pretzler* takes a closer look at Aineias' narrative strategies in the stories he provides as examples, usually in the form of short narrative passages or historical vignettes. While these passages are often not very useful as a source for detailed historical information, they do give us a clear insight into Aineias' approach to his work, and the goals he wanted to achieve when he included them. Moreover, these historical vignettes show how recent events were remembered and discussed, probably before there was a widely known historiographical treatment of the period which would have provided a framework for such stories; they also add to the range of experiments with historical writing well beyond historiography which we can detect in Greek literature of the fourth century BC.

The conference organisers were particularly keen to include a discussion of Aineias' language, which is unique in classical Greek literature and, apart

9 Xen. *Hell.* 3.6.7; mentioned in relation to events in 367BC; cf. Whitehead (1990) 4–13.

from Herodotos, represents a unique sample of classical prose writing not using standard Attic Greek. *José Vela Tejada* had been doing a lot of work on the language of Aineias,[10] and he provides a detailed analysis of Aineias' Greek here, particularly focusing on what it might tell us about the creation of *koine*, the Greek dialect widely used in the literature of the Hellenistic period. This is a chapter where we have refrained from translating or transliterating all Greek, since the discussion of linguistic details will mainly be of use for those with a knowledge of the language. However, we have provided translations of phrases in the text where more general arguments are presented, and we hope that this chapter will allow all readers to appreciate the significance of the *Poliorketika* as a source for the history and development of the Greek language, while also providing plenty of detailed discussion for the experts.

Aineias himself has a lot to say about writing and communication in his period, yet commentators have mainly been interested in his comments on secret messages and cryptography. *Peter Liddel* shows that the *Poliorketika* tells us a lot more about how Greeks in the fourth century BC may have thought about information exchange, and especially about spreading messages within communities. Aineias' interest in secret messaging provides fascinating thoughts on different means of communication (verbal/non-verbal, oral/written) and their comparative practicality. Comparisons with epigraphical evidence provide a context for Aineias' thinking, while also clearly showing that there were aspects of communication within Greek *poleis* which have no interest for him and are therefore absent from his work.

The *Polioretika* assumes a generic Greek *polis* under pressure, and one of Aineias' major concerns is therefore the danger of civil conflict, sabotage, treason and sedition. *Maria Pretzler* revisits the question of Aineias as a source for *stasis*, civil unrest in the Greek world of the fourth century. A comparison with the discussions of *stasis* by Thucydides and by Aristotle in the *Politics* emphasises Aineias' focus on the desperate individual and the seeds of disloyalty which can, in the extreme case, lead to serious internal conflict or collaboration with an enemy and the loss of many lives. This chapter shows how Aineias surveys the weaknesses and resources of his community, in this case less material than psychological, and how he justifies his approach of constant vigilance combined with an understanding of human weakness in the face of extreme pressure.

With *Marco Bettalli*'s chapter we finally reach the central topic of warfare in Aineias's work. As a general discussion of how Aineias' depiction of warfare

10 Vela (1991), Vela (1994).

relates to developments in the fourth century this chapter also provides an introduction for the more specialised discussions which follow. Bettalli shows how Aineias' depiction of warfare and his observations about soldiers and their communities relate to contemporary depictions of war. Aineias offers unique insights and practical advice and Bettalli also shows that at least some of these ideas remained relevant even after technology and the routine availability of larger armies shifted the focus of war (at least as we see it in our literary sources) away from the small Greek *polis*.

Nick Barley sees in Aineias a welcome antidote to the usual emphasis in both ancient Greek historiography and modern scholarship on pitched battles between hoplites. What we see in the *Poliorketika* is a much more varied image of warfare, with a strong emphasis on light infantry, high quality command and reliable communications. This chapter also discusses what Aineias actually means by 'siege': in the *Poliorketika* an actual assault on the walls of a city is an extreme event—and before this comes to pass, if it happens at all, the enemy's attack affects the territory of the *polis*, and every possible measure should be taken to never let it get any further. Barley also discusses Aineias' emphasis on the internal defences of a city, both in terms of morale and the prevention of revolutions or betrayal, and demonstrates how these aspects, too, form an integral part in what the ancient author understands as a siege on a city.

James Roy's discussion of mercenaries in Aineias provides an opportunity to compare some of Aineias' advice with the reality of warfare in the fourth century. The picture which emerges here shows that Aineias, despite being known for his attention to mundane detail, is not covering this subject as fully as might be expected. This discussion raises questions about the practicalities of employing soldiers, including payment, maintenance and camp-followers which make the picture more complicated than the advice given in the *Poliorketika* suggests. What we do see, however, is what assumptions about mercenaries and their role in a Greek city's military activities any local commander could apparently take for granted in the mid-fourth century.

One aspect of the *Poliorketika* that is often overlooked are the many references to harbours, shores and ships. *Philip de Souza* discusses what we can learn from Aineias about this aspect of Greek warfare, despite the fact that the tantalizing last sentence suggests that more advice on naval warfare was to come in a part of the work which is no longer preserved. What we can glean from the work we have is an alternative to the activities of big fleets which dominate the accounts of warfare in historical accounts; the use of ships as part of other operations and on a small scale. Aineias also teaches us to appreciate the vulnerability of a city by the sea, and the risks inherent in day-to-day shipping activities and the infrastructure necessary in any port.

When planning the conference, we realised that the conversation needed to include an archaeologist's perspective on small Greek cities in the mid-fourth century BC. Stymphalos in northern Arcadia seemed the ideal site for inclusion in our discussion. The tentative connection with Aineias himself is of course tempting and makes a discussion of the site more poignant as a reference point, although we do not want the focus on this particular city to be read as an editors' endorsement of the idea that it was, in fact, the author's home. The site of ancient Stymphalos is, in fact, a particularly good example for a relatively small city planned and fortified in Aineias' life time, and deserved inclusion on those grounds alone.

As one of the excavators of Stymphalos, *Ben Gourley* was therefore invited to discuss the site in context of Aineias' work. In terms of date, modern exploration and geographical context Stymphalos represents an excellent illustration of the kind of place Aineias had in mind when he talks about his generic *polis*. The city is only missing one crucial ingredient, namely access to the coast, but its well-explored fourth century walls and compact territory make it easier to get a more concrete sense of Aineias' conception of warfare and provide further insights into how a small community needed to prepare for exactly the kind of eventuality discussed in the *Poliorketika*.

Tracey Rihll investigates what Aineias can tell us about the use of technology in classical Greece. Readers have often been disappointed that a work on siege warfare (as the title promises) has little to say about complex siege engines and eye-catching technological innovations. The *Poliorketika* offers a much more down-to-earth insight into every day technology applied to extraordinary circumstances, as this chapter demonstrates, and invites us to think about the development of technology well beyond spectacular inventions.

Although the Greek city continued to flourish for centuries, warfare became increasingly a matter of larger entities, be it the Hellenistic kingdoms or the Roman Empire. One problem for the survival of Aineias' work was therefore that the generic small city at its centre not only no longer existed, but was perhaps even difficult to imagine for later readers, not least experts in warfare now used to operating within large, well organised and well funded military bodies.

Philip Rance's chapter therefore tells a gripping story of survival, following a trail which has become very faint indeed, yet also showing how it was possible that at least a considerable part of Aineias' work survived into the Renaissance, when the publication of printed copies finally secured its transmission into our own age. This chapter represents a considerable amount of original research which shows just how much can be gained from investigating the history of texts and ideas through the Byzantine period. Without Philip Rance's work the story of Aineias would not be complete, and it would be more difficult to understand its unique position within the genre of ancient Greek military writing.

As editors we hope that this collection of papers reflects the stimulating discussion we had at the conference, and provides new insights into a work which we both find informative and immensely enjoyable. Our thanks are due to all the contributors, particularly for their patience as this volume finally took shape, and to all those who attended and contributed to the lively discussion both during formal sessions and those spontaneous, informal, sessions that make conferences such stimulating environments. We would also like to thank the Callaghan Centre for the Study of Conflict at Swansea University and its former director, Prof Nicola Cooper, for its generous financial contribution to the conference, and Swansea University's Research Institute for Arts and Humanities (RIAH) for its support. Our thanks are also due to Colin Hough whose help during the conference was crucial to its success, and to the hardworking staff of Gregynog Hall. Bogdan Burliga and Nicholas Sekunda delivered papers to the conference and made significant contributions to the discussion, but were unable to provide a chapter for this volume. Finally the editors would like to thank all those students and colleagues with whom we have shared and developed our enthusiasm and understanding of Aineias; his is a text which rewards all comers, and we hope that the following chapters do him justice.

Swansea, September 2016

Bibliography

Delebecque, É. (1950) *De l'Art équestre: Texte et traduction avec une introduction et des notes*. Société d'Édition les Belles Lettres, Paris.

Gray, V. (1989) *The Character of Xenophon's Hellenica*. Duckworth, London.

Haase, F. (1835), 'Über die griechischen und lateinischen Kriegsschriftsteller', *Jahns Jahrbücher für Philologie und Paedagogik* 14: 88–118.

Hansen, M.H. (2007), 'Aineias the Tactician', in Hansen, M.H. (ed.), *The Return of the Polis. The Use and Meanings of the Word* Polis *in Archaic and Classical Sources*. (Papers from the Copenhagen Polis Centre 8. *Historia* Einzelschriften, 198). Franz Steiner Verlag, Stuttgart. 243–246.

Hunter, L.W. and Handford, S.A. (1927), *Αἰνείου Πολιορκητικά. Aineias on Siegecraft*. Clarendon Press, Oxford.

Lane Fox, R. (Ed.) (2004) *The Long March. Xenophon and the Ten Thousand*. Yale University Press, New Haven and London.

Marchant, E. (1968) *Xenophon: Scripta Minora* (Loeb). Havard University Press, Cambridge, MA.

Vela Tejada, J. (1991) *Estudio sobre la lengua de la poliorcética de Eneas el Táctico.* Departamento de Ciencias de la Antigüedad de la Universidad Zaragoza, Zaragoza.

Vela Tejada, J. (1994) 'La *Poliorcética* de Eneas el Táctico: nuevas aportaciones para el estudio de la *koiné*', in: *Actas del VIII congreso español de estudios clásicos (Madrid, 23–28 de septiembre de 1991)*. 323–330.

Whitehead, D. (1990) *Aineias the Tactician, How to Survive under Siege. Translated with Introduction and Commentary*. Clarendon Ancient History Series, Oxford.

Editions, Commentaries and Translations (in Chronological Order)

Casaubon, I. (1609) *Æneæ, vetustissimi Tactici, Commentarius de toleranda obsidione* (appended to Casaubon's edition of Polybius), Marnius, Frankfurt.
Editio princeps; Greek text, Latin translation.

Orellius, J.C. (1818) *Aeneae Tactici Commentarius de toleranda Obsidione*. In Libraria Weidmannia, Leipzig.
Greek text, Latin translation, notes.

Köchly, H., Rüstow, W. (1853) *Griechische Kriegsschriftsteller*. W. Engelmann, Leipzig (Aeneas: vol. 1 1–183, vol. 2, 336–351).
Greek text, German translation, introduction, notes.

Hercher, R. (1870) *Aeneae Commentarius poliorceticus*. Weidmann, Berlin (with a second, corrected edition, also Weidmann, Berlin 1870).
Greek text (with new readings).

Hug, A. (1874) *Aeneae Commentarius poliorceticus*, Teubner, Leipzig.
Greek text and critical notes.

Schöne, R. (1911) *Aeneae Tactici de obsidione toleranda commentarius*, Teubner, Leipzig.
Greek text (completely new edition; basis for the Loeb edition).

Oldfather, W.A. (1923) *Aeneas Tacticus—Asclepiodotus—Onasander: with an English translation by members of the Illinois Greek Club*. William Heinemann and G.P. Putnam's Sons, London and New York.

Loeb Classical Library: Greek text, English translation, introduction.

Hunter, L.W. and Handford, S.A. (1927), *Αἰνείου Πολιορκητικά. Aineias on Siegecraft*. Clarendon Press, Oxford.
Greek text, English translation, introduction, commentary.

Belyaev, V.F. (1965) 'Aeneas Tacticus: De obsidione toleranda', *Vestnik drevnej istorii* 91: 239–268 and 92, 219–243.
Russian translation, introduction.

Dain, A. and Bon, A.M. (1967) *Énée le Tacticien. Poliorcétique*. Les Belles Lettres, Paris.
Budé edition; Greek text, French translation, introduction, commentary

Marek, P.V. (1977) 'Aieneias Taktikos. O obraně měst', in: P.M. Marek, J. Kalivoda (eds), *Antické válečné umění*. Svoboda, Prague.
Czech translation, introduction, commentary.

Whitehead, D. (1990) *Aineias the Tactician, How to Survive under Siege. Translated with Introduction and Commentary*. Clarendon Ancient History Series, Oxford. [Updated edition Bristol Classical Press (2001) = Duckworth (2002)]
English translation, introduction, historical commentary.

Bettalli, M. (1990) *Enea Tattico. La difesa di una città assediata (Poliorketika). Introduzione, traduzione e commento*. ETS editrice, Pisa.
Greek Text, Italian translation, introduction, commentary.

Vela Tejada, J. (1991) *Eneas el Táctico. Poliorcética. La estrategia militar griega en el siglo IV a.C.* Ministerio de Defesa, Madrid.
Spanish translation, introduction

Vela Tejada, J. and Martín García, F. (1991) *Eneas el Táctico, Poliorcética. Polieno, Estratagemas*. Gredos, Madrid.
Greek text, Spanish translation, introduction, commentary.

Battistini, O., et al. (1994) *La Guerre: trois tacticiens grecs: Énée, Asclépiodote, Onasandre*. NiL Ed, Paris.
French translation, introduction, commentary.

Burliga, B. (2007) *Eneasz Taktyk. Obrona oblezonego miasta*. Proszynski i S-ka, Warsaw.
Polish translation, introduction, notes.

Brodersen, K. (2017) *Aineias/Aeneas Tacticus. Stadtverteidigung/Poliorketika* (Tusculum). De Gruyter, Berlin.

CHAPTER 2

'The Other Aineias'*

David Whitehead

The typography of my title is precise. 'The other Aineias' looks like a quotation because that is what it is; and readers of the present volume may well know, or can recall, the provenance. Almost from its inception in the late 1880s, the *Classical Review* adopted the pleasant convention of prefacing its reviews with a suitable "headline" word or phrase to give notice of what the review would be about, just as the title of any of the journal's short articles and notes flagged up that piece's contents (despite its name, *Classical 'Review'* began life as an all-purpose periodical). Thus over the years a harmless incidental pleasure of reading *CR* has been to observe the skills, and the thought-processes, of its successive editors in devising these headers. And when *CR* 1991 published a review (by John Lazenby) of my own treatment of the work to which this present book is devoted, the then editors, Alex Garvie and Harry Hine, introduced it with the phrase I have now revived: 'The other Aineias'.[1]

Other than whom? A question nobody will have asked, even rhetorically. The subtext was perfectly clear then and remains so now. If the name Aineias meant anything to people in general, and to *CR* readers in particular, it would be because of time spent with Vergil.

Thanks to Latin orthography, Vergil's hero is (always and inevitably) AENEAS. My own book deliberately showcased 'AINEIAS the Tactician', and I noted in my Preface that his name was being transliterated directly from the Greek 'to evade intrusive Vergilian overtones'.[2] What Garvie and Hine did served as a reminder (if a reminder were needed) that any and every bearer of this name cannot but be, for us, 'other' than the son of Anchises and Aphrodite. Even so, it is curiously satisfying that the Trojan has been immortalised not as the Aineias he of course is (in Homer and, by and large, in the rest of Greek literature, too) but as Vergil's

* What follows here began life as the opening 'keynote address' at the Gregynog conference (September 2010) which was the genesis of this present volume. Besides eliminating the informalities and kindred features suitable to that genre but inappropriate in print, I have allowed myself a limited amount of hindsight-driven allusions (almost all of them in footnotes) to the contributions that came after mine and are now, likewise, chapters in this book.

1 Lazenby (1991).

2 Whitehead (1990) viii.

 | DOI: 10.1163/9789004352858_003

Aeneas. That leaves the Aineias version of the name free to be assigned to our man, who can reasonably be claimed to be its second most important known holder.[3]

The Greek adjective *taktikos*, when applied to a person rather than to an impersonal idea or area of knowledge, is first found attested in Xenophon's *Cyropaedia* (8.5.15). Xenophon ascribes there, to his fanciful recreation of King Cyrus the Great of Persia, an appreciation of the qualifications needed by a *taktikos anêr*. And in Greek authors the term *taktikos* continues to be used—when used in that sense—of individuals who, evidently, have some practical expertise in military matters but are themselves not (or at least not necessarily) writers on the subject. Other instances can be found in Strabo, Plutarch, Josephus, Diogenes Laertius, and elsewhere.[4] However, make the simple change from Greek *taktikos* to Latin *tacticus* (a word not present in Lewis and Short, I was surprised to discover), and one is instantly transported to a world of scribes, scholars, bibliophiles and, especially, librarians. Here are the ancient and medieval precursors of Melvil Dewey, keen to assign authors to genres, and to preserve their writings in a suitably rational and orderly manner.

Some of these genres, and the classifications associated with them, are clearcut and self-defining. No competent ancient librarian (or bookseller) will have located, say, Plato under Epic Poetry, or Aristophanes with the Medical Writers. But the *tactici*—a category still found useful by the TLG, as will be seen—make up a significantly more nebulous grouping, whether they are viewed in ancient terms or in modern.

The ancient view of who the ancient Greeks' military writers were—or to be more precise, what the writings produced by such writers were—is best appreciated if one examines the contents of those large Byzantine and early-medieval manuscripts which aimed (so one must suppose) to isolate precisely

---

3 Progress in this regard can be seen in, for instance, a comparison between the second (1970) and third (1996) editions of the *Oxford Classical Dictionary*. The former proffers 'Aeneas(1)' and 'Aeneas(2)'. In the latter this becomes (unnumbered) 'Aeneas' and 'Aeneas (Aineias) Tacticus'. The second of these 1996 articles appeared above the initials D.W., but no matter: it was the editors, Simon Hornblower and Tony Spawforth, who saw the larger picture.

4 Strab. 10.4.10, 12.3.33; Plut. *Mor.* 618C, 650F; Joseph. *AJ* 13.340; Diog.Laert. 4.15. (Diog.Laert. 2.6 identifies Xenophon's particular qualities as "a lover of horses and hunting, and a *taktikos*, as is clear from his writings" ("φίλιππος καὶ φιλοκύνηγος καὶ τακτικός, ὡς ἐκ τῶν συγγραμμάτων"), but none of these three terms is being used in upper case, so to speak.)

such a *corpus*. This is what I recently did, in a paper published in the *Acta* of the Hungarian Academy, and I will take the liberty of quoting from it here:[5]

> In the ninth and tenth centuries, Byzantine Greek scribes were busy transcribing the uncial scipt of earlier manuscripts (virtually all of them now lost) into cursive or miniscule Greek. Cursive is quicker and easier to write; it is also smaller (allowing more words per page); and this, combined with moves towards collection and classification more generally, had the natural result of bringing together works of similar type in compendious codex-manuscripts. For my present purposes, four specimens of this are especially notable.
>
> (a) The mid-tenth century *Laurentianus* LV–4 (preserved in the Biblioteca Laurenziana, the library of the Medici in Florence) contains military handbooks, and a group of ancient ones occupy the central position. There are six of them, from five different writers—arranged there in an order not chronological (which would have been very helpful) but quasi-alphabetical. The writers are: Asklepiodotos, Aelian, Aineias, Arrianos (two treatises), Onasandros.
>
> (b–c) Two of the other great repositories of Byzantine manuscripts, the Bibliothèque Nationale and the Vatican, contain, respectively, the (probably eleventh-century) *Parisinus Graecus* 2442 and *Vaticanus Graecus* 1164. They each include not only two of the five authors from the Florence manuscript (Ailianos and Onasandros) but five others—here in no discernible order: Athenaios, Biton, Heron (two treatises), Apollodoros, and Philon (two treatises).
>
> (d) *Parisinus inter supplementa Graeca* 607. This comprises several manuscripts bound together in the late-fifteenth or early-sixteenth century. The binding's decoration and a short Latin inscription in Greek letters inside the back cover show that it was once in the library of King Matthias Corvinus at Buda, destroyed by the Turks in 1526; it was brought to Paris in 1843, and appropriated by the Bibliothèque Nationale (then Impériale) twenty years later. In its central section, folios 16–104 (which Wescher dated to the late-ninth or early-tenth century, though others would say later), there is a miscellany of 25 items on a discernibly military theme. Some are extracts, from historians (e.g. Thucydides on the siege of Plataia), alongside complete works by, again, Athenaios, Biton, Apollodoros, and (three times) Heron.

5 Whitehead (2008) 140–141. See also Rance in this volume, 290–311.

> Summarily, then, we have ten named Greek *Kriegschrifsteller* (by no means a Canon of Ten, merely an accidental decet), and fourteen works. Together they span the period between the mid-fourth century BC and the mid-second century AD.

The term *Kriegschriftsteller* is one I borrowed from Hermann Köchly and Friedrich Rustow's study of that name, published in the 1850s. Yet it is noteworthy that their pioneering attempt to define and bring together such 'war-writers' in the modern era found room for only five of my ten individuals: namely (in chronological order) Aineias, Philon, Asklepiodotos, Heron, and Aelian. Carle Wescher, whom I mention there, entitled his own compendium (published in Paris in 1867) *Poliorcétique des Grecs*, and that narrower compass reasonably excluded Asklepiodotos and Aelian; 'reasonably' because neither Asklepiodotos nor Aelian have anything to say about sieges. On the other hand, by concentrating on the *Parisinus inter supplementa Graeca* 607 manuscript, which does not contain him, Wescher found no room for Philon either; and Philon certainly did cover sieges.

To be sure, in the modern TLG, Philon is classified not as a Tact(icus) but as a Mech(anicus), yet that fact raises more questions of classification than it solves, because Biton, Heron, Athenaios and Apollodoros too—in sum, five of my *de facto* ten—are similarly labelled, as against four so-called *tactici* (Aineias, Asklepiodotos, Onasandros, and Aelian), and also Arrian, who anyway, given the totality of his output, is dubbed 'Hist. et Phil.' And mentioning Arrian inevitably brings up the strange—for our purposes—matter of his predecessor and exemplar Xenophon. Both are writers who stubbornly resist confinement to a single genre. In Xenophon's case, one can argue the relative merits of his present-day classification as an historian, as against the ancient and medieval view of him as primarily a philosopher. But both Xenophon and Arrian were known also for works—and surviving works, at that—of a specifically military kind; and whereas, as we have seen, Arrian's *Taktika* and the smaller and more narrowly-focused *Ektaxis kata Alanôn* were included in the *Laurentianus* LV–4 manuscript (a.k.a. M, the *Mediceus*), Xenophon's *Hipparchikos* and *Peri Hippikês* were not. One wonders why.

In recent years I myself have become more absorbed than I used to be in these questions of genre, as they pertain to military writers. I have also become very much more absorbed in what might be called its Mech. as well as its Tact. side. In 2004 I co-authored, with P.H. Blyth, the first English translation of, and commentary on, Athenaios Mechanicus;[6] and 2010 saw a similar volume on

6 Whitehead and Blyth (2004). As with buses, one can wait interminably for a commentary

Apollodoros Mechanicus, a.k.a. Apollodoros of Damascus.[7] Though the Apollodoros book is sole-authored, by me, I am in fact no less indebted in it to the acuity of Henry Blyth. He it was, twenty years ago, who realised that Apollodoros' *Poliorketika*—his entirely practical suggestions for siege-machinery suitable for the mountainous terrain and conditions to be anticipated by Trajan in Dacia in the early first century AD—is, as transmitted, very heavily contaminated by later 'armchair' material, much of which cheerfully breaches the laws of physics and of sensible military campaigning alike.[8] Yet even without that extra layer of complication in Apollodoros, it is self-evident that the mere fact of being a military (or 'military') writer does not mean that either Apollodoros or Athenaios before him live in anything like the same conceptual world as Aineias, so as to be fruitfully comparable and contrastable with him. For that exercise, the man we need is Philo(n)—Philo Mechanicus, Philon of Byzantion (second half of the third century BC)—and I will return to him from time to time as I proceed.

The classification Mechanicus—of which it could be said that the *mechanicus militaris* is a more or less distinct sub-species—is in fact a category with defining characteristics that are usually quite easy to see and to endorse. Something is being designed and constructed, out of durable material; wood and/or metal. Precise mathematics and physics are involved, whether up-front or in the background, so as to produce an artefact of exact dimensions. (Indeed, very exact, when it comes to that particular, rebarbative group of treatises which deal with various kinds of artillery.)

In a military context, as I have noted elsewhere,[9] this is far more often set in a scenario of siege than one of battle. And such a state of affairs is then underscored by the detached, abstract character of those treatises that do, however remotely, envisage combat between opposing forces on open ground, rather than in the attack vs. defence roles of a siege or blockade. Of course, Asklepiodotos, Onasandros and the others have to be labelled *tactici* inasmuch as it would be unhelpful (and unjustifiable) to label them *mechanici*. It was, though, a *taktikon genos* that, by the time they were making their contributions to it, had been well and truly hi-jacked by philosophy and rhetoric.

---

on Athenaios only to have two come along in close succession; so see now, also, Gatto (2010).

7 Whitehead (2010).

8 Blyth (1992).

9 Whitehead (2008) 141.

So far as good—were it not for the fact that Aineias had already, in the mid-fourth century BC, muddied the waters. He too has to be called a *tacticus* in that he is manifestly not a *mechanicus*, or any other categorisation that one could readily think of and apply to him. Yet he is writing about real, blood-and-guts life, not about philosophical or rhetorical abstractions; and the incipient military crisis which underlies everything that he discusses is—at least by the latter stages of the treatise—siege rather than battle. He is (and the phrase is unavoidable, I think) *sui generis*.

At this juncture it is necessary to remind ourselves that he survives only in part. Somewhat like Philon in a later age and context, Aineias appears to have written a cluster of short (or shortish) works on different facets of his general, overall theme. Even if the totality of this fell well short of the 'encyclopédie militaire' or *Stratêgika* that some have seen in it,[10] it certainly included a *Paraskeuastikê biblos* (*On Preparations*) and a *Poristikê biblos* (*On Procurement*), together with as many as six other possible items.[11]

Needless to say, the loss of what has been lost is most regrettable, and any fortuitous reappearance of it, whether in another manuscript or (less likely) on papyrus, will be a cause for rejoicing. But unless and until that happens, students of Aineias have to remain content with gratitude to a succession of people for the fact that one can read and use even the single treatise that is extant. It was probably, in my view, called the *Poliorketikê biblos* (*On Sieges*), and subsequently—with an eye to greater precision—subtitled *pôs chrê poliork-oumenous antechein, How to hold out when under siege*.[12]

Some of these predecessors of ours must remain anonymous. Under this head, pride of place obviously goes to the meticulous calligrapher who, somewhere around the year 950, was responsible for *Laurentianus* LV–4, the *Mediceus*, the only manuscript of its kind (and period) which includes anything by Aineias. But as regards those individuals to whom one can put a name, here are some who would get into my personal Hall of Fame.

First, in respect of antiquity itself, material of this kind would have sunk without trace, surely, if influential readers had not found it useful enough to read and mention (and thereby to perpetuate). Kineas the Thessalian evidently did so, in either the late fourth or the early third century BC, when, for his

10 For these characterisations, by C. Graux and Y. Garlan respectively, see Whitehead (1990) 16 with n. 43.

11 On the lost works see Whitehead (1990) 13–16; Bettalli (1990) 10–12.

12 Περὶ τοῦ πῶς χρὴ πολιορκουμένους ἀντέχειν. On the title see Whitehead (1990) 16–17; Bettalli (1990) 6–7.

master King Pyrrhos of Epeiros, he epitomised what Aelian (*Tactica* 1.2) called, in a tantalising phrase, Aineias' "considerable number of military books".[13] It is one that seems bound to mean more than just the *On Sieges*. Similarly, as is well-known, Polybios (10.44) not only mentions Aineias and calls him 'the writer of the military treatises' (plural);[14] he also cites him, at length, on fire-signalling. This topic, though briefly alluded to in the surviving treatise (7.4.), was covered more fully in one of the lost ones, the *Paraskeuastikê*.

Polybius' treatment of Aineias' ideas in this area is rather condescending; and, as a case like that of Timaios of Tauromenion shows (*mutatis mutandis*), to come under the harsh critical spotlight of Polybius was no guarantee of survival *in extenso*. All the same, it is probably reasonable to think that the chances of any of Aineias's writings managing to last into the Christian era were increased, rather than diminished, by Polybius' interest in him.

And Polybius is in any case the main bridge between the second century BC and the seventeenth century AD. Long before that lower terminus—namely, at some (indeterminable) time after Aelian but before the age of Constantine Porphyrogenitus—Aineias' 'considerable number of military books' shrunk to just the one. And it is entirely possible that that 'one', once its mere survival was sufficiently assured by the existence of a family of manuscripts descended from the *Mediceus*, would have enjoyed the same restricted circulation, and the same recognition, as the products of Philon, Biton, Heron and the rest. It would have been, that is to say, something familiar to specialists in the field but largely unknown outside it. However, thanks to one man—or two if, as I say, we count Polybius again—Aineias' future was to be different. In 1609 Isaac Casaubon, at the height of his powers in Paris, published his great edition of Polybius, and an appendix to it comprised the *editio princeps* of Aineias: text,[15] Latin translation, and notes.

To have attracted the attention, and justified the labour, of one of the most outstanding scholars of his time lifted Aineias onto a plane of celebrity never attained by any of the other *tactici* (let alone the *mechanici*), before or since. Furthermore, this new, superior status of his can only have been enhanced by Casaubon's presentation of him as an author of the classical period, no less; a man to be identified with the Aineias of Stymphalos, general of the Arkadian

13 στρατηγικὰ βιβλία ἱκανά.

14 "Αἰνείας ... ὁ τὰ περὶ τῶν Στρατηγικῶν ὑπομνήματα συντεταγμένος, τοῦ γε μὴν δέοντος ἀκμὴν πάμπολυ τὸ κατὰ τὴν ἐπίνοιαν ἀπελείφθη". The Loeb translator of Polybios, W.R. Paton, obscures this point with his 'the author of the work on strategy'.

15 Based not on the *Mediceus*, but on *Parisinus* 2443, a manuscript decended from it at two removes.

League in 367 BC, who makes a fleeting appearance in Xenophon's *Hellenica* (7.3.1). This celebrated identification has been challenged from time to time, since Casaubon, and differing opinions about it are still current. My own, I find, has not changed since 1990. Given the relative rareness of the name, and (especially) in the light of the other, quite independent, reasons for dating the treatise at the very mid-point of the fourth century BC, I continue to think it highly probable that Aineias of Stymphalos is our man.[16]

Post-Casaubon, the roll-call of individuals who have done important work on Aineias—initially, again, as a corollary of studying Polybius; then, from Orelli[17] onwards, addressing him in his own right—could be rather a long one. However, since I am not engaged in a bibliographical *tour d'horizon* for its own sake, I will take as read these fine specimens of eighteenth- and especially nineteenth-century Continental scholarship, and fast-forward us to the key decade, at least as far as Anglophone readers are concerned: the 1920s.

The Clarendon Press edition of 1927, originally begun by the precociously-gifted Leslie Hunter and completed after his premature death (during the First World War) by S.A. Handford, was not only seminal itself but offered the first detailed introduction to, and line-by-line commentary on, Aineias in English. Naturally, it went out of print long ago. But still happily *in* print, after nearly ninety years, is a work which, given its user-friendly format of introduction, Greek text with facing translation, and sporadic footnotes, has always reached a much wider audience. I refer to the Loeb edition: *Aeneas Tacticus—Asclepiodotus—Onasander: with an English translation by members of the Illinois Greek Club*, first issued in the Loeb Classical Library in 1923. Its preface memorably declares:[18]

> Early in 1917, *Marte iam diu furente*, the attention of the Faculty Greek Club of the University of Illinois was turned toward the art of war, in which, as in so many other fields of scientific and humanistic interest, the Greeks achieved results of more than transitory value. The military manual of Aeneas, styled the Tactician, suggested itself as a monograph in this field well suited for discussion by such a club, and portions of

16 Whitehead (1990) 4–13, esp. 10–12. See also Bettalli (1990) 1–6, and more recently (e.g.) Hornblower (2004) 182–184. For extreme scepticism on this point see most recently Schellenberg (2016) section II, esp. n. 5.

17 J.C. Orellius [J.K. Orelli], *Aeneae Tactici commentarius de toleranda obsidione* (Leipzig 1818).

18 Oldfather et al. (1923) ix–x.

> this treatise were accordingly translated by the following members. [17 names are then listed.] Of the versions thus produced a number were discussed and criticized at a series of meetings, and all were subsequently revised and edited by Messrs. W.A. Oldfather, A.S. Pease, C.M. Moss, and H.V. Canter. An introduction, critical apparatus, notes and index have been added to make the work conform to the general plan of the Loeb Classical Library.

The key name there is of course that of William Abbott Oldfather (1880–1945), and it is appropriate to say something about him here.

After a first BA from Hanover College, Indiana, and a second one (and an MA) from Harvard, Oldfather took a doctorate, between 1906 and 1908, at the University of Munich; more of that below. On subsequently arriving at Urbana-Champaign, in 1909 (here I quote his biographer John Buckler), Oldfather

> wished to institute at Illinois the German concept of large, joint academic endeavors. He organized the Illinois Greek Club and launched numerous bibliographical programs, including word indices to Seneca's tragedies, Apuleius' works, and Cicero's *Letters* and *Rhetoric* [...] Early in 1917 the war prompted him and some colleagues who constituted the Illinois Greek Club to examine Greek military writers, some of whom were readily available only in the German translation of [...] *Griechische Kriegschriftsteller* [...] As usual, Oldfather began with exhaustive textual studies of the manuscripts before he and his colleagues produced their translation for the Loeb Classical Library (1923).[19]

Aineias, presumably, would not have been enough, in and of himself, to justify a Loeb volume; so, as we see, the Greek Club ventured considerably further off the beaten track, to include Asklepiodotos and Onasandros. Here in short, prompted by Oldfather and his colleagues and students, the Loeb Library was proffering what amounted to *Select Tactici* (from three different eras). And consequently, what I think is so striking about Aineias-studies since the 1920s is the major shift in direction that they have taken. While Aineias has never stopped entirely being regarded as a *tacticus*, and duly assessed in terms of the expressly military circumstances before, during and after his day, diachronic evaluation has for several decades now been at some risk of being elbowed aside by synchronic. Study of Greek warfare itself has (rightly) become embed-

19 Buckler (1990) 347, 350.

ded in study of Greek society at large, and as a corollary of that, the old notion of Aineias the narrowly military writer has yielded ground to that of Aineias the socio-economic commentator, especially valuable for his insights into "typical", unstable, struggling poleis of the mid-fourth century.

There is a connection here which I noticed only recently, and which is worth brief mention in case one believes (as I am tempted to do) that it might be causal. In his postgraduate work in Munich Oldfather studied Greek with Otto Crusius, Latin with Friedrich Vollmer, archaeology with Adolf Furtwängler, and ancient history with Robert von Pöhlmann. For present purposes the last of these four names is the significant one. Here is what I wrote about him in 1990, when essaying a sort of *Forschungsbericht* on what was then recent work on Aineias, and with particular regard to his personal beliefs and convictions.[20]

> Much of this modern work, since the early 1960s, in in fact a delayed response—the delay attributable to the failure of Aineias's treatise, until then, to command any sort of consistent place in Greek history's academic agenda—to a particularly provocative statement of the question made long ago, in an extraordinary book by an extraordinary scholar, both of them more familiar nowadays to historiographers than to historians. The scholar was Robert von Pöhlmann: the book, his two-volume *Geschichte des antiken Kommunismus und Sozialismus* (Munich 1893–1901), retitled for the second (1912) and posthumous third (1925) editions *Geschichte der sozialen Frage und des Sozialismus in der antiken Welt*. Von Pöhlmann (1852–1914) was a conservative libertarian who familiarized himself with Marxist and socialist literature in order to play devil's advocate with it; that is, he shared with Marxists a hypersensitivity to 'class' conflicts but his personal sympathies therein were the diametric opposite of theirs. For him, class conflict (and its outcome) in antiquity was intended to serve as a dire warning to his contemporaries, a demonstration of the evils which communism and socialism entailed.

Reading Buckler on *Oldfather* suggests a more complex psychological profile than this, and it would be wrong to suppose that he was simply intent on propagating von Pöhlmann's views about Aineias. On the contrary: in the place where he could most appropriately have done so, the closing couple of pages of the Loeb Introduction, Oldfather not only forebears to cite von Pöhlmann by name;

20 Whitehead (1990) 30.

he makes a scrupulously polite but nonetheless polemical statement which clearly has him, amongst others, in mind. Having quoted Plato on rich vs. poor *stasis*, Oldfather writes: 'Aeneas [...] seems to have held aloof from the partisanships of politics so that some who have failed to grasp his true attitude have regarded him as a moderate democrat and others as a supporter of oligarchy'. At the time of writing, this latter view was championed by no-one more vigorously than by Robert von Pöhlmann. So, as I say, no Munich-inspired reading of Aineias by Oldfather there. Insofar as Oldfather had overt political convictions of his own, they have been characterised by another present-day scholar who has probed his biography, Michael Armstrong, as those of a 'socialist'.[21] Be that as it may (and terms like 'socialist' do mean significantly different things on different sides of the Atlantic), the von Pöhlmann connection may well explain, on a much simpler level, why Oldfather had decided to spend time and effort on Aineias at all.[22]

Nowadays Aineias is a much less marginal figure than he was when the Illinois Greek Club set about their business. As I have commented already, it was precisely the product of their work, the Loeb edition, which deserves the lion's share of the credit for bringing him into the mainstream; but of course there are other reasons too. And some of them are very specific reasons—by which I mean that some very particular aspects of what Aineias wrote about have gained and then maintained a reputation of their own, not merely amongst scholars but also amongst a considerably wider and in some instances non-academic readership.

Here one thinks, inevitably, of chapter 31 in particular. This is the long and rightly famous treatment of the sending of secret messages, whether by steganographic methods (which aim to hide the actual existence of the message) or by cryptography proper (which can leave the message visible but seeks to render it unintelligible).[23]

---

21 Armstrong (1993).

22 Oldfather's actual Munich PhD dissertation was not on Aineias but on the early history and mythology of Lokris: *Lokrika, Sagengeschichtliche Untersuchungen*. (Buckler misleading lists this as 'Diss., Tübingen, 1908'. Rather, it was a Munich dissertation published—such publication being the norm at this time—in Tübingen (Laupp), as well as being issued as *Philologus* 67 (1908) 412–472.) Furthermore, as this project came under Greek rather than ancient history, Oldfather's supervisor was not von Pöhlmann but Crusius. Even so, consideration of the Lokrian "maiden-tribute" will have brought the young American up against the evidence of Aineias (31.24).

23 See most recently on this Debidour (2006). Philon, it may be noted, devoted an entire

The first edition of my own Aineias commentary was written while I was on leave at the Center for Hellenic Studies in Washington DC. The Center's administrator at that time was a young woman called Rose Mary Sheldon. Nowadays she is Colonel Sheldon, Professor of History at the Virginia Military Institute, and the author of several books and articles on ancient military history, with particular reference to espionage and intelligence. But even in 1988/89 she had much to tell me—since I was starting from a position of such ignorance—about the history of cryptography, and Aineias's place in it. I seem to recall her saying that ch. 31 of Aineias still received honourable mention in introductory courses on cryptography for CIA trainees and the like. My memory of this has perhaps embellished itself over the intervening years. But what is unquestionably true is that such mention (the world's 'first instructional text on communications security') occurs in the most comprehensive and authoritative history of the subject, David Kahn's *The Codebreakers*.[24]

Furthermore, surf the web for just a while, as I recently did, and one finds that the situation remains much the same. Not only the authors of traditional hard-copy work on cryptography but also the wider and, in some instances, wackier world of websites and blogs still pay the necessary homage to Aineias in this regard—even if a firm grip on him can sometimes be lost or garbled in the process. I have not seen the piece for myself, only the summary of it on www.informaworld.com (the website of the Taylor and Francis publishing group), but I did enjoy coming across what is reported of a 1999 article about Maple.[25] Maple is a symbolic computation language used to simulate the performance of rotor machines, the time-varying substitution cipher devices most famously exemplified by the German ENIGMA machines of World War II. In the course of this 1999 study, it seems, mention is made of the military treatise by 'a Roman called Aeneas Tacticus'. The curse of Vergil strikes again?

Naturally, one must not be too hard on such mistakes, made by specialists in one field when they venture out of it into another, utterly different one. And to the best of my knowledge, ch. 31 of Aineias is unique in the degree to which it has burst the bounds of its own discipline(s). Otherwise, most people who

treatise to secret messaging, but it has not survived, and in what we have (D76–83 of the *Poliorketika*) he gives only samples. See further Rance in this volume, pp. 347–354.

24 Kahn (1967) 82–83 (retained unchanged in the 'revised and updated' New York (Scribner) edition of 1996).

25 McKeown (1999).

encounter Aineias, nowadays just as in previous times, do so by design rather than by accident, and the main way in which present-day readers and users are better placed than their predecessors is the ever-increasing ease of gaining access both to the treatise itself and to relevant analysis of it.

Periodically, via the Liverpool 'Classicists' mailing list, Graham Shipley used to ask whether anyone knew of an on-line translation of Aineias. And until quite recent times the answer was No—not least because (for mysterious reasons) the Illinois Greek Club's Loeb volume was not one of the ones available on Perseus. Nowadays it is. The Loeb translation can also be found on Bill Thayer's Lacus Curtius website.[26] Furthermore, the out-of-print Hunter/Handford translation has been given a welcome new lease of life by its inclusion on Maria Pretzler's dedicated Aineias site.[27]

Besides that, in the thankfully not-yet-obsolete world of hard copy, there is continuing good news also, in the shape of the steady growth of modern-language translations, with some degree of commentary. My own effort of 1990 can claim the distinction, such as it is, of being the first English translation since Oldfather et al., and the first full(ish) commentary since Hunter/Handford. However, I was only a matter of months—and no time at all, in terms of a publication-year—before Marco Bettalli produced his Italian edition: introduction, text and translation, commentary.[28] Written initially (and obviously) for the benefit of Italian readers, Bettalli's is in fact, I would assert, the finest treatment of Aineias, in the translation-and-commentary format, currently available in any language at all. And besides English and Italian, Aineias can nowadays be read and studied in Czech,[29] in Spanish,[30] in French,[31] in Polish,[32] and in German.[33]

Access to Aineias is one thing, further front-line research on him another. What is left to do? The short answer to that question is: plenty. Aineias-studies are, if not (yet) positively booming, at least in a buoyant state. Much good work has already been done, especially over the last half-century or so, but scope

26 http://penelope.uchicago.edu/Thayer/E/Roman/Texts/Aeneas_Tacticus/home.html.

27 www.aeneastacticus.net.

28 Bettalli (1990).

29 Marek and Kalivoda (1977).

30 Vela Tejada (1991); Vela Tejada and Garcia (1991).

31 Besides the 1967 Budé edition of Alphonse Dain and Anne-Marie Bon, see now Battistini et al. (1994).

32 Burliga (2007).

33 Brodersen (2017).

for more remains. What that 'more' is will, in part, emerge from the rest of this present volume, but here are some preliminary (and selective) thoughts, under three heads.

(a) Literary and historiographical perspectives. Aineias' treatise could so easily have been a dull and dreary read, if he had confined himself austerely to the Do's and Don'ts of pragmatic didacticism. Instead he made an inspired decision. He would embellish and corroborate his precepts, positive and negative alike, with real historical examples—thereby not only furnishing us with an invaluable set of pointers to his date but also providing all those miniature narratives. Some of them, to be sure, are so very miniature that they come and go in an instant, but collectively they would repay further study.[34] Possible comparators could be not only Herodotos (recently addressed by Burliga)[35] and other *bona fide* historians but also, as another narrator on a miniature scale, Polyainos.

(b) Viewed as a *tacticus*, Aineias needs to be placed even more precisely in the overall scheme of ancient military writings.[36] Important early specimens of it are lost, but Xenophon survives; and on the relationship (on all levels) between Aineias and Xenophon there is probably still more to be said.

And as regards the post-Aineias material, let me again mention Philon. An in-depth Aineias/Philon comparison would be greatly to the benefit of our appreciation of both of them. True, Philon is on the face of things a drabber writer. This is in part because he (unlike Aineias) proffers only flat didacticism, without the leavening of concrete paradigmatic examples.[37] Nevertheless he was a real individual living in a real historical era and catering for a real geographical location. To judge from the casual frequency of the mention of date-palms, that location has been deemed likely to be the border-lands fought over by the Ptolemies and the Seleucids. But be that as it may, Philon's city is an analytical construct which needs to be brought into focus from several perspectives, and then compared and contrasted with Aineias's. In an article, thirty years old now, entitled 'Cités, armées et stratégie à l'époque hellénistique

34 See now Pretzler in this volume, 68–95.

35 Burliga (2009).

36 See meanwhile Vela Tejada (1993) and Vela Tejada (2004).

37 I cannot refrain from noting here my mystification at the fact that Peter Fraser, in his magisterial *Ptolemaic Alexandria*, got this the wrong way round. He writes of Philon's 'vivid style'—specifically in the siegecraft material—and he asserts that it is Aineias who comes off worst in the comparison: Fraser (1972) 433.

d'après l'oeuvre de Philon de Byzance'—in effect, a set of partial prolegomena for his translation and commentary on Philon in *La Poliorcétique Grecque*, published the following year—Yvon Garlan made an excellent start on Philon *per se* in this regard.[38] He did not, though, undertake a detailed Philon/Aineias comparison, so that is one of the analyses included in my own recent translation (the first into English) of, and commentary on, Philon.[39]

On a military level, some things are very different indeed, especially the dominant role played by technology in the siege warfare of Philon's day. But other things are, or at least seem, hardly different at all. Most notably, Philon's city, like Aineias's, is every bit as likely to succumb to stasis and/or corruption as to main force.[40] And the society underlying the military subject-matter? Here is one example, in brief. Readers of Aineias are made to realise that, in the sort of small or medium-sized mid-fourth-century poleis that he was catering for, foreigners—be they metics, slaves, allies or mercenaries—pose a particular threat to security.[41] So does he have any good ideas about how to lessen that threat? Apparently not. He is all stick and no carrot.[42] But here is Philon on the subject (C45–48): 'if any of the foreigners [*xenoi*: he means mercenaries] are wounded, let all their needs be attended to, and for those of them who do not have carers, let them be placed in the houses of the citizens. Those of them who show bravery ['become *andres agathoi*' is the phrase] should be promoted and given a command and garlanded. If any of them die, give them the most lavish

38 Garlan (1973).

39 Whitehead (2016).

40 Striking comments in this regard come in passages of the section where Philon adopts the attackers' perspective. D65: "Do not stint on money, whether for bribery or for other (*sc.* related) expenses, as you will get it back many times over once you have taken the city". D83: If secret letters to would-be collaborators fail, "… send others to the leading men, promising them great rewards and money; when they come to light they will make some people seditious ⟨and others …⟩". D96–97: "Try too, at the outset, to corrupt the generals or commanders, by giving money and promising rewards, for, if you decide thus, victory occurs and there is no other stratagem like it".

41 See Whitehead (1992); Boëldieu-Trevet and Mataranga (2006); and on mercenaries Roy in this volume, pp. 206–213.

42 Such a verdict, whether concerning *xenoi* or any other issue, must of course be tempered by the fact that even more of Aineias than of Philon is lost. Note in particular 40.8: "As regards grainless rations, shortages during a siege, and the means of making water fit to drink, these are matters discussed in my book *Preparations*". Philon's *Paraskeuastika* does address, at some length, that kind of topic (B1–48); and it may be that Aineias also had things to say about e.g. medical care in his besieged city, as Philon certainly does (B47, C72–73).

of burials at public expense, and if they leave their children or wives, let them be treated with unstinting consideration; for this is an especially good way of making them, if they are well-disposed [*eunooi*] to the generals and to the citizens, the best combatants'. Food for thought there, clearly.[43] Aineias' well-recognised *homonoia* and Philon's *eunoia* are two words for much the same thing, surely—so, where *xenoi* are concerned, why the difference?

I began this point with the assertion that 'Viewed as a *tacticus*, Aineias needs to be placed even more precisely in the overall scheme of ancient military writings'. Yet time and again one discovers that addressing what purports to be the military content of a writer like him (or Philon) is not in fact something that can, or should, usefully take place in a vacuum. On the contrary, it leads inevitably on to:

(c) Aineias's broad historical insights into his period. Scholars nowadays do not talk and write, so freely as they did half a century ago, of the 'crisis' or (worse) the 'decline' of fourth-century Greece. Much of that sort of analysis, it came to be realised, had two major and interconnected flaws in it: extrapolation from Athens (that most un-Aineian of poleis) to the Greek world as a whole, and an over-interpretation of certain military, political, and psychological landmarks, notably Sparta's victory in the Peloponnesian War and, especially, Philip's victory at Chaironeia. The Macedonians, as is often remarked, cast little or no discernible shadow over Aineias and his work. But for the immediately pre-Macedonian era it remains true that life in a "typical" Greek community is nowhere better illuminated.

Thanks to the Great Dane Mogens Herman Hansen and his colleagues, we all have on our bookshelves now the superb *Inventory of Poleis*, and we wonder how academic life was tolerable without it.[44] No maps in it, but never mind. The *Barrington Atlas* (another latter-day marvel),[45] to which the Copenhagen Inventory makes systematic cross-reference, can tell us exactly where—let me take a non-random example—Stymphalos is. The *Inventory* (item 296 in this instance, by Thomas Heine Nielsen) summarises the surviving evidence, from and about Stymphalos. But how much can it reveal of what it was like *to live*

43 For comparison see e.g. the mid-third-century agreement between Eumenes I and his mercenaries (OGIS 266, *Staatsverträge* 481), one succinct clause of which appears to deal with any orphaned children.

44 Hansen and Nielsen (2004). On Aineias, see specifically Hansen in Hansen (2007) 243–245.

45 Talbert et al. (2000).

there, in Stymphalos or in any of the hundreds of other one-horse, run-of the-mill places like it? For that, we must turn to Aineias.

Sir Kenneth Dover once wrote as follows, of Thucydides: '[i]t is the common experience of people who study Thucydides intensively over a long period of time that one goes on indefinitely noticing things in him that one has not noticed before. This could be said of other authors too, but in most cases the returns diminish; in the case of Thucydides there always seems to remain the possibility that something really important is waiting to be noticed'.[46] It is not, I venture to say, required of students of Aineias to pretend that he is a Thucydides in this respect. A writer of bottomless profundity? No. But nor is he—not yet, at any rate—a writer so limited in his scope and insights that the law of diminishing returns, invoked by Dover, can already be seen to have set in. Rather, Aineias's work—written when it was, and written how it was—is multi-dimensional. Hence the Gregynog conference, and this volume which has flowed from it.

Bibliography

Armstrong, M.S. (1993) 'A German Scholar and Socialist in Illinois: the Career of William Abbott Oldfather', *Classical Journal* 88: 235–253.

Battistini, O., et al. (1994) *La Guerre: trois tacticiens grecs—Énée, Asclépiodote, Onasandre*. Editions Nil, Paris.

Bettalli, M. (1990) *Enea Tattico, La difesa di una città assediata (Poliorketika). Introduzione, traduzione e commento*. ETS Editrice, Pisa.

Blyth, P.H. (1992) 'Apollodorus of Damascus and the *Poliorcetica*', *Greek, Roman, and Byzantine Studies* 33: 127–158.

Boëldieu-Trevet, J. and Mataranga, K. (2006) 'Étrangers et citoyens: le maintien de l'ordre dans une cité assiégée selon Énée le Tacticien,' in M. Molin (ed.), *Les régulations sociales dans l'Antiquité: actes du colloque d'Angers, 23–24 mai 2003, sous la direction de M. Molin*. Presses Universitaires de Rennes, Rennes. 21–41.

Brodersen, K. (2017) *Aineias/Aeneas Tacticus, Stadtverteidigung/Poliorketika. Griechisch-deutsch. Herausgegeben und übersetzt*. Walter de Gruyter, Berlin and Boston.

Buckler, J. (1990) 'William Abbott Oldfather', in W.W. Briggs and W.M. Calder III (eds), *Classical Scholarship: a biographical encyclopedia*. Garland, New York and London. 346–352.

46 Dover (1973) 44.

Burliga, B. (2007) *Eneasz Taktyk. Obrona oblezonego miasta*. Proszynski i S-ka, Warsaw.

Burliga, B. (2009) 'Aeneas Tacticus Between History and Sophistry: the Emergence of the Military Handbook' in J. Pigoń (ed.), *The Children of Herodotus: Greek and Roman Historiography and Related Genres*. Cambridge Scholars Publishing, Newcastle. 92–101.

Debidour, M. (2006) 'Le secret et les messages secrets dans la Poliorcétique d'Énée le Tacticien', in Olivier, H., Giovannelli-Jouanna, P. and Bérard, F. (eds), *Ruses, secrets et mensonges chez les historiens grecs et latins*. Université Jean Moulin Lyon 3, Lyon. 213–241.

Dover, K.J. (1973) *Thucydides*. Clarendon Press, Oxford.

Fraser, P.M. (1972) *Ptolemaic Alexandria*, vol. I. Clarendon Press, Oxford.

Garlan, Y. (1973) 'Cités, armées et stratégie à l'époque hellénistique d'après l'oeuvre de Philon de Byzance'. *Historia* 22: 16–33.

Gatto, M. (2010) *Il Περὶ μηχανημάτων di Ateneo Meccanico. Edizione critica, traduzione, commento e note*. Aracne editrice, Rome.

Hansen, M.H. (ed.) (2007) *The Return of the Polis. The Use and Meanings of the Word* Polis *in Archaic and Classical Sources*. (Papers from the Copenhagen Polis Centre 8. *Historia* Einzelschriften, 198). Franz Steiner Verlag, Stuttgart.

Hansen, M.H. and Nielsen, T.H. (2004) *An Inventory of Archaic and Classical Poleis*. Clarendon Press, Oxford.

Hornblower, S. (2004) *Thucydides and Pindar: Historical Narrative and the World of Epinikian Poetry*. Clarendon Press, Oxford.

Kahn, D. (1967) *The Codebreakers: the Story of Secret Writing*. Weidenfeld and Nicolson, London.

Lazenby, J.F. (1991) 'Review of Whitehead 1990', *Classical Review* 41: 34–35.

McKeown, G.P. (1999) 'Specifying and Simulating Rotor Machines Using Maple', *International Journal of Mathematical Education in Science and Technology* 30: 867–887

Marek, V. and Kalivoda, J. (1977) 'Aineias Taktikos. O obrane mest', in Marek, V. And Kalivoda, J. (eds), *Antické Válecné Umení*. Svoboda, Prague.

Oldfather, W.A. (1923) *Aeneas Tacticus—Asclepiodotus—Onasander: with an English translation by members of the Illinois Greek Club* (Loeb Classical Library). William Heinemann and G.P. Putnam's Sons, London and New York.

Schellenberg, H.M. (2016) 'Reflections on the Military Views of the "Military Writer" Aeneas Tacticus', in Rance, P.R. and Sekunda, N.V. (eds), *Greek* Taktika. *Ancient Military Writing and its Heritage (Proceedings of the International Conference on Greek* Taktika *held at the University of Toruń, 7–11 April 2005)*, Akanthina Monographs 13. University of Gdańsk, Gdańsk, 81–93.

Talbert, R.J.A. et al. (2000) *Barrington Atlas of the Greek and Roman World*. Princeton University Press, Princeton and Oxford.

Vela Tejada, J. (1991) *Eneas el Tactico, Poliorcética. La Estrategia Militar Griega en el Siglo IV a.C.* Ministerio de Defensa, Madrid.

Vela Tejada, J. (1993) 'Tradición y originalidad en la obra de Eneas el Táctico: la génesis de la historiografía militar', *Minerva* 7: 79–92.

Vela Tejada, J. (2004) 'Warfare, History and Literature in the Archaic and Classical Periods: the Development of Greek Military Treatises', *Historia* 53: 129–146.

Vela Tejada, J. and Martin Garcia, F. (1991) *Eneas el Tactico, Poliorcética. Polieno, Estratagemas*. Gredos, Madrid.

Whitehead, D. (1990) *Aineias the Tactician, How to Survive under Siege. Translated with Introduction and Commentary*. Clarendon Ancient History Series, Oxford. [Updated edition Bristol Classical Press (2001) = Duckworth (2002)]

Whitehead, D. (1992) 'L'image de l'étranger dans la *Poliorcétique* d'Énée le Tacticien', in R. Lonis (ed.) *L'Étranger dans le monde grec II: actes du colloque organisé par l'Institut d'Études Anciennes Nancy, septembre 1991* (Nancy 1992) 315–331.

Whitehead, D. (2008) 'Fact and Fantasy in Greek Military Writers', *Acta Antiqua Academiae Scientiarum* 48: 139–155.

Whitehead, D. (2010) *Apollodorus Mechanicus, Siege-Matters. Translated with Introduction and Commentary*. (*Historia* Einzelschriften 216). Franz Steiner Verlag, Stuttgart.

Whitehead, D. (2016) *Philo Mechanicus, On Sieges. Translated with Introduction and Commentary* (*Historia* Einzelschriften 243). Franz Steiner Verlag, Stuttgart.

Whitehead, D. and Blyth, P.H. (2004) *Athenaeus Mechanicus, On Machines. Translated with Introduction and Commentary* (*Historia* Einzelschriften 182). Franz Steiner Verlag, Stuttgart.

CHAPTER 3

Aineias the Author: Who, Where and When?

Robin Lane Fox

Aineias Tacticus gives us the best surviving evidence for life in fourth-century *poleis* outside Athens and Sparta. Fully commented on by scholars, his text is still not exhausted for its historical, social and cultural potential. One reason for its being so rewarding is that continuing epigraphic finds throw a new light on words or phrases which Aineias uses in passing. When he describes a coup in Argos, he uses a turn of phrase which has helped to explain a similar turn in the recently found law of the Eretrians in 341/0 BC against coups by oligarchs or tyrants.[1] The shifting emphases of historians also encourage us to reread Aineias. His comment on the presence of Lokrian maidens at Ilion has long exercised scholars. He also assumes that there will be a theatre in any *polis* to which his advice is directed. He reminds us, therefore, how purpose-built theatres, the great Athenian legacy, had spread widely through the Greek world by the mid-fourth century.[2]

There are gains, too, for historians of animals in Greek history, a subject which has yet to find its apt historian: Aineias is explicit (unlike Thucydides) about dogs in a *polis* and their noise. Aineias also makes us think more clearly about such matters as gates, debtors and travelling textiles. From one modern point of view the rural and urban territory of a Greek *polis* is to be seen as a whole, a 'variously-peopled landscape'.[3] Aineias' detailed concern for the city-gate and its key alerts us to contrary evidence, the importance of locking people in and out of the urban part of this landscape every night. The first person attested (but nowadays, underemphasized) as locked out of his city and thereby losing control of it to a coup is Gorgos, king of Cypriot Salamis, in 499 BC.[4] Control of a city's gate-key is a major theme for Aineias and remained so, symbolically, in subsequent Greek history. If we read ahead into Hellenistic history with Aineias in mind we can see how keys continued to matter in a continuing age of polis-commitment. In 243 BC, Aratos seized Corinth from Antigonos Gonatas, greeted the crowds and, significantly, gave them back the

---

1 Ain. Tac. 11.8; Knoepfler (2002) 168–169; see Liddel in this volume, p. 133.

2 Ain. Tac. 31.24; Hornblower (2015) 406–412; Ain. Tac. 1.9, 3.5; Diod. Sic. 15.40.2 for a theatre in Phigaleia.

3 Osborne (1987) 26.

4 Hdt. 5.104.

 | DOI: 10.1163/9789004352858_004

keys of their gates for the first time since the era of Philip II.[5] More people throughout Greek history were locked nightly into their *polis* than we readily envision. Hellenistic Corinth can be matched to Chalcidic cities in the time of Thucydides as well as to nearby Sikyon in Aratos' lifetime. Aineias' concerns for the gate and security find practical illustration in the Greek historians, in bits of text which seem typical if read in the light of his manual. The best bits are near the end of Thucydides book four and the start of Plutarch's *Life of Aratos*. Brasidas' penetration of Amphipolis and later of Torone exemplify some of Aineias' fears, while book four ends on a note in keeping with Aineias' tone, the bell which one of the night-guards of the walls of Poteideia had 'passed on' and which, mysteriously, he had not yet 'come back to'.[6] Above all, Aratos' capture of Sikyon in 251 BC reminds us how Aineias' fears were not out of date even a century later. Here we see an attempt on a walled city through the memoirs of the 'besieger', Aratos himself, who is Plutarch's source.[7] Once again, the bell for the night-watch is significant and nowhere in all Greek history is there a more vivid fear of dogs. The fierce 'little dogs' of a gardener who lives by the walls terrify Aratos' conspirators. In turn, a huge hunting-dog, a *thēratikos*, terrifies them, and as a result the 'entire place is filled with lots of barking'. The hazards in the enterprise are Aineias' hazards, seen from outside, not inside.

As for the debtors in a *polis*, no other author gives us such a graphic image of them as Aineias: they are a very serious threat lying in wait (πολύ γε φοβερώτατοι ἔφεδροι).[8] Here and elsewhere in Aineias' text, de Sainte Croix, in 1981, detected 'some interesting evidence of the fear by the propertied class of revolution by the burden of debt'.[9] The force of Aineias' word ἔφεδροι needs to be emphasized here too, people 'waiting to pounce'. For de Sainte Croix's 'revolution' I would sometimes read treachery, but the fear was real enough. For Hunter, writing in or before 1914, 'the rich, as usual in Aineias, have to suffer'.[10] However, Aineias is not quite so one-sided. He advises using the well-off (εὔποροι) in responsible jobs, especially the job of gatekeeper which is not a role for 'ordinary men'. The richer classes are to serve, not suffer. Aineias does not have a dislike or fear of the rich as such: during sieges they are particularly useful because they have most at stake in the city's survival.[11] Nor does he look down on riff-raff. A

5 Plut. *Aratos* 23.4.

6 Thuc. 4.103.4–5, 4.110, 4.135; Villing (2002) 223–295.

7 Plut. *Aratos* 5.3–9.1; 5.5, 6.3, 7.3–4 on the dogs, with Ain. Tac. 22.14 and 24.18.

8 Ain. Tac. 14.1.

9 De Sainte Croix (1981) 298; Ain. Tac. 14.1 and Hunter and Handford (1927) 146.

10 Hunter and Handford (1927).

11 Ain. Tac. 5.1; see also Pretzler in this volume, p. 154–157.

general, he advises, should praise and blame those who are fighting on the city-walls but not go angrily after the ordinary men, οἱ τυχόντες. Otherwise, they may lose heart, a loss of morale to which Aineias, the hardened military captain, is sensitive elsewhere. He would thoroughly disapprove of Homer's Odysseus, speaking mildly to the important heroes in the *Iliad* and threatening to thump the ordinary warrior. Aineias has progressed beyond Homeric values.[12]

Indirectly, he gives us an invaluable insight into the objects which passed into any *polis* with a harbour. In 1973, in his *Ancient Economy*, M.I. Finley questioned whether there was significant 'export of manufactured goods' between Greek *poleis*. 'How much "must there have been" in the way of manufacturing for export despite its non-recognition in the available sources?',[13] he asked ironically. He noted a few possible counter-examples, but minimized them, including trade in clothes and shoes. For the clothing trade, he cited the 'summer mantles, made we do not know where, which an Athenian brought to Kyrene once a year in such small supply that Bishop Synesius was impatient lest he miss the opportunity to buy'. He also cited the 'high-grade linen garments' from Tarsus which 'brought the weavers of that city an apparently steady livelihood, but on so low a level …'. One of his examples comes from a fifth-century AD source, the other from the first century AD.[14] But what about Aineias, centuries earlier, when describing the plot to capture an (unnamed) city, one of goods being brought in (τὰ ἐπὶ πράξει γεγενημένα)?[15] Linen breastplates, στολίδια (jerkins) and weaponry were made ready in chests used for transporting merchandise (ἐν κιβωτοῖς ὡς φορταγωγοῖς). The harbour-officials, ἐλλιμενισταί, opened the chests and saw ὡς ἱμάτια μόνον, 'only clothing', and 'put a mark on' (the chests, I presume), until the importers should give a value. They missed the weapons underneath, but they missed them because they assumed the κιβωτοὶ φορταγωγοί (transport containers) were filled with clothing-imports. To the supervisors of harbour-dues, therefore, imports of clothing by traders were a very familiar fact of life. Aineias does not name the city, though Meinecke, unconvincingly, tried to identify it as Amphipolis in 424 BC.[16]

The question I want to answer in a new way is: where and when was the author of these vivid details writing? In his seminal article of 1962 Bengtson raised this question, noted the hypothesis that Aineias was an Arkadian but advised that the question was best left alone: 'man wird die Frage des Verfassers

12 Ain. Tac. 38.4; contrast Homer, *Il.* 2.188–206.

13 Finley (1973) 134–136.

14 Synes. Epist. 54; Dio Chrys. 34.21–23.

15 Ain. Tac. 29.3–12.

16 Meineke (1867) 174–190, at 183.

besser auf sich beruhen lassen'.[17] Nonetheless, de Sainte Croix in 1981 noted 'in my opinion, the work was most probably written in the early 350s' and David Whitehead considered it 'probable (or even certain) that Aineias was indeed the Aineias of Stymphalos who is independently known to us in 367 BC'.[18] In some ways the historical value of Aineias' text does not depend on further precision about the author. The examples which it cites belong on any view before c. 340 BC, making the text into a mid-fourth century source. Many of them contain personal names and *poleis* which tie up with external evidence and locate them precisely too. There is, however, an author behind them who continues to engage his readers. I wish to place him in a context more detailed, and novel, than commentators have so far recognized. My arguments draw on an underemphasized item in his vocabulary, an unnoticed context for one of his most distinctive bits of advice and a general pattern in the distribution of his historical examples. His text is not autobiographical and nowhere does the author use a significant first-person singular. Nonetheless, I believe we can infer some of his biography by attending to the range of examples which he gives. An analogy will make my approach clearer. Suppose we had an anonymous text, evidently by an eighteenth-century author, who was identified at the head of its only manuscript as 'E. G.'. Suppose, too, that its historical examples were located in Hampshire, Oxford, Paris and Lausanne, with a few drawn from seventeenth-century Britain but nowhere else. The inference would be secure that the author was Edward Gibbon. For Aineias, we have no such securely attested life to whose external evidence we can match the distribution and dating of his anecdotes. However, they suggest one, which can be linked in turn to a credible life for the author. On that credibility, and the consequent enlightenment given by its lower date, I rest my case.

The external evidence for various Arkadian Aineiases of Stymphalos is varied and well known. In 472/468, the Aineias in Pindar *Olympian* 6 is (surely) the *chorodidaskalos* of this ode, sung in honour of his fellow Arkadian Hagesias, the winner of the mule-chariot race (no more) at Olympia. This Aineias is from Stymphalos, as is Hagesias through his mother's family, but Hagesias has citizenship of Syracuse too. In my view (opinions vary) Aineias is implied by Pindar to be going west to Syracuse with his choir.[19]

The next Aineias to be attested is the most memorable. He was present on Xenophon's Anabasis in 401 BC and in 400/399 BC, among the rock-rolling

17 Bengtson (1962) 461.

18 Whitehead (1990) 13.

19 Pind, *Ol.* 6.103–105, with Hutchinson (2001) 424; also 6.88.

Taochians he was remembered for a spectacular death. When the Taochians' stronghold was captured they made a 'terrible sight': the women 'began to thrown their little children down off the rocks, then themselves, and their men did likewise'. Aineias was a commander and noticed a man running to throw himself off who was wearing a fine robe. He grabbed him in order to stop him, but was pulled along and both of them rolled down the rocks to their death.[20] Would Aineias have tried to stop one of the women rather than this man of distinction who seemed such a waste? I suspect he would not.

In Xenophon's *Hellenica* we then meet the Aineias of Stymphalos who was 'general of the Arkadians' in 366 BC. He took his army into the acropolis of Sikyon and summoned the 'most powerful' (κρατίστους) of Sikyonians in the city and 'sent for those who had been exiled (by the populist Euphron) without decree': the latter, probably, were pro-Theban oligarchs.[21] The initiative did not come to much and before long Euphron was back, Aineias having disappeared from Xenophon's narrative.

In a dense and detailed discussion of the evidence and other texts which he relates to it, Simon Hornblower has concluded in his *Thucydides and Pindar* that 'chorus-leaders, seers and military commanders and adventurers, the various men called Aineias of Stymphalos and their relatives have one foot in the world of Pindar and one in the text of Thucydides'.[22] For 'Thucydides', I would substitute Xenophon. What we have, across more than a century, are Arkadians with links to Sicily who have a tragic exploit to their credit in Anatolia, a history of military command and a short-lived capture of Sikyon's acropolis which had more than a hint of wishing to drive populist troublemakers out of the *polis*.

With this well-known background in mind I wish to look at the distribution in time and space of the historical examples in Aineias' text.[23] By 1914, Hunter had drawn up a helpful table which arranged them zone by zone in the Greek world and here, I propose only to simplify his divisions, keeping 'Peloponnese and central Greece' but amalgamating 'western Greece' with 'Sicily' and his 'Hellespontine district' with his 'coast of Asia Minor'.[24]

The anecdotes from 'Peloponnese and central Greece' include one (but only one) culled directly from Thucydides: the siege of Plataia at 2.3. The others, however, are almost all well known too, although one (the 'Achaian city' at

20 Xen. *Anab.* 4.7.13–14.

21 Xen. *Hell.* 7.3.1–3; Griffin (1982) 72–74. The date is not certain, and others opt for 367 BC, but not Roy (2000) 321–324.

22 Hornblower (2004) 184.

23 For an overview see the Pretzler's table in this volume, 89–93.

24 Hunter and Handford (1927) xxxvi–xxxvii; for the book's date, ii and v.

18.8) is textually defective (not 'reticent') and one (gates on Aigina) only occurs in a remark in passing. What about their dates? Like David in *AJP* 1986, I place the 'second oligarchical revolt' in Argos (11.7) back before 370 BC.[25] Like Casaubon in 1612, but few since, I place the Theban invasion of Sparta (2.2) in 369, not 362.[26] I do not put the assault of Chalkis in 357, because Diodoros 16.7.2, our only source, contains nothing specific about Chalkis in this way. There is, however, an exact allusion to the capture of the harbour at Sikyon which belongs in the 360s and is usually credited to Pammenes in 369. What is striking is the absence of anything datable to the 350s, or (on my view of 2.2) later than c. 366 BC. However, those who take 2.2 to refer to the Theban invasion in 362 BC, as I do not, will allow that it was so famous that its presence in Aineias' texts implies nothing about his own whereabouts. He could have heard it anywhere.

What is striking, as Hunter observed, is his knowledge of a detail about what happened in the port of Sikyon, much to the detriment of the Sikyonians. All modern commentators compare the anecdote in Polyainos 5.16.3 whereby Pammenes captured Sikyon by pretending that his boatload of hoplites was only a boat of some merchants (ἐμπορικοί τινες). Hunter inferred that 'the occurrence in our book of an allusion to an event definitely known to have taken place at Sikyon in 369 does certainly give support to the view that the author was Aineias of Stymphalos', the man who seized Sikyon's acropolis shortly afterwards.[27] Hunter went on to argue that aspects of the *polis* for which Aineias' text is 'legislating' fit Sikyon very well. He even suggested that 'when Aineias' official command came to an end he stayed on at Sikyon, either at his own wish or at the request of the grateful inhabitants'. He acted, Hunter suggested, as 'their chief military adviser' and 'incidentally' employed 'his leisure in the composition of his *magnum opus*'.

The generalized *polis* for which Aineias legislates should not be narrowed down to Sikyon only, and Hunter's reasons for narrowing it are not convincing. However, his point about Aineias of Stymphalos and Sikyon is a good one and can, perhaps, be strengthened. Is Aineias really alluding to an event of 369 BC? In that year, certainly, Thebans and their allies went west from Corinth to take Sikyon and Pellene but Xenophon, the only source for these events, says nothing about any attack from the seaward side. The impression, rather,

25 David (1986) 343–349, opting for a date between 415 and 370.

26 Ain. Tac. 2.2, with Cartledge (1987) 232–236. The Spartans' (successful) barricading of their *polisma* could as well fit 370/69 as 362 BC, in my tentative opinion.

27 Hunter and Handford (1927) xxv, with xxvi.

is of a land campaign.[28] In 366, Aineias took Sikyon's acropolis and Xenophon mentions no Theban governor in the city. Soon afterwards Euphron returned to the lower city and then a Theban governor reappears in the course of events.[29] Was it in this context that Pammenes' Thebans had deceived the Sikyonians in the port, the last episode (perhaps) in a series of captures of their harbour? If so, Aineias is alluding to a story connected with his own, a Theban recapture of Sikyon's port in 366 BC. The likelihood that he is drawing here on personal knowledge then becomes even stronger.

The 'Hellespontine district and the coast of Asia Minor' are quite another matter. There are far more anecdotes here, sixteen in all, and several of them give named participants about whom we know next to nothing, or nothing, in any other source. Python of Klazomenai, Temenos of Rhodes (a name with good Argive–Rhodian roots), the fall of Astyanax at Lampsakos: these are, for us, very rare items. None comes from a known written source. A story about Datames' siege of Sinope contains details which are absent from Nepos' *Life* (using Deinon) and although Sekunda has suggested a date of c. 368, tentatively, for Aineias' siege, it could be a second siege, later in the 360s (Datames died c. 360).[30]

External evidence suggests a date for six of the sixteen anecdotes in this region. Only one, about Nikokles on [Na]xos, belongs earlier, in 376 BC, and only if the text is correctly supplemented by modern editors. The other five all fit in the years from 365 to c. 360 BC, culminating in the 'inside story' of Charidemos' capture of Ilion and the passwords used.[31] In this geographical division I am also struck by three local observations which are added to their respective stories. The *polis* of Teos is παμμεγέθης, 'very big indeed'; Herakleia on the Pontus had three tribes each with four *hekatostyes*, which were then changed to sixty *hekatostyes*. Above all, 'when the Chiotes celebrate the Dionysia ... they occupy in advance the streets which lead to the agora with guards and many forces'.[32] These observations are of considerable interest in their own right, about demography, civic structures and festival-practice. They are most unlikely to have been passed to Aineias by oral tradition with its 'tyrannical exclusion of the commonplace' when retelling stories about coups and precautions at the places involved.[33] They read like the personal observations of the

28 Xen. *Hell.* 7.1.18 and 22.

29 Xen. *Hell.* 7.2.11, 7.3.1–2, 7.3.4–5.

30 Ain. Tac. 28.5, 18.13–19, 31.33; 40.4–5 with Sekunda (1988) 35–53.

31 Ain. Tac. 22.20, with Whitehead (1990) 157–158; Ain. Tac. 24.

32 Ain. Tac. 18.13; 11.10–11; 17.5.

33 Cadbury (1927) 34.

narrator himself, someone who knew Chios and Teos personally and perhaps Herakleia too.

Could Aineias, then, be a man from the east Aegean? The name 'Aineias' is well attested there, and not only (of course) in the post-Homeric Troad.[34] Was he writing his text, even, inside the Persian empire? That allusion to rare events at Sikyon tells against this possibility because it was such an unlikely piece of local knowledge to be picked up by an author in Asia minor. There is, however, support for Aineias' first-hand Asian knowledge in an under-exploited detail of his vocabulary.

In chapter 33, on burning (ἔμπρησις), Aineias discusses protections 'If there are any wooden towers in the city, or if any part of the wall is made of wood' (ἄν τινες ὦσι τῆς πόλεως ξύλινοι μόσυνες ἢ τοῦ τείχους τι.)[35] Μόσυνες is an exceptionally rare word, which is only attested in prose of the classical period for the wooden towers, or battlements, of the Mossynoikoi as observed by Xenophon and immortalized in his *Anabasis*.[36] The word entered Greek only from, and for, the wooden-towered settlements in this one region by the Black Sea. In his fine work on Greek fortifications, F.E. Winter struggled to apply the term μόσσυν more widely to city-fortifications in fourth-century Greece. As he admits, forts and frontier-posts would not be wooden μόσσυνες because they would be so easily torchable by an enemy.[37] Were there wooden sections in the curtains, perhaps, of *polis*-walls on the Greek mainland? Winter cites Thucydides for the Plataians' wooden additions to their walls, but these wooden extras were made only in an emergency.[38] He then, wisely, hesitates. 'In view of the scarcity of timber in Greece it is unlikely that the whole upper structure was of wood.' However, Aineias' ξύλινοι μόσυνες are exactly that. Winter backtracks, but wonders, splendidly, in a final note whether some 'round structures at Halieis may not have been the underpinning for something similar to Aineias' *mosynes*'.[39]

What Winter rightly sees is that 'certainly, such structures must have existed in [Aineias'] day'. Were they due to 'hasty building', he wonders, the 'need for economy', perhaps, or the conversion of 'old-fashioned bastions into genuine two-storeyed towers'?[40] But we know very well where they existed: up on the Black Sea. Aineias has not picked up this exotic word casually from Xenophon's

34 Hunter and Handford (1927) xv, to which many more can be added.
35 Ain. Tac. 33.3. (μόσυνες); cf. μόσσυνες in Callim. *Aetia* fr. 43, line 68.
36 Xen. *Anab.* 5.4.
37 Winter (1971) 75–77.
38 Thuc. 2.75.4–5.
39 Winter (1971) 77.
40 Winter (1971) 77.

text and applied it, at random, to Greek walling elsewhere. He has picked it up 'live', from first-hand reports or personal observation of μόσυνες in their home territory. Only in one other Greek text within a century of Aineias is the word also used: I will return to it.

Aineias knows a rare story about Python at Klazomenai. He knows another about Iphiades at Parion, a figure whom A. Wilhelm brilliantly recognized as the man of Abydos who captured Sestos across the Hellespont in 360/59 BC. This coup is known from Demosthenes: Iphiades' assault on Parion is highly likely to belong c. 365–360 BC too.[41] Aineias also knows invaluable facts about the Lokrian maidens at Ilion and the very passwords used there in Charidemos's siege. Manifestly he served in it in c. 360 BC or talked to participants, maybe Charidemos himself. He also knows about μόσυνες which could only be encountered near the Pontic coast. These items of knowledge imply personal acquaintance with the 'Hellespontine district and the coast of Asia minor'. The *poleis* of Pontus were just as vulnerable as those in Greece to dangers which Aineias describes. In the illuminating treaty between Sinope and Herakleia, probably dating to the late 350s, more or less Aineias' era,[42] the two *poleis* swear to help each other if attacked, except by the Persian king. The possibility of help by an external city is not discussed in Aineias' text: like other *polis*-theorists, including Aristotle, he thinks of his 'imagined *polis*' in isolation. In lines 28 ff., however, of the Sinope–Herakleia treaty there are provisions if the *dēmos* of Sinope or the tyrants of Herakleia are overthrown. These possibilities are closer to the warnings about *stasis* in Aineias' text.

In his text, local detail about Sikyon coexists with first-hand detail about the east Aegean and Pontus: can they be related to one Aineias? In 366 BC Aineias of Stymphalos' occupation of Sikyon was quickly challenged again by Euphron. As Xenophon shows, it was a time of exiling in Arkadia: was Aineias exiled by his fellow-Arkadians for his short-lived interference at Sikyon? Long before Xenophon's *Anabasis*, Greeks from the mainland were no strangers to mercenary service in Asia. Aineias knew the options for an Arkadian under political threat,[43] so 'Et ego ex Arcadia' (correcting the title of Fields' recent article),[44] he took his military talents east. Surely he remembered his forbear, perhaps the very grandfather whose name he bore, the Aineias who had fallen so stirringly to his death off a Taochian cliff? Did he, deep down, want to go and see the site for himself? He saw Teos and Chios, Herakleia Pontica and saw

41 Ain. Tac. 28.5, 28.6–7 with Wilhelm (1974) 23–29.

42 Ain. Tac. 31.24, 24; Whitehead (1990) 8; French (2004) no. 1.

43 Roy (2004) 264–288.

44 Fields (2001) 102–130, whose title misplaces the 'et', an error with a long prehistory.

(or heard about) the Pontic μόσσυνες in which the Mossynoikoi (north-west of the Taochians) still lived. He served with Charidemos at Ilion in 360 and learned about those maidens or talked with others who knew. If he migrated from Arkadia in 366/5 BC, it is easy to understand why so many of his eastern anecdotes in the 'Hellespontine district and coast of Asia minor' cluster from 364 BC (Klearchos) to c. 360 BC (Charidemos).

There remains the division of 'Western Greece and Sicily'. The anecdotes here are far fewer, but one has a relevance to Aineias himself which has never been detected. He tells a story about Chares and Korkyra (probably 361/0), another about Dionysios I and Leptines (which belongs in the mid- to later 380s) and one about Dionysios and an unnamed city (certainly before 369 BC).[45] The most illuminating one is 31.30, the 'secret messages' involving Dionysios and Herakleidas. Here, scholars have gone astray. Manuscript M of Aineias' text reads ΔNCCKΛC, which scholars since H. Schöne have filled out as 'Dionysios Kolos', meaning 'Dionysios is Knackered'. This reading of the letters suits the universal modern assumption that the context of this 'secret message' is an exchange from Sicily (perhaps from Dion) to Herakleidas (back in Greece) urging him to come west because Dionysios' tyranny is on its last legs.[46] No source refers to any such message, but if it existed, its context would be 357/6 BC.

The modern reading of the message prefers a strange adjective (*kolos*) to an obvious alternative (*kalos*). It prefers *kolos* because it cannot find any occasion when *kalos*, sent by Herakleidas, makes historical sense. What has been missed is Plutarch, *Dion* 48.7. At that point Syracuse, except Ortygia, had been liberated in 356 BC; Herakleidas had supported a proposal in Syracuse that both land and houses should be redistributed; he then went off to Messina where 'he began to make agreements secretly through Pharax the Spartan' who had come as a military captain to Sicily.[47] These secret agreements needed secret communications, Aineias' exact subject. Herakleidas needed to be polite in his approaches to Dionysios: 'Dionysios kalos' fits perfectly. Dionysios needed Herakleidas; hence 'Let Herakleidas come', preserved by Aineias in the Doric dialect which Dionysios the Syracusan would have used.

For Debidour in 2006, the question of the message's wording 'n'importe guère pour apprécier le procédé technique d'Enée.'[48] Indeed it does not, but it matters greatly for approaching his date and milieu. Aineias did not avoid

---

45 Ain. Tac. 11.13–15, 10.21–22, 40.2–3.

46 Bengston (1962) 460 for the history of scholarship here.

47 Plut. *Dion* 48–49; Fuks (1968) 207–233.

48 Debidour (2006).

examples from the distant past on principle. Just before this 'secret message', he discussed tricks for secretly conveying messages and retold Herodotos' old story about Histiaios of Miletus' slave with a shaven and tattoed head.[49] Secret messages, however, were not (by definition) to be found in detail in previous historians, so when he wished to illustrate one he had to illustrate one which he himself knew. He did not reach back into the past. To take a modern British analogy for an author writing in 2010/11: he did not go back across 36 years to find a secret coalition-message, 'Jeremy Thorpe is *kalos*', 'Let Ted Heath come', hilarious though their undertones would now seem. Rather, he cited the most recent: 'Clegg is *kalos*', 'Let Cameron come', secret exchanges in 2010 which he himself knew.

How ever did he know this secret lettering? Pharax the Spartan knew it, but the boldest and simplest answer is that Aineias knew it because he was on the spot when these messages were sent and received. When his anecdotes from the east Aegean and Asia are datable, none is later than 360 BC. The reason, I suggest, is that Aineias returned to Greece soon afterwards. He went to the Peloponnese, heard of Dion's inspiring ambitions, met him, like other military men and set off west to liberate Syracuse.[50] He was not the first Arkadian by any means to go west: once again, there was a family precedent, Aineias of Stymphalos the *chorodidaskalos* who had gone west with a choir to Syracuse for his fellow-citizen Hagesias. No doubt the family had remained Syracusan *xenoi* since c. 472 BC. According to Plutarch, *Dion* 22, Dion's mercenaries were all 'known for their many great campaigns' and were 'far and away the most excellent in experience and daring'. Who better to join than Aineias the veteran of the seizure of Sikyon, who had known the east Aegean from 365–360 and even understood a wooden μόσσυν?

Here, I observe in passing Aineias' anecdote about Korkyra (Korkyra was indeed a stopping-point on Dion's westward voyage), I note evidence which he cites specifically from Thessaly at 31.32 (a Thessalian diviner, Miltas, was present with Dion's mercenaries on the voyage west).[51] I even note Aineias' reports (*legetai*) about the horse-drawn chariots (not the 'supporting waggons' detected here by Peter Greenhalgh)[52] which 'are said' to have been used by the men of Kyrene and Barke (Dion's expeditionary force was blown off course and ended up briefly on the coast of north Africa).[53] Of course such stories

49 Ain. Tac. 31.28; Hdt. 5.35.3; Polyaen. Strat. 1.24.

50 Plut. *Dion* 23.

51 Plut. *Dion* 24.1–2.

52 Greenhalgh (1973) 16, rightly opposed by Whitehead (1990) 144.

53 Plut. *Dion* 25.

might have been told wherever mercenaries met and talked: my point is simply that ones told about the West coincide with places Dion's men (and Aineias?) visited. I cannot resist remarking that Herakleidas' secret message to Dionysios was sent from Messina in Sicily.[54] Messina is the site of the only other μόσσυνες which are attested in Greek before 200 BC, the 'wooden towers' which feature in Kallimachos' verses about Zancle–Messina and which the first founders of the city, Euboeans both of them, had had built round the site's great treasure, the sickle, or δρέπανον, with which Kronos had once castrated Heaven, the 'Great Castrator', whose progress I have followed in my book *Travelling Heroes*.[55] If Aineias was at Messina when Herakleidas exchanged these messages, he could see the only wooden μόσσυνες attested outside the Mossynoikoi. He knew the prototypes: did he first use the word to describe Messina's similar battlements of wood?

More solidly, I emphasize the dating. Herakleidas' and Dionysios' 'secret messages' belonged in 355 BC. When Aineias wrote, these messages were fresh for him, the ones he could best cite. In a passing reference in his big commentary on Arrian, Bosworth cited Aineias 11.3, the magistrate's betrayal of Chios, and connected it with the Karian Idrieus' takeover of the island, which is generally dated between 351 and 346.[56] The anecdote needs no Karian behind it, and mentions no Karian in it, and even if such a takeover occurred, it is not Aineias' reference. His text does not belong so late. Nothing in it belongs later than 355 BC, the date of the secret messages, so I put the work c. 354/3 BC. This dating helps to explain two important tenses which it contains.

At 5.2, Aineias tells us how Leukon, *tyrannos* of the Bosporos (he is not called *archon*, as in Athenian decrees) used to sack any of his guards who fell into debt through gambling and so forth. Hunter remarks 'the use of the imperfect need not cause us any difficulty', but on each reading I find it very difficult if Leukon was still alive and well.[57] Diodoros 14.93 gives Leukon 'forty years'' rule, under 394/3, but the numbers of years for the Bosporan *tyrannoi* are notoriously problematic, even if they derive ultimately from Diodoros' chronographic source. In 1982, Tuplin remarked that 'prolonged consideration encourages doubts as to whether anything valuable can be said to emerge from any part of it', Diodoros' catalogue.[58] But Diodoros 16.37.6 is another matter. It synchronizes the fall of Methone (354 BC), Dion's death (autumn 354) and

54 Plut. *Dion* 48.7.

55 Callim. *Aetia* fr. 43 line 68; Lane Fox (2008) 286–288.

56 Bosworth (1980) 178.

57 Ain. Tac. 5.2.

58 Tuplin (1982) 126.

Leukon's death 'after forty years'. For an Aineias, writing in 354/3, Leukon was indeed past history, and the imperfect tense refers to him with the right nuance.

Iphikrates, too, attracts an imperfect tense at 24.16: 'Iphikrates used to order (ἐκέλευε) that the patroller and the sentry should not have the same password'. Ἐκέλευε, but when? When Aineias served with him personally in the east Aegean? Or during Iphikrates' career, now ended? Iphikrates' career ended in 355/4 BC, and he died soon after, so the latter is accommodated if the text was written in 354/3 BC.[59]

Where did Aineias write? Was he still in the West, even, perhaps, with Dionysios after Dion's death in 354? He is not, I think, writing for a Western public or showing any wider familiarity with Sicily itself. He refers at 10.22 to a 'city named Himera' (πόλιν ὄνομα Ἱμέραν), as if Himera was unfamiliar both to readers and to himself. It may reassure my own readers that I can see no answer to his whereabouts when writing: anywhere from south Italy to Asia Minor is possible.

What my itinerary for Aineias of Stymphalos offers instead is a little on why he wrote as he did. He liked to categorize and subdivide and use cumbersome classificatory adjectives, ποριστικός, παρασκευαστικός and so forth.[60] Back in 424 BC Aristophanes had brilliantly lampooned the trendies in contemporary Athens who liked to use adjectives ending in -ικός: they should all be packed off, he proposed, to go hunting.[61] Seventy years later, Aineias was still an -ικός-man, grinding on with a sort of theoretical subdividing which had been fashionable in Athens a lifetime earlier. Had it survived in him as a military habit of mind? The nearest ancient comparison with him and his treatises are Sex. Julius Frontinus' *Strategemata*, especially Book 3. There, Frontinus begins by setting out the 'species' of stratagems in siege-warfare. He solemnly lists eleven distinct 'species' concerning 'besieging' and then seven more 'species' 'concerning protection of the besieged'.[62] The latter are the nearest parallels in antiquity to Aineias' text. Like Aineias, Frontinus was a proven military commander, Gaul his Arkadia, Wales his Asia Minor and Germany (perhaps) his Sicily.[63] Introductory subdividing appealed to his tidy mind.

Unlike Frontinus, Aineias generalized from his experiences and addressed a generic *polis* as an ideal type. He did so, I believe, on the strength of his far-flung experience, just as Aristotle was to generalize about the nature of the *polis*

59 Nepos, *Iphicr.* 3.3 and Kirchner (1901) 514, under no. 7737 Iphikrates.

60 Ain. Tac. 7.4, 14.2, 21.2; note also ἐνεδρευτικός at 1.2.

61 Aristoph. *Equ.* 1375–1383.

62 Front, *Strat.* 3 pref.; compare his stated aims in 1 pref.

63 Richmond (1967) 206–207 for Frontinus' career and connections.

on the strength of his collection of constitutions of 158 different Greek states. Military services had made Aineias a 'global Greek', from Arkadia to Pontus to Syracuse. The experience inclined him to generalize. He is followed by other such 'global Greeks', the Phokian mercenaries with Phalaikos who fought in central Greece, withdrew to the Peloponnese, planned to serve in south Italy and then went off to bother Crete[64] and above all, Demaratos of Corinth, who was to serve Corinth, Timoleon in Sicily, Philip in Macedon and Alexander as far as Susa and beyond, dying in Bactria some twenty-six years after Aineias wrote.[65]

There was one unfortunate fact. What Aineias wrote in 354/3 paid no attention to the major event of that year, Philip's capture by siege of Methone after his capture by siege of Amphipolis (357) and Poteidaia (356). Just as Xenophon's text on cavalry-service took no account of the new 'shock and awe' of the Macedonian cavalry-charging, evident from 359 onwards, so Aineias, on one major issue, was out of date as soon as he appeared. He ignored the news from the North, a zone which none of his anecdotes addresses, and showed no awareness of the rising importance of machines and artillery with which Philip and Alexander transformed the art of besieging.[66] Yet machines were only ever one part of the story. To read Plutarch's *Aratos* (based on Aratos' memoirs) is still to engage with a world in which tricks, not artillery, took cities in ways which Aineias had understood.[67] When Frontinus set out to discuss sieges, he too saw no point in discussing machines, because their 'invention has long since been completed' and he could see no further scope for the subject. What he thought useful, instead, was to list tricks and stratagems, just like Aineias, which had lost none of their relevance by AD 100 and about which people still needed to know more.[68]

Like Frontinus, Aineias omitted machines, but continued to be read in antiquity. One other omission was well advised. Like Thucydides, Aineias 'regards μάντεις, diviners, as particularly troublesome people.'[69] 'Evidence of any belief in the higher powers is conspicuously absent'.[70] Unlike Xenophon, but like Hieronymos (as seen through Diodoros),[71] Aineias knew through widespread

64 Diod. Sic. 16.61; Parke (1933) 141–142.

65 Berve (1926) no. 253.

66 Marsden (1977) 211–223.

67 Plut. *Aratos* 5–9 and 17.

68 Front *Strat.* 3 pref.

69 Hunter and Handford (1927) xxxiv.

70 Hunter and Handford (1927) xxxiv.

71 Hornblower (1981) 35; cf. Diod. 19.55.8.

military experience that the gods are best left out of practical instructions on how to survive in war.

Bibliography

Bengtson, H. (1962) 'Die griechische Polis bei Aeneas Tacticus', *Historia* 11: 458–468.

Berve, H. (1926) *Das Alexanderreich auf prosopographischer Grundlage*. Beck, Munich.

Bosworth, A.B. (1980) *A Historical Commentary on Arrian's History of Alexander*, vol. I. Clarendon Press, Oxford.

Cadbury, H.J. (1927) *The Making of Luke–Acts*. Macmillan, New York.

Cartledge, P. (1987) *Agesilaos and the Crisis of Sparta*. Duckworth, London.

David, E. (1986) 'Aeneas Tacticus 11.7–10 and the Argive Revolution of 370 BC', *American Journal of Philology* 107: 343–349.

Debidour, M. (2006) 'Le secret et les messages secrets dans la Poliorcétique d'Énée le Tacticien', in Olivier, H., Giovannelli-Jouanna, P. and Bérard, F. (eds), *Ruses, secrets et mensonges chez les historiens grecs et latins*. Université Jean Moulin Lyon 3, Lyon. 213–241.

Fields, N. (2001) 'Et Ex Arcadia Ego', *Ancient History Bulletin* 15: 102–130.

Finley, M.I. (1973) *The Ancient Economy*. University of California Press, Berkeley and Los Angeles.

French, D.H. (ed.) (2004) *The Inscriptions of Sinope. Part I*. R. Habelt, Bonn.

Fuks, A. (1968) 'Redistribution of Land and Houses in Syracuse in 356 BC and its Ideological Aspects', *Classical Quarterly* 18: 207–233.

Greenhalgh, P.A.L. (1973) *Early Greek Warfare: Horsemen and Chariots in the Homeric and Archaic Ages*. Cambridge University Press, Cambridge.

Griffin, A. (1982) *Sikyon*. Clarendon Press, Oxford.

Hornblower, J. (1981) *Hieronymus of Cardia*. Oxford University Press, Oxford.

Hornblower, S. (2004) *Thucydides and Pindar*. Oxford University Press, Oxford.

Hornblower, S. (2015) *Lykophron: Alexandra*. Oxford University Press, Oxford.

Hunter, L.W. and Handford, S.A. (1927), *Αἰνείου Πολιορκητικά. Aineias on Siegecraft*. Clarendon Press, Oxford.

Hutchinson, G.O. (2001) *Greek Lyric Poetry*. Oxford University Press, Oxford.

Kirchner, J. (1901) *Prosopographia Attica*, vol. I. Georg Reimer, Berlin.

Knoepfler, D. (2002) 'Loi d'Érétrie contre la tyrannie et l'oligarchie', *Bulletin de Correspondance Hellénique* 126: 149–204.

Lane Fox, R. (2008) *Travelling Heroes*. Allen Lane, London.

Marsden, E.W. (1977), 'Macedonian Military Machinery and its Designers under Philip and Alexander', *Archaia Makedonia* 2: 211–223.

Meineke, A. (1867) 'Coniectanea in Aenean Tacticum', *Hermes* 2: 174–190.

Ogilvie, R.M. and Richmond, I. (eds) (1967) *Tacitus, Agricola*. Clarendon Press, Oxford.
Osborne, R. (1987) *Classical Landscape with Figures*. George Philip, London.
Parke, H.W. (1933) *Greek Mercenary Soldiers*. Clarendon Press, Oxford.
Roy, J. (2000) 'Problems of Democracy in the Arcadian Confederacy, 370–362 BC', in R. Brock, S. Hodkinson (eds) *Alternatives to Athens*. Oxford University Press, Oxford: 308–326.
Roy, J. (2004), 'The Ambitions of a Mercenary', in Lane Fox, R. (Ed.) (2004) *The Long March. Xenophon and the Ten Thousand*. Yale University Press, New Haven and London: 264–288.
de Sainte Croix, G.E.M. (1981) *The Class Struggle in the Ancient Greek World*. London
Sekunda, N. (1988) 'Some Notes on the Life of Datames', *Iran* 26: 35–53.
Tuplin, C. (1982) 'Satyros and Athens: *IG* ii^2 212 and Isokrates 17. 57', *Zeitschrift für Papyrologie und Epigraphik* 49: 121–128.
Villing, A. (2002) 'For Whom the Bell Tolls in Ancient Greece. Archaic and classical Greek bells at Sparta and beyond', *Annual of the British School at Athens* 97: 223–295.
Whitehead, D. (1990) *Aineias the Tactician, How to Survive under Siege. Translated with introduction and commentary*. Clarendon Ancient History Series, Oxford.
Wilhelm, A. (1974) *Akademieschriften zur griechischen Inschriftenkunde* II. Zentralantiquariat der Deutschen Demokratischen Republik, Leipzig.
Winter, F.E. (1971), *Greek Fortifications*. Routledge, London.

CHAPTER 4

Aineias Tacticus in His Intellectual Context*

D. Graham J. Shipley

Much effort has been expended in the attempt to identify the author of the late classical treatise known as *Poliorketika*, generally by deduction from the meagre evidence in the text itself. As far as this goes, the likeliest candidate remains the younger Aineias of Stymphalos, militarily active in the 360s.[1] This, remains unconfirmed, however, as the range of views in this volume shows. Accordingly, I sometimes refer to our author as 'Aineias' (in inverted commas). Whoever the author was, the work seems likely to belong to the mid- or late 350s.[2]

Aside from questions of name and date, what persona does 'Aineias' convey through his text? What identity, or identities, does he intentionally assert or unintentionally reveal? In the light of this, what can we further ascertain about the cultural and intellectual context within which the work was composed, and about the author's aims? To what extent, indeed, were those aims military, scientific, political—or even rhetorical and literary? As we shall see, recent advances in the understanding of the *polis*,[3] of Arkadian politics and identities,[4] and of the nature of oligarchy[5] allow us to fine-tune our appreciation of his self-presentation.

---

* I thank Maria Pretzler for inviting me to contribute to the 2010 conference *in absentia* and to this volume, as well as for copies of publications and for editorial guidance; the anonymous referee for percipient suggestions; and the editors and other contributors for sight of their drafts. My paper has its genesis in the interchange of ideas about Aineias Tacticus with students at the University of Leicester, especially the members of my department's Greek Club and of successive source study groups. On specific points I thank Philip de Souza, Peter Rhodes, and Aimee Schofield for useful advice. Particular thanks for helpful comments are also due to Tony Barratt, Séraphine Harman, and Sarah Scott. The invaluable assistance of the library staff at the University of Leicester and the Institute of Classical Studies, London, is, as ever, appreciated. *Special abbreviation*: A. = Aineias Tacticus.

1 Whitehead (2002) 10–13. For the elder Aineias, see Xen. *Anab*. 4. 7. 13.

2 Whitehead (2002) 8–10.

3 See esp. Hansen and Nielsen (2004) and the many preceding volumes from the Copenhagen Polis Centre (main ideas summarized by Shipley 2003; Hansen 2006).

4 e.g. Nielsen (1999); Roy (2000); Nielsen (2002); Nielsen (2004a); Pretzler (2009).

5 See esp. Ostwald (2000a), also Ostwald (2000b).

 | DOI: 10.1163/9789004352858_005

Ethnic and Political Identities

At the broadest scale, 'Aineias' writes as a Hellene. The observation may seem trivially true; but it does help define the work, for it is noticeable that—despite being familiar at some level with the work of Herodotos[6]—he draws only a tiny number of historical illustrations from non-Greek history; the story of Glous (31. 35) seems the only straightforward case. His focus is thus almost exclusively on the Greek world, and he has little interest in warfare and other violence in other places. I say 'warfare' rather than 'siege warfare', because examples of actual sieges in the text are few: in the final part of the work (chapters 38–40), specifically devoted to what happens once a siege is under way,[7] he uses only three historical examples as against 61 in the preceding 31 chapters.[8] He scarcely even takes any examples from conflicts involving both Hellenes and non-Hellenes. 'Barbarians' as a general category are once, perhaps, envisaged as possible enemies (24. 17, on secret signals), while in specific terms we hear only of Abdera's encounter with the Triballians (15. 8–10), a Spartan governor in Thrace (27. 7),[9] Timoxenos and Artabazos (31. 25–27), Histiaios's ruse in the Ionian revolt (31. 28), the siege of Barke by Amasis (37. 6–7), and Sinope's conflict with Datames (40. 4–5). Other Greek communities with barbarian neighbours are discussed without their enemies being named.[10] In discussing passwords he imagines an army of 'mixed' men from a variety of '*poleis* or *ethne*' (24. 1); whether the latter are conceived of as Greek or non-Greek, he expects all the soldiers to speak Greek (24. 1–2 and 15–16). While his focus is strongly Hellenocentric, however, the allusion in the preface (§ 4) to certain Hellenes who have suffered calamity and subsequently recovered is a striking, perhaps unique, refocalization (to which we shall return presently); a momentary stepping outside the Greek world; perhaps an attempt to universalize his observations to the whole of that world.[11]

---

6 For A.'s use of Herodotos, see 31. 9 with Brown (1981) and Whitehead (2002) 38 n. 114 (with references), 185–186; see also A. 31. 14 and 25–29, 37. 6–7.

7 I follow Whitehead (2002) 20, who identifies a tripartite structure in the work.

8 See 37. 6–7, 40. 2–3 and 4–5. At 39. 6 he refers to a technique that has been used, but seems to have no specific incident in mind. On A.'s illustrative stories, see Pretzler, this volume.

9 Probably an 'area harmost' rather than in a Greek *polis* (Whitehead 2002, 176).

10 Bosporan kingdom, 5. 2; Herakleia, 11. 10a and 12. 5; Apollonia, 20. 4. Kyrene and Barke, 16. 14.

11 *Contra* Hunter and Handford (1927), xxv–xxvi, 104. the work is unlikely to have been intended for a single commissioning patron. See Bettalli (1990) vii, 15–16, revising the view he stated at Bettalli (1986) and now concurring with Whitehead (2002), 39–40, as to the

To what extent, if at all, does he write as a Peloponnesian,[12] an Arkadian, or indeed a citizen of any particular *polis*, be it Stymphalos or another? Only once does he hint at local knowledge, when he identifies *paneia* ('panic') as a Peloponnesian and especially Arkadian word (27. 1).[13] This supports the notion that he himself was from the Peloponnese, perhaps from Arkadia; but he is unlikely to have been writing for a purely Arkadian audience, who would have known that the term *paneia* was characteristically their own. Since Arkadia is landlocked,[14] the point about a wider intended readership is supported by the references to coastal locations (e.g. 16. 13, a coastal road; 16. 21, a fleet; 17. 1, the docking of ships; 29. 12, dockyards) and by the fact that the closing lines of the text as we have it lead up to a discussion of naval matters.[15] Likewise, recommendations such as that about removing domestic animals to a neighbouring *chora* (10. 1–2) presume a continuous landscape divided into *poleis*. His theoretical landscape may be a re-imagined Peloponnese, but it could equally be in central Greece, western Asia Minor, Sicily, or Magna Graecia; or could be a sort of amalgam of these, for in none of them is the typical *polis* far from its neighbours. (None of this, I think, bears upon the question of his own civic affiliation, which must be judged on other grounds.) He frames his prescriptions and observations so as to give them validity for the Greek world in the widest sense. Only in respect of islands is his vision somewhat impoverished: he never uses the word 'island', he names few island *poleis*,[16] and when he does he never notes that that is what they are.

Those possible identities, therefore, are recessive in the work. 'Aineias' does not even name the exemplary people or peoples who were restored to viability after calamity (above): they could as well be Athenians as Spartans or others.[17] Clearly he wants his work to have general applicability across the Greek world,[18] and therefore chooses to veil his allegiances almost completely. But

universalizing character of the work. A. envisages a range of possible circumstances and responses.

12 On Peloponnesian identity and whether it was a significant concept, see Vlassopoulos n.d. [2008]. Whitehead (2002), 39, considers the possibility that, if the author is not Aineias of Stymphalos, he is at least Peloponnesian.

13 See Whitehead (2002) 10–11.

14 Leaving aside coastal Lepreon in Triphylia, incorporated into Arkadia in the 4th cent.: Ps.-Skylax 44 with Shipley 2011 ad loc.; Nielsen (2004b), 543–544.

15 See Whitehead (2002), 24 and 114, for A.'s assumption that the sea is close by.

16 Only Chios (11. 3; 17. 5), Kerkyra (11. 13–15), Aigina (20. 5), Naxos (22. 20, if the text is correct: see Whitehead (2002), ad loc.), and Mytilene (31. 34).

17 Hunter and Handford (1927) 102 (Sparta); Bettalli (1990) 212–213 (Athens).

18 At other points (e.g. 23. 3–5 and 7–11) editors have thought that he conceals the name of

his Greek world is a kind of finite universe with almost nothing outside it. To put it another way, what lies outside it is almost of no interest to him, and his prescriptions (written, of course, in Greek) are framed for use by Greeks alone—not humankind in its entirety, as one could say of Thucydides. This differs, too, from Aristotle in the *Politics*, whose arguments explicitly depend on the differences between Hellenes and others (1. 1, women and barbarians as natural slaves; 1. 9, the barter economy of barbarians; the whole notion of the *polis* as *telos*, e.g. 1. 2), and whose field of view includes Carthage (e.g. 2. 11); and in the opposite sense from Xenophon in his treatise *Hipparchikos* (*Cavalry Commander* or *Mag. eq.*), which mentions only Attic places (1. 26, 3. 1–14 *passim*, 7. 1, 7. 4), but is in some other respects comparable to that of 'Aineias'.

More importantly, 'Aineias' is a man of the *polis*. He mentions a great many Greek city-states,[19] and seems to assume that warfare will involve a city on one side or both; a city, indeed, organized in the characteristic manner of a Greek *polis*. Its people is divided into political tribes (*phylai*, e.g. 3. 2). It has an *agora* where they gather (e.g. 1. 9, 3. 5, 10. 15), as well as civic offices (*archeia*, 22. 2 and 4), a *prytaneion* (10. 4), magistrates (e.g. 1. 4, 4. 6), a council (10. 4), a theatre (1. 9, 3. 5, 22. 4), and so on. The population is based in the urban centre, not in the countryside.[20] The distinction between principal streets (*diodoi*) and alleys (*rhymai*, 2. 5) is consistent with, though it does not entail, the kind of grid-plan layout that was typical of Greek towns founded or rebuilt in the archaic and classical periods, and was indeed adopted at Stymphalos in the second quarter of the fourth century.[21]

Does 'Aineias' prefer, or identify with, a particular kind of *polis*? Previous detailed investigations have suggested that he offers no grounds for inferring that he favours a particular political system such as democracy or oligarchy;[22] rather, he is a defender of the *status quo*. In the light of recent studies of the classical *polis*, however, as well as of the nature of Greek oligarchy and the political history of the Peloponnese, we can nuance this view somewhat. It is true that his recommended managerial techniques include several that are nowadays thought to have been characteristic of democracies, such as selection by drawing lots (3. 1 and 6, 18. 1, 20. 2). He remarks that rich men should be selected for certain roles (5. 1, 22. 15), that the well-off should subsidize the city's

---

a Peloponnesian community for reasons of discretion or politics (Hunter and Handford 1927, xvii, on 29. 3–10); but see Whitehead (2002) 180.

19 For his importance to the study of the *polis*, see Hansen (2007).

20 Bettalli (1990) 17.

21 Shipley (2005b) 361.

22 See esp. Bettalli (1990) 19–20; Whitehead (2002) 29–32; Pretzler, this volume, 153–157.

military needs (13. 1–4), and that one should make an example of them if they fall short in performing their duty (38. 5). At first sight these provisions have a democratic flavour. The lot was the cornerstone of the Athenian constitution, a way of ensuring that a large sample of citizens exercised civic power in turn. At Athens the same selectivity towards the rich was shown in filling key positions such as that of treasurer (*tamias*, e.g. Arist. *Ath. pol.* 43.1). A man with a great deal at stake in the economy, one who had more to lose, was thought more reliable than a poor man who might be open to bribery.

But the lot was almost certainly a time-hallowed Greek practice and was also consistent with oligarchy, the commonest constitutional form in the Peloponnese,[23] at least before the 360s. The crucial difference in a democracy was that the lot was used to select from among all eligible citizens, not from a limited number.[24] Again, oligarchies, no less than democracies, deployed members of the elite in key positions.[25] In fact, 'Aineias' envisages that the financial contributions of the rich will be refunded in the fullness of time (13. 1–4), which was certainly not the case with liturgies at Athens, or citizen benefactions elsewhere. He even alludes to his own discussion, in another book, of how to adjust the terms of debts and interest rates 'equitably and painlessly for the rich' (14. 2). From the last two illustrations we may deduce that 'Aineias', while not actively anti-democratic, is not a radical democrat of Athenian type.[26] In this connection it may be recalling that, while the historical Aineias grew up in an oligarchic world under Spartan domination, the Arkadian league after 369 promoted democracy,[27] so that his role as a commander was precisely to sustain democracies (see below on his actions at Sikyon).

Other evidence of the writer's political leanings might be sought among the seemingly illiberal steps that he sometimes recommends, especially in chapter 10. Here and elsewhere he advises the commander to issue proclamations on many matters, imposing tight controls upon citizens and foreigners (including impoverished ones, *talaipeirioi*, 10. 10),[28] limitations on free speech, and

23 Gehrke (1986) *passim*; cf. Shipley (2005a) 325–326.

24 Headlam-Morley (1933) 78–87, esp. 80 and n. 1, citing Schömann (1861) 154, who in turn cites Anaximenes, *Rhetorica ad Alexandrum* (4th c. BC), 2. 14; but the key passage is 2. 18. 1424a40–b2, where it is explicitly said that in *oligarchies* most magistrates should appointed by lot.

25 Bettalli (1990) 19–20.

26 Whitehead (2002) 31, rightly castigates Hunter and Handford (1927) 233, for protesting that A. would victimize the well-off.

27 Roy (2000); Nielsen (2002) 485–490.

28 Unless the word simply means 'without lodgings': Whitehead (2002) 122.

restrictions on the use of public spaces; there are even 'street commanders' (*rhymarchai*, 3. 4–5). Cattle are to be removed to the next *polis*, if necessary (10. 1). We may wonder what kind of military officer or civic official could, in reality, legally close inns (10. 10), set a premium on imports (10. 12), announce rewards to be paid presumably by the *polis* (10. 16), control the movements of hostages' relatives (10. 23), abolish interest rates and loans (14. 1), and compel the mouths of domestic animals to be cauterized so as to silence them (23. 2). These measures have been seen as totalitarian and troublingly anti-democratic.[29] They may also be purely fanciful.

In some of these cases, however, 'Aineias' may be tacitly assuming that normal decision-making processes will be followed: the commander persuading the council and people to adopt the measures (such as selecting as ambassador a man he wants out of the way, 10. 20).[30] In at least one instance he refers to what must surely be a longer-term strategy on the part of the *polis*, not to hire too many mercenaries (12. 4). He gives no hint of whether such decision-making processes are democratic or oligarchic; either would involve council and assembly.[31] On the other hand, in some cases it seems clear that he assumes that the military commander will have executive authority over non-soldiers.

Even tyranny, though our author's focalizing language generally distances us from it, is not depicted pejoratively. Tyrants, like other governments, may adopt clever ruses (5. 2), though they can also make mistakes (31. 33). Some military techniques are recommended specifically where one is defending 'a tyrant's citadel' (22. 19).[32] Tyranny is simply another kind of regime (10. 11), albeit one that may be brought about by the failure on the part of a *polis* to take the right steps to protect itself (12. 5). Is our author's formal neutrality towards tyranny evidence of a preference on his part for autocratic over populist politics, or only of a desire to universalize his recommendations about the conduct of war? The latter seems likely, in view of what we have already seen; so it may be only through incompleteness of coverage that tyranny is the only form of government that he presents as 'other' (apart from, perhaps, Persian kingship).

If the author is the younger Aineias of Stymphalos, the notice he takes of tyrannies may relate to his own military intervention at Sikyon against the tyrant Euphron (Xen. *Hell.* 7. 3. 1),[33] when he enlisted the support of the

29 See Jenkins (2006) 51–59; also Whitehead (2002) 40–41.

30 On this passage, see Whitehead (2002) 27 and n. 81.

31 Ostwald (2000a) *passim.*

32 See Whitehead (2002) 157, on this passage, with his p. 32 and n. 92.

33 Here I take a slightly different line from Whitehead (2002) 157 (on 22. 19).

'leading men' (called *kratistoi* and *beltistoi* by Xenophon, thus not necessarily hereditary aristocrats) who had been expelled without a *dogma*, a decree of the people. The mobilization of influential men does not necessarily mean Aineias was not acting to support democracy. Indeed, the situation at Sikyon was unusually complicated, involving a *demos* seemingly devoted to a leader who had become tyrannical, but also rich exiles who were pro-Spartan rather than pro-Arkadian.[34] In any case, the episode tells us nothing directly of Aineias's personal preferences: he was a general of the Arkadian league with a job to do. We must simply assume he would not have been elected *strategos* if there had been any doubt about his commitment to democracy, Arkadian style.

Thus far we have seen that, although it is possible to detect slight differences between the author's vision of a well-run *polis* and the fullest form of democracy, as well as some hints of an antipathy towards tyranny, for the most part he scrupulously veils his own political preferences. Although he cannot be said to be either a democrat or an oligarch in a simple sense, in his writing he is an instinctive defender of Hellenism and an active defender of the constitutional *polis*. We may take this neutral pose as reflecting his generalizing impulses, his efforts to map out his subject.

There remains the question whether the author's more dictatorial recommendations could be applied in real life.[35] Is he writing theoretically, even impractically, advocating measures that citizens of no actual *polis* would contemplate even in a crisis? Were such steps seen as incompatible with the values of a citizen-state? Were they actually taken, or at least contemplated, in the classical period? And, if so, were they taken by generals without reference to other authorities?

Sources for classical history offer little support for a positive answer to the last two questions. Athenian generals could call an assembly (e.g. Thuc. 2. 22. 1; 2. 59. 3), but we have no evidence that they had authority to micro-manage the city.[36] When Spartan generals, for example, summon assemblies (Brasidas at Torone and Skione, Thuc. 4. 114 and 120; Gylippos at Syracuse, Thuc. 7. 21), it is quite possible that they have gone through the usual channels, ask-

34 See Roy (2000) 323–324, for the convoluted nature of Sikyonian politics.

35 At 26. 12 he appears refers 'the *politarchos*' (πολίταρχος) checking on the guards. The word is unparalleled; Whitehead (2002) 174, leaves it open whether a military or civil office is meant, but the etymological meaning should be 'commander of the *citizens*' (πολῖται), presumably citizen *soldiers*; the word may simply be a synonym for *strategos*, implying nothing about either civil magistrates or generals ordering non-soldiers about.

36 On the limited implication of their power to call assemblies, see Hamel (1998) 6–12.

ing the magistrates to convene the meeting. In such cases—and indeed in the circumstances set forth by 'Aineias'—we may be dealing with informal, *de facto* authority rather than the exercise of defined powers. Brasidas probably behaved autocratically in various ways when defending Sparta's allies in northern Greece against the Athenians; but if one examines his individual decisions one sees that they were taken in pursuit of military objectives and/or with the implied consent of the community, such as when he evacuated citizens' wives and children to safety (Thuc. 4. 123).

Nor do Greek *strategoi* of the fifth and fourth centuries, when present in an allied *polis* to provide military leadership, seem to have thrown their weight around in the way that Aineias envisages. Neither is it clear that when a man was appointed *strategos autokrator*, 'self-empowering general', he had authority to govern by decree. We possess only vague indications of the powers of such men.[37] There is no sign, for example, that Dionysios of Syracuse (appointed στρατηγὸς αὐτοκράτωρ at Diod. 13. 95. 1) exercised the kind of authority over civilians that Aineias describes.

Some of the treatise's thoroughgoing prescriptions recall indirectly the improvisations of classical commanders in the field rather than inside a *polis*. For example, stratagems attributed by Polyainos (writing in the second century AD) to the fourth-century BC Athenian commander Timotheos include melting down coins in order to strike new ones (3. 10. 14), normally the prerogative of a *polis* or ruler. But nothing in Polyainos's copious stories of Iphikrates, Timotheos, and Chabrias (3. 9–11) matches our author's prescriptions for the internal management of a *polis*. Rather, it seems correct to say that he is 'studiously vague' about the formal position of his imagined commander.[38] It begins to look as though, even if his recommendations may be both practical in principle and grounded in experience, they do not conform to what we know of late classical warfare and might have required a new kind of military organization. It is true that we have little evidence for the conduct of generalship in the typical middling and small Greek *poleis*; but if real commanders' powers were not so autocratic as those Aineias portrays, we may choose to regard his work in this respect as theoretical, not only practical.

---

37 For the term and its limited implications, see Hamel (1998) 201–204.

38 Whitehead (2002) 40–41.

Literary and Intellectual Context

Our author, whether Arkadian or not, is writing in a very specific historical and political–military context: one that influences his selection and presentation. Can we go further and identify the 'research context' in which he is working? Is he simply generating his ideas from his own knowledge and what he has personally read or heard, and then elaborating them? Or does he mean us (as I shall argue) to see his work as situated in a wider tradition of reflection and theorization?

On one level his response to his predecessors is to use them as sources of information. It has long been known, for example, that he retails material from Herodotos[39] and less often Thucydides,[40] though he does not name them and often adds important details to their accounts.[41] Awareness of Homer, Xenophon, and other writers is confirmed by examination of his vocabulary,[42] though it is less certain that he uses Xenophon's *Anabasis*[43] or Ephoros.[44]

On a deeper level, it is uncertain whether he follows earlier writers as models for organizing his work. We are told that the sophists had written 'many' books (Xen. *Cyn.* 13. 1–8 *passim*): among them Simon of Athens, whose work on horsemanship survives in part and who is cited by Xenophon in *Hippike* or *Horsemanship* (1. 1 and 3; 11. 6).[45] As a general model for a mid-fourth-century

39 Cases listed at Whitehead (2002) 38 n. 114.

40 Thuc. known to A.: Hornblower (1995) 53. A. uses Thuc. at 2. 3–6 for Plataea (cf. T. 2. 2–6) and at 38. 2 for Brasidas (T. 5. 9. 8, also the view of Hornblower (1996) 444; Whitehead (2002) 201 ad loc.; cf. Bettalli (1990) 330 for 38. 1 τὸ γὰρ ἐπιόν as a possible verbal echo of T.). Referring to the book as a whole, Hunter and Handford (1927) lxxxi, convict A. of having 'few of the virtues and many of the vices of a conscious literary style', both the virtues and the vices reflecting admiration of Thuc.; while certain phrases (e.g. at 10. 21, 16. 4) 'have an unmistakably Thucydidean ring' (lxxxi n. 1). Yet they resist the notion that A. consciously imitates Thuc. (lxxi).—For 4th-cent. knowledge of T., see generally Hornblower (1995). Cf. also Marchant and Bowersock (1968), 469, for T.'s influence on the 'Old Oligarch'; while Bowersock dates O.O. to the mid-5th cent., Marr and Rhodes (2008), 3–6 with 31–32, date him 431–424; Hornblower (2000) to the 4th cent.

41 Brown (1981), 366, 388; (Burliga) 2009, 101.

42 Hunter and Handford (1927), lii–lxiv; cf. (Bettalli) 1990, 10 n. 25.

43 Bettalli (1990), l.c. and 295 on 27. 11, leaving open the possibility that A. is citing Xen. inaccurately from memory. Cf. however the use of μόσυνες at 33. 3, a word specific to the Mossynoikoi of the E. Black Sea (cf. Ps.-Skylax 85–86) previously used only by Xen. *Anab.* 5. 4. 26 (with σσ).

44 Bettalli (1990) 10 n. 25, *contra* Bengtson (1962) 460.

45 Text in Widdra (1964) 41–44.

man embarking on the writing of a treatise, one could also invoke Plato's earlier dialogues, including the *Laches*, an inconclusive discussion of whether young men can be taught courage. Indeed, it would be unwise to underestimate the impact of the sophists, and of Plato's Academy, upon political elites across Greece. A literate man in a fairly important *polis* outside Athens, no doubt from a prosperous family, will surely have known about the philosophical innovations occurring at Athens, and may even have spent time there—or at least read Plato.[46] Indeed, we know that a man of Stymphalos, one Sophainetos (*FGrHist* 109), wrote about the expedition of Cyrus;[47] like Xenophon, he must have known the elder Aineias of Stymphalos, who died on that march (Xen. *Anab.* 4. 7. 13) and whose son or nephew may be our author. What we do not have is any evidence that any writer before 'Aineias' composed a technical treatise on warfare; and previous studies have pointed to him as the originator of the model,[48] a 'pioneer'.[49]

There can be little doubt that 'Aineias', writing in the mid- or later 350s,[50] responds in certain respects to the work of his contemporary (and possible family friend) Xenophon.[51] The latter's *Hipparchikos* (above) is earlier than the *Hippike*, which cites it (12. 14); both are earlier than the *Poliorketika*.[52] Xenophon's *Kynegetikos* or *On Hunting* (*Cyn.*) is earlier still.[53] The relationship of our text to Xenophon's *Poroi* or *Resources* (355 BC or shortly after) is less clear, since they were written around the same date. Be that as it may, the two writers could easily have corresponded or met each other, even if our author was not

46 Cf. the outline suggestion of Slater (2010) that A. may have been connected with current philosophical trends.

47 Whitehead (2002) 37 n. 111.

48 For earlier Sophistic discussions of the soldier's art as practical and orally based rather than theorized in writing, see Bettalli (1990) 14–15. A. himself sometimes hints at earlier areas of (oral?) debate (2.7; 22. 13; 24. 16; 26. 6 and 12; 27. 1): see Whitehead (2002) 36–37.

49 Bettalli (1990) 13–15, against the idea that A. is following a predecessor who wrote *about this topic*; similarly Whitehead (2002) 34–35 and n. 101.

50 Whitehead (2002) 8–10.

51 Whitehead (2002) 36, suggests they exchanged ideas.

52 Marchant and Bowersock (1968), xxviii, says *Hipp.* is 365; p. xxxii, *Peri Hipp.* soon after. Anderson (1974) 184, dates *Peri Hipp.* to Xen.'s old age, and *Hipp.* (which *Peri Hipp.* cites at 12. 14) 'some years' earlier. Bettalli (1990) 15: they are closely contemporaneous. Bettalli (1990), 14 n. 48, cites Delebecque (1973) (*non vidi*) as revising his earlier view, set out in Delebecque (1957) 430 and 444 (*non vidi*), that Aineias's work was earlier than *Hipp.*

53 Marchant (in Marchant and Bowersock 1968, xl) dates it early in Xen.'s *oeuvre*; Anderson (1985) 161–162 n. 4, suggests a date after Xen. moved to Skillous, perhaps post-387.

the historical Aineias of Stymphalos. If he was Aineias, of course, it is even likelier that they knew each other, because Xenophon knew the elder Aineias and because, between about the late 390s and the early 360s, Xenophon lived at Skillous near Olympia—perhaps a long day's ride, no more than two, from Stymphalos.[54] Nor is the window for the direct exchange of ideas limited to the years of his residency there: Xenophon and our author—whether or not he is Aineias of Stymphalos—could have corresponded by letter or met face to face before he left Skillous, or in Athens after his recall from exile.

Xenophon's *Hippike* provided our author with a model for a theoretical preface, if he needed one; but he chose to write a very different opening from Xenophon's pious disquisition on the role of the gods and heroes in the origins of hunting.[55] Closer in time to our work, though possibly written just after it, the *Poroi* offers a closer parallel for an abstract preface, though it states a 'research question' in a way that Aineias does not:[56]

> For my part I have always held that the constitution of a state reflects the character of the leading politicians. But some of the leading men at Athens have stated that they recognize justice as clearly as other men; 'but', they have said, 'owing to the poverty of the masses, we are forced to be somewhat unjust in our treatment of the cities'. This set me thinking whether by any means the citizens might obtain food entirely from their own soil, which would certainly be the fairest way. I felt that, were this so, they would be relieved of their poverty, and also of the suspicion with which they are regarded by the Greek world.
>
> 1. 1, trans. MARCHANT 1925, 193

In other respects, however, the treatises of Xenophon are very different in content from that of 'Aineias'. This is partly because of their different intended readerships: *Hipparchikos* is aimed at Athenian commanders, *Hippike* at unnamed novices in horsemanship; neither cites any historical exempla to support its recommendations. Whitehead points out, however, that what unites 'Aineias' and Xenophon is a belief in the value of practical experience as

54 Arrival at Skillous: by c. 393, Anderson (1974) 165; 392/1 or later, Tuplin (2004) 252. Departure: Xen. recalled from exile c. 369, Anderson (1974) 184 (despite Paus. 5. 6. 4–7, on which see Tuplin (2004) 270–272).

55 Gray (1985) shows that the preface and technical section of *Cyn.*, like the rest of the treatise, are original.

56 On ancient historiographical prefaces see cf. Earl (1972), with Bettalli (1990) 211 ad loc.

opposed to the sophists' over-theorization.[57] Their treatises also reflect the evolution of military thinking in the new geopolitical circumstances of the first half of the fourth century.[58]

Other studies, then, have shown that our author responds to a tradition of reflection upon military practice that was, by the mid-fourth century, being translated out of an oral tradition into a written one; and that in the matter of military writing he is probably a pioneer. In the remainder of this paper I aim to go further: arguing, on the basis of one passage in particular, that he consciously attaches himself to a literary heritage and its rhetorical forms as a way of claiming expert authority and thus promoting the adoption of his ideas.[59] That he does so only briefly is less significant than that he does so at all.

The passage in which our author seems most acutely conscious of his place in a literary culture is the preface, which has not received the attention it deserves. It may be a 'bolt-on', added as a justification of the work after its completion. The first two subsections, in particular, seem to disclose the author's literary awareness, particularly when compared with a speech of Perikles as recalled and re-narrated by Thucydides.

'Aineias', preface

(1) Ὅσοις τῶν ἀνθρώπων ἐκ τῆς αὑτῶν ὁρμωμένοις χώρας ὑπερόριοί τε ἀγῶνες καὶ κίνδυνοι συμβαίνουσιν, ἄν τι σφάλμα γένηται κατὰ γῆν ἢ κατὰ θάλασσαν, ὑπολείπεται τοῖς περιγιγνομένοις αὐτῶν οἰκεία τε χώρα καὶ πόλις καὶ πατρίς, ὥστε οὐκ ἂν ἄρδην πάντες ἀναιρεθείησαν. (2) Τοῖς δὲ ὑπὲρ τῶν μεγίστων μέλλουσι κινδυνεύειν, ἱερῶν καὶ πατρίδος καὶ γονέων καὶ τέκνων καὶ τῶν ἄλλων, οὐκ ἴσος οὐδὲ ὅμοιος ἀγών ἐστιν, ἀλλὰ σωθεῖσι μὲν καὶ καλῶς ἀμυναμένοις τοὺς πολεμίους φοβεροὺς τοῖς ἐναντίοις καὶ δυσεπιθέτους εἰς τὸν λοιπὸν χρόνον εἶναι, κακῶς δὲ προσενεχθεῖσι πρὸς τοὺς κινδύνους οὐδεμία ἐλπὶς σωτηρίας ὑπάρξει.

(3) Τοὺς οὖν ὑπὲρ τοσούτων καὶ τοιούτων μέλλοντας ἀγωνίζεσθαι οὐδεμιᾶς παρασκευῆς καὶ προθυμίας ἐλλιπεῖς εἶναι δεῖ, ἀλλὰ πολλῶν καὶ παντοίων ἔργων πρόνοιαν ἑκτέον, ὅπως διά γε αὑτοὺς μηδὲν φανῶσι σφαλέντες· (4) ἂν δὲ ἄρα τι σύμπτωμα γένηται, ἀλλ' οἵ γε λοιποὶ τὰ ὑπάρχοντα εἰς ταὐτό ποτε καταστήσαιεν ⟨ἄν⟩, καθάπερ τινὲς τῶν Ἑλλήνων εἰς τὸ ἔσχατον ἀφικόμενοι πάλιν ἀνέλαβον ἑαυτούς.

---

57 Whitehead (2002) 34–36, 37, citing Xen. *Mem.* 3. 1. 1ff., Pl. *Euthd.* 273c, and Pl. *La.* for criticisms of sophists.

58 Bettalli (1990) 15.

59 This relates to the practicability issue discussed above; cf. the portrayal of A. as an essentially practical thinker by Whitehead (2002) 35–37.

1. When men leave their country and engage in warfare and encounter perils beyond their own frontiers, and disaster occurs by land or sea, the survivors still have their own country and city and fatherland between them and utter destruction; 2. but for those who have to fight for all that is most dear to them, for temples and fatherland, for parents and children and all they posses, the struggle is of a wholly different kind: a successful and stout resistance to the enemy will make them dreaded by their foes and more secure from future invasion, while any weakness in meeting the peril will leave them no hope for safety.

3. When, therefore, men have to fight for all these precious stakes, they must omit no preparation and no personal effort: rather they must think beforehand of every possible precaution, that the world may never see them beaten through their own fault, 4. and if disaster does come, the survivors may be able later on to make good the loss, like certain of the Hellenes who have been reduced to the direst straits and yet have recovered.

trans. HUNTER 1927, 3

Thucydides, book 2

43. (3) ἀνδρῶν γὰρ ἐπιφανῶν πᾶσα γῆ τάφος, καὶ οὐ στηλῶν μόνον ἐν τῇ οἰκείᾳ σημαίνει ἐπιγραφή, ἀλλὰ καὶ ἐν τῇ μὴ προσηκούσῃ ἄγραφος μνήμη παρ' ἑκάστῳ τῆς γνώμης μᾶλλον ἢ τοῦ ἔργου ἐνδιαιτᾶται. (4) οὓς νῦν ὑμεῖς ζηλώσαντες καὶ τὸ εὔδαιμον τὸ ἐλεύθερον, τὸ δ' ἐλεύθερον τὸ εὔψυχον κρίναντες μὴ περιορᾶσθε τοὺς πολεμικοὺς κινδύνους. (5) οὐ γὰρ οἱ κακοπραγοῦντες δικαιότερον ἀφειδοῖεν ἂν τοῦ βίου, οἷς ἐλπὶς οὐκ ἔστιν ἀγαθοῦ, ἀλλ' οἷς ἡ ἐναντία μεταβολὴ ἐν τῷ ζῆν ἔτι κινδυνεύεται καὶ ἐν οἷς μάλιστα μεγάλα τὰ διαφέροντα, ἤν τι πταίσωσιν. (6) ἀλγεινοτέρα γὰρ ἀνδρί γε φρόνημα ἔχοντι ἡ μετὰ τοῦ μαλακισθῆναι κάκωσις ἢ ὁ μετὰ ῥώμης καὶ κοινῆς ἐλπίδος ἅμα γιγνόμενος ἀναίσθητος θάνατος.

(3) For the whole earth is the grave of distinguished men: they are commemorated not only by the inscription on the tombstone in their own land, but even in foreign territory there lives in every man's heart an unwritten memorial, of their purpose rather than their accomplishment. (4) You now must emulate them, judging that happiness depends on freedom, and freedom on a good spirit, and not looking anxious at the dangers of war. (5) It is not the victims of misfortune, men with no hope of a good outcome, who are most justified in being generous with their lives, but those who risk a great downfall if their life continues, and the greatest reversal of fortune if they fail. (6) For a man of spirit the arrival of misfor-

> tune attended by cowardice is more distressing than a barely perceived death attended by firmness and hope for one's country.
>
> trans. RHODES 1988, 91

Both passages exhort the citizens to show resolve and highlight the motivation of knowing what precious things they stand to lose; though whereas Aineias stresses the importance of defensive warfare at home, Thucydides emphasizes that the sacrifice of those fighting abroad is no less glorious. 'Aineias' and Thucydides use some of the same nouns,[60] employ similar evaluative language,[61] and choose words connoting calamity.[62] This is not surprising, since both are discussing the psychology of warfare;[63] but the resonances go deeper. Both authors attempt to maintain a philosophically or rhetorically persuasive tone by the repeated use of simple contraries[64] and more complex oppositions,[65] some approaching paradox.[66] On a word-by-word level, the effect seems reinforced by the syntax (e.g. dative of relative pronoun plus verb;[67] conditional statements;[68] clauses begun with relative pronouns)[69] and by generally complex periods, especially in 'Aineias' (his sentences have 40–51 words, those of Thucydides only 20–34). Even the authors' grammatical choices add rhetorical and philosophical colour: compound optatives,[70] present participles of compound, middle, or deponent verbs,[71] aorist participles in the middle or passive voice, and so on.[72]

---

60 γῆ; ἐλπίς; and words related to ἐνάντιος, κίνδυνος, and πόλεμος.

61 A. κακῶς, T. κακοπραγοῦντες, κάκωσις.

62 A. ἀναιρεθείησαν, T. θάνατος.

63 Cf. A. ἀγῶνες—ἀγών, φοβερούς, perhaps δυσεπιθέτους (a *hapax*, Hunter and Handford 1927, 102); T. μνήμη, κρίναντες, μὴ περιορᾶσθε, ἀλγεινοτέρα, φρόνημα ἔχοντι, ῥώμης, ἀναίσθητος; plus ἐλπίς, above.

64 A. ἐκ τῆς αὑτῶν χώρας ... οἰκεια χώρα, T. ἐν τῇ οἰκείᾳ ... ἐν τῇ μὴ προσηκούσῃ; also A. κατὰ γῆν ἢ κατὰ θάλασσαν.

65 A. καλῶς ἀμυναμένοις ... κακῶς προσενεχθεῖσι; T. οὐ στηλῶν μόνον ἐπιγραφή ... ἄγραφος μνήμη, τῆς γνώμης ... τοῦ ἔργου; see also A.'s §1. 1 versus §1. 2; and T.'s §43. 3, 5, and 6, each expounding a contrast.

66 A. οὐκ ἴσος οὐδὲ ὅμοιος ὁ ἀγών ἐστιν, T. οὐ γάρ οἱ ... ἀλλ' οἷς. The striking, chiastic abstraction of T.'s τὸ εὔδαιμον τὸ ἐλεύθερον, τὸ δ' ἐλεύθερον τὸ εὔψυχον, is not, however, matched—at least in this passage—by A.'s bland ὑπὲρ τῶν μεγίστων.

67 A. τοῖς δέ ... οὐδεμία ἐλπὶς ὑπάρξει, T. οἷς ... ἐλπὶς οὐκ ἔστιν.

68 A. ἄν τι σφάλμα γένηται, T. ἤν τι πταίσωσιν.

69 A. ὅσοις, T. οὕς.

70 A. ἀναιρεθείησαν, T. ἀφειδοῖεν.

71 A. ὁρμωμένοις, περιγιγνομένοις, T. προσηκούσῃ, κακοπραγοῦντες, γιγνόμενος.

72 A. σωθεῖσι, ἀμυναμένοις, προσενεχθεῖσι; T. ζηλώσαντες, κρίναντες.

It has scarcely been noticed in modern commentaries that 'Aineias' here may be imitating the style of Thucydides.[73] The whole preface (including §§3–4) stands in stark contrast with most of the rest of the work. What is 'Aineias' doing? Is he simply originating a rhetoric *ex nihilo*—not very skilfully but in awareness that complexity of expression may create distance and authority?

The clear resonances of Thucydides in the preface, far more than the factual recollections in other parts of the *Poliorketika*, point to a conscious attempt—albeit in a 'bolted-on' passage—to claim a place in an intellectual tradition dominated by Athenian writers. In order to persuade us that his practical expertise is theoretically grounded, he adopts a rhetorical style that evokes philosophical, even 'scientific' writing. To this extent the aim of his treatise, for all its grounding in experience and its technical detail, is more than merely practical.

There are, in fact, features of the rest of the work that suggest that 'Aineias' is trying to do more than simply record and communicate technical and psychological devices for maximizing the survivability of war. Recent work has highlighted his narrative subtlety.[74] One may also highlight small-scale rhetorical devices in the text that are designed to enhance his claim to expertise: for example, the withholding of further expert knowledge at 19 ('I could mention ...'), or the seven or eight references he makes to his other writings.[75] The very fact that he supports his recommendations with named, specific examples from the experiences of other Greek observers, not merely his own experiences or *a priori* theorization, puts him in the same ambit of 'scientific' writing as Aristotle's *Politics*. In this respect he departs from the Xenophontic model that was surely present to his mind, for exempla drawn from outside Athens are rare in Xenophon's treatises on horses and hunting and, when they do occur, are largely non-Greek.[76]

---

73 Hunter and Handford (1927) note the careful and rhythmical periods of the preface (p. 102, ad loc.), and see A. as 'grappl[ing]' with abstract ideas (lxxx). Whitehead 2002, 38, merely mentions 'a brief but noticeable attempt at a polished preface'. Bettalli (1990) 13 and 211, notes its careful composition and use of rhetorical devices; he comments that the treatment of general themes lent itself to such a format, and that A. had many models to follow for a preface. Both he and Whitehead (2002) 98, compare Thuc. 4. 92. 4 and cite Lammert 1938, 397, for comparison with e.g. Vitruvius 10. 10. 1; Vegetius 4. 31.

74 Burliga (2009).

75 See discussion at Whitehead (2002), 13–16.

76 Xen. *Peri Hipp.* 8. 6, Persians and Odrysians riding downhill safely; more allusively, the Boiotian style of helmet, 12. 3; the Persian sabre, 12. 12; mounting 'in the Persian fashion', *Hipp.*

Comparison with the *Periplous* of Pseudo-Skylax (c. 337–c. 335 BC) is perhaps relevant. The author of that technical treatise, working within the ambit of the post-Platonic Academy or the early Peripatos,[77] adopts only rudimentary narrative techniques and rhetorical devices to enliven a text that remains, for the most part, a rather dry enumeration of coasts and peoples. Yet a generation earlier 'Aineias', perhaps without the advantage of having worked in Athens, makes greater efforts to establish his credentials, though almost exclusively in a preface that may well be an afterthought.

Conclusion

This study has contended that 'Aineias' presents his recommendations as universally applicable, but implicitly only within the world of the *polis*; and that since he is pioneering, mapping out a new field of written discourse, he responds (at least in one passage) to the Greek literary tradition—more directly than scholars have realized—in order to enhance his credibility and promote the adoption of his views, which are in a sense both theoretical and practical. The realization that he engages, to some degree, with the literary tradition enriches our understanding of the scientific context within which he is working, and thus of his aims in writing.

Bibliography

Anderson, J.K. (1974) *Xenophon*. Duckworth, London.

Anderson, J.K. (1985) *Hunting in the Ancient World*. University of California Press, Berkeley and Los Angeles.

Bengtson, H. (1962) 'Die griechische Polis bei Aeneas Tacticus', *Historia* 11: 458–468.

Bettalli, M. (1986) 'Enea Tattico e l'insegnamento dell'arte militare', *Annali della Facoltà di Lettere e Filosofia dell'Università di Siena* 7: 73–89.

Bettalli, M. (1990) *Enea Tattico. La difesa di una città assediata (Poliorketika). Introduzione, traduzione e commento*. ETS editrice, Pisa.

Brown, T.S. (1981) 'Aeneas Tacticus, Herodotus and the Ionian revolt', *Historia* 30: 385–393.

1. 17 and *Peri Hipp.* 6. 12. In *Cyn.* direct references are more numerous: Phasian or Carthaginian flax, 1. 4; Indian hounds, 9. 1; other hounds, 10. 1 and 4; hunting in foreign lands, 11. 1–4.

77 Shipley (2012); Shipley (2011), 15–18.

Burliga, B. (2009) 'Aeneas Tacticus Between History and Sophistry: the Emergence of the Military Handbook', in Pigoń, J. (ed.), *The Children of Herodotus: Greek and Roman Historiography and Related Genres*. Cambridge Scholars Publishing, Newcastle upon Tyne. 92–101.

Delebecque, É. (1957) *Essai sur la vie de Xénophon*. Klincksieck, Paris.

Delebecque, É. (ed. 1973) *Xénophon, Le Commandant de la cavalerie. Texte établi et traduit* (Budé). Les Belles Lettres, Paris.

Earl, D.C. (1972) 'Prologue-form in Ancient Historiography', *Aufstieg und Niedergang der römischen Welt* 1 (2): 842–856.

Gehrke, H.-J. (1986) *Jenseits von Athen und Sparta: das dritte Griechenland und seine Staatenwelt*. Beck, München.

Gray, V.J. (1985) 'Xenophon's Cynegeticus', *Hermes* 133 (2): 156–172.

Hamel, D. (1998) *Athenian Generals: Military Authority in the Classical Period* (*Mnemosyne* Supplement 182). Brill, Leiden.

Hansen, M.H. (2006) *Polis: An Introduction to the Ancient Greek City-state*. Oxford University Press, Oxford.

Hansen, M.H. (2007) 'Aineias the Tactician', in Hansen, M.H. (ed.), *The Return of the Polis. The Use and Meanings of the Word Polis in Archaic and Classical Sources*. (Papers from the Copenhagen Polis Centre 8. *Historia* Einzelschriften, 198). Franz Steiner Verlag, Stuttgart. 243–246.

Hansen, M.H., and Nielsen, T.H. (2004) *An Inventory of Archaic and Classical Poleis* (*An Investigation conducted by the Copenhagen Polis Centre for the Danish National Research Foundation*). Oxford University Press, Oxford.

Headlam-Morley, J.W. (1933) with D.C. Macgregor, *Election by Lot at Athens*, second edition. First published 1891. Cambridge University Press, Cambridge.

Hornblower, S. (1995) 'The Fourth-Century and Hellenistic Reception of Thucydides', *Journal of Hellenic Studies* 115: 47–68.

Hornblower, S. (1996) *A Commentary on Thucydides*, vol. II: *Books IV–V. 24*. Clarendon Press, Oxford.

Hornblower, S. (2000) 'The Old Oligarch (Pseudo-Xenophon's *Athenaion Politeia*) and Thucydides: a Fourth-Century Date for the Old Oligarch?' in Flensted-Jensen, P., Nielsen, T.H. and Rubinstein, L. (eds), *Polis and Politics: Studies in Ancient Greek History presented to Mogens Herman Hansen*. Museum Tusculanum Press and University of Copenhagen, Copenhagen. 363–384.

Hunter, L.W. and Handford, S.A. (1927), *Αἰνείου Πολιορκητικά. Aineias on Siegecraft*. Clarendon Press, Oxford.

Jenkins, T.E. (2006) *Intercepted Letters: Epistolary and Narrative in Greek and Roman Literature* (Roman Studies: Interdisciplinary Approaches). Lexington Books/Rowman and Littlefield, Lanham, MD.

Lammert, F. (1938) 'Die antike Poliorketik und ihr Weiterwirken', *Klio* 31 (n.s. 13): 389–411.

Marchant, E.C. (ed. 1925) *Xenophon, Scripta Minora*, first edition (Loeb Classical Library, 183; Xenophon, 7). Harvard University Press and Heinemann, Cambridge, MA and London.

Marchant, E.C., and Bowersock, G.W. (eds 1968) *Xenophon, Scripta Minora; Pseudo-Xenophon, Constitution of the Athenians*, second edition (Loeb Classical Library, 183; Xenophon, 7). Harvard University Press and Heinemann, Cambridge, MA and London.

Marr, J.L., and Rhodes, P.J. (eds 2008) *The 'Old Oligarch': The Constitution of the Athenians attributed to Xenophon* (Aris & Phillips Classical Texts). Aris & Phillips and Oxbow, Oxford.

Nielsen, T.H. (1999) 'The Concept of Arkadia: the People, their Land, and their Organisation', in T.H. Nielsen, T.H. and Roy, J. (eds), *Defining Ancient Arkadia* (Acts of the Copenhagen Polis Centre, 6; Historisk-filosofiske meddelelser, 78). Kongelige Danske Videnskabernes Selskab, Copenhagen. 16–79.

Nielsen, T.H. (2002) *Arkadia and its Poleis in the Archaic and Classical Periods* (Hypomnemata, 140). Vandenhoeck & Ruprecht, Göttingen.

Nielsen, T.H. (2004a) 'Arkadia', in Hansen, M.H. and Nielsen, T.H. (eds), *An Inventory of Archaic and Classical Poleis*. Oxford University Press, Oxford. 505–539.

Nielsen, T.H. (2004b) 'Triphylia', in Hansen, M.H. and Nielsen, T.H. (eds), *An Inventory of Archaic and Classical Poleis*. Oxford University Press, Oxford. 540–546.

Ostwald, M. (2000a) *Oligarchia: The Development of a Constitutional Form in Ancient Greece* (*Historia* Einzelschriften, 144). Franz Steiner Verlag, Stuttgart.

Ostwald, M. (2000b) 'Oligarchy and Oligarchs in Ancient Greece', in Flensted-Jensen, P., Nielsen, T.H. and Rubinstein, L. (eds), *Polis and Politics: Studies in Ancient Greek History Presented to Mogens Herman Hansen*. Museum Tusculanum Press and University of Copenhagen, Copenhagen. 385–396.

Pretzler, M. (2009) 'Arcadia: Ethnicity and Politics in the Fifth and Fourth centuries BCE', in Funke, P. and Luraghi, N. (eds), *The Politics of Ethnicity and the Crisis of the Peloponnesian League* (Hellenic Studies, 32). Center for Hellenic Studies, Trustees for Harvard University/Harvard University Press, Washington, DC/Cambridge, MA/London. 86–109.

Rhodes, P.J. (ed. 1988) *Thucydides, Histories II* (Aris & Phillips Classical Texts). Aris & Phillips, Warminster.

Roy, J. (2000) 'Problems of Democracy in the Arcadian Confederacy 370–362 BC', in Brock, R. and Hodkinson, S. (eds), *Alternatives to Athens: Varieties of Political Organization and Community in Ancient Greece*. Oxford University Press, Oxford. 308–326.

Schömann, G.F. (1861) *Griechische Staatsalterthümer*, 2nd edn, i: *Das Staatswesen*. Weidmann, Berlin.

Shipley, D.G.J. (2003) 'Review of M.H. Hansen and others, *Acts and Papers of the Copen-*

hagen Polis Centre; A Comparative Study of Thirty City-state Cultures', in *Journal of Hellenic Studies* 123: 234–237.

Shipley, D.G.J. (2005a) 'Between Macedonia and Rome: Political Landscapes and Social Change in Southern Greece in the Early Hellenistic Period', *Annual of the British School at Athens* 100: 315–330.

Shipley, D.G.J. (2005b) 'Little Boxes on the Hillside: Greek Town Planning, Hippodamos, and Polis Ideology', in Hansen, M.H. (ed.), *The Imaginary Polis* (Acts of the Copenhagen Polis Centre, 7; Historisk-filosofiske Meddelelser, 91). Kongelige Danske Videnskabernes Selskab, Copenhagen. 335–403.

Shipley, D.G.J. (2011) *Pseudo-Skylax's Periplous: The Circumnavigation of the Inhabited World (Text, Translation, and Commentary)*. Bristol Phoenix Press and The Exeter Press, Exeter.

Shipley, D.G.J. (2012) 'Pseudo-Skylax and the Natural Philosophers', *Journal of Hellenic Studies* 132: 121–138.

Slater, N.W. (2010), 'A Horse is a Horse? Pursuing Plato in Aeneas Tacticus 31.15', Abstract of a paper given at the Annual Meeting of the Classical Association of the Middle West and South, 24–27th March 2010, at https://camws.org/meeting/2010/program/abstracts/03E1.Slater.pdf (last accessed 19/09/16).

Tuplin, C.J. (2004) 'Xenophon, Artemis and Scillus', in Figueira, T.J. (ed.), *Spartan Society*. Classical Press of Wales, Swansea. 251–281.

Vlassopoulos, K. (n.d. [2008]), 'The Regional Identity of the Peloponnese', in *Being Peloponnesian* (Nottingham Centre for Spartan and Peloponnesian Studies), published online at https://www.nottingham.ac.uk/csps/documents/beingpeloponnesian/kostas.pdf (last accessed 19/9/2016).

Whitehead, D. (2002), *Aineias the Tactician, How to Survive under Siege. Translated with Introduction and Commentary*, updated edition. Bristol Classical Press and Duckworth, Bristol and London.

Widdra, K. (ed. 1964) *Xenophontis de re equestri*. Teubner, Leipzig.

CHAPTER 5

Aineias and History: The Purpose and Context of Historical Narrative in the *Poliorketika*

Maria Pretzler

Aineias Tacticus' work includes a lot of historical material, usually short accounts which are meant to illustrate and support his advice on how to cope with a siege. Anybody who has ever tried to extract specific historical information from the *Poliorketika* will know how problematic these passages usually are. They offer tantalising glimpses of historical information, even some details which are not known from other sources, but it is very difficult indeed to contextualise this information and to put it to good use as historical source material. From a historian's perspective, Aineias' approach to historiography (if we can call it that) can cause more problems than it solves; in fact, questions concerning the identification and date of specific events mentioned in the *Poliorketika* take up considerable space in commentaries, and yet many problems remain. But, as Burliga has recently demonstrated,[1] the value of these historical passages must not be seen merely in the information they convey to us. In this article, I investigate what Aineias can tell us about attitudes to history in the mid-fourth century BC and about the role such narrative material could play outside the narrow confines of 'proper' historiography.

Small, paragraph-sized historical narratives are a common feature in ancient Greek literature. We might call such passages 'historical vignettes': the term can remind us that, just as in looking at a miniature painting, we are invited to notice the care with which these short narratives have been shaped and consider the thought processes involved in reducing the story without losing its essential features. Historical vignettes are particularly familiar from texts connected to the Second Sophistic, the Greek cultural revival of the Roman imperial period. Such works, for example Pausanias' *Periegesis* and, particularly relevant in this context, Polyainos' *Strategemata*, have shaped our approach and expectations in handling historical vignettes, particularly when if comes to mining ancient texts for historical information.[2] These later texts will there-

1 Burliga (2008).

2 Cf. Pretzler (2010).

 | DOI: 10.1163/9789004352858_006

fore provide comparative material for this discussion, especially to illustrate how different Aineias' pioneering approach is, in spite of some superficial similarities, from the well-honed literary habits of the Second Sophistic some five hundred years later.

In spite of their widespread use in ancient literature and their importance as historical source material, historical vignettes are hardly ever considered for their own merit as a literary feature. The standard question is whether we can extract historical information from them, and how accurate this material is likely to be. There is no question that this kind of analysis is important, but if we want to appreciate the full potential of these bite-sized histories, we have to consider their purpose within the text and the intentions of the author: most historical vignettes were clearly not created with the prime purpose of preserving unique historical details for later historians.

This chapter starts with an evaluation of Aineias' approach to these historical vignettes. I focus first on the historical context he provides in part of his narrative passages, namely chronological (in the widest sense) context, geographical information and references to historical characters. This is followed by a discussion of Aineias' sources and an investigation of how he adapts his material for the *Poliorketika*. The second part of the chapter looks at narrative passages which have little specific historical context, and uses these to analyse Aineias' aims as a didactic writer. Finally, I focus on historical vignettes as a means of enhancing the author's credibility.

Even in Aineias' time, small, paragraph-sized narratives which digress from the main text were nothing new: they are an integral part of Homeric storytelling, and by the classical period, they were also firmly established in prose texts. Herodotos' numerous digressions include many which might qualify as historical vignettes, and by the fourth century, historical examples were a standard ingredient of rhetoric, particularly in a political context.[3] Nevertheless, in this as in several other aspects of ancient literature Aineias shows himself as an innovative author: assuming that he was not following a template which is now entirely lost to us, we see him experimenting with an entirely new kind of didactic literature and with ways in which historical examples could be used to enhance practical or technical advice. The historical vignettes, inserted to illustrate and instruct, are therefore a crucial aspect of his innovation.

3 Attic orators: Worthington (1994), Harding (1987), Perlman (1961), Pearson (1941); cf. Baragwanath (2008) 17–22, Gould (1989) 55–58.

At this point we need to pause and reconsider our own perspective. Most comments on Aineias' historical vignettes have been focusing on the expectations of historians mining the text for information. For this purpose, it is crucial to identify historical events accurately, so that Aineias' reports can be compared with alternative source material; we need to assess the accuracy of his accounts and, if possible, identify the sources of Aineias' historical material. These issues are important and deserve extensive discussion, but this strictly utilitarian historian's approach does not do justice to Aineias' original aims: how did *he* want these passages to be read? How did Aineias approach his historical material, and how did he assess its purpose within the work? We have to remember that through a long history of the reception of ancient literature, modern scholarly approaches are significantly informed by ancient literary trends which were perhaps just beginning in Aineias' time, but which became dominant only decades later and continued to define ancient Greek literature for centuries. I am referring to developments of the Hellenistic period, particularly linked to intellectual work at the Library of Alexandria, which meant that writers were increasingly aware of the heritage of Greek history and literature and began to find ever more creative ways of using it in their works. Postclassical authors often employed partial or covert references to historical events as a kind of elaborate game for the highly educated.[4] They could rely on their readers' knowledge of a shared literary canon which would allow them to contextualise events immediately. These authors interact with the literary tradition, and since they themselves were used to mining texts for information, the details offered in historical vignettes were carefully crafted for an audience wanting to do the same.[5]

For example, the stratagems in Polyainos' work are presented in a roughly historical/ethnographical framework, rather than grouping together stratagems which illustrate a particular area of military expertise.[6] The *Strategemata* are therefore arguably of more use to those wanting to discover and consider historical details within a new context, rather than to those in need of military instruction. Since postclassical authors and their readers were themselves so familiar with the process of gathering historical information from a variety of texts, their approach to writing historical vignettes often responds well to a reader's need to identify sources and contexts, and these texts—apart from

4 Bowie (1974), Swain (1996) 79–87, Whitmarsh (2001) 26–29.

5 Bowie (1974) 170–174, Anderson (1993) 69–85, Swain (1996) 89–100; Specific examples: Pretzler (2007) 73–90 (Pausanias), Pretzler (2008) (Polyainos).

6 Krenz and Wheeler (1994) vii–viii, Schettino (1998) 39–43, 191–278.

Polyainos particularly the works of Plutarch and Pausanias—have served modern scholars well in providing historical information through such passages. This learned and consciously secondary approach to historical material made the works of many postclassical authors eminently suitable for the purposes of modern scholarship, and their works therefore had a particularly strong influence on what scholars today expect of (apparently) factual narratives embedded in ancient literary texts.

Aineias, however, wrote before Greek education and intellectual activity became so focused on the literary tradition. If Aineias wants to increase his readers' knowledge of specific events, he does not do so for the sake of their literary or historical education, but in order to make up for their lack of military experience. This may sound self-evident, but in the light of the large volume of post-classical literature, and the strong influence it has had on the reading habits of classical scholars, it is worth spelling out just how different Aineias' basic assumptions are. Despite superficial similarities of form, we need to interpret historical vignettes in the *Poliorketika* in their own context, independent of what later literary tradition has taught us to expect.

Before tackling these more comprehensive questions we need to take a closer look at Aineias' way of writing historical vignettes: what did he consider necessary in order to present short historical narratives effectively? The table at the end of this chapter provides an overview of all passages which have been considered for this purpose: there are 64 in all, and the table also indicates why any given passage has been included. The lines between historical vignette and general advice are sometimes blurred, and I have been inclusive wherever possible. As a basic requirement for inclusion, a passage had to convey a sense, however slight, that Aineias is dealing with specific events or circumstances.

The minimum one would expect from any historical account is that it provides at least a few specific details which allow the audience to recognise the passage as a historical event and ideally also to locate the story in a specific context. Quite a few of the passages in the *Poliorketika* do not actually offer sufficient detail, and I shall return to these later. In the first instance, we need to observe how Aineias introduces information that might help us to locate a story in history, focusing, in turn, on chronology, geography and finally on named participants. Polyainos, for example, collects hundreds of historical vignettes for his collection of stratagems, and he is remarkably consistent in providing these crucial details for all but a few exceptional passages.[7] Aineias is a lot less

7 Schettino (1998) 97–115, Pretzler (2010) 95–96.

systematic or reliable in supplying specifics. In fact, a number of his historical vignettes include so little information that even the best informed ancient readers would not have been able to reconstruct the historical context, unless they already knew the story in question and managed to identify it by specific details in the narrative; readers today, with only fragmentary information at our disposal, are in an even more difficult position: perhaps, in some cases, he wanted readers to understand more than we can know now.

For example, Aineias tells the famous story of Histiaios' message to Aristagoras which was tattooed onto the shaved head of a slave and concealed once the hair had grown back.[8] Aineias never mentions that the secret message was sent on a journey which took about three months,[9] all the way from Susa to Miletos, although this particular method would be of little use for communications which could not wait for such a long period; presumably he was counting on his readers' previous knowledge of Herodotos' story and merely decided to jog their memory with a short reference among the many examples provided in chapter 31. On the whole, such extreme brevity in summarising the narrative and a focus on the most relevant details is typical of Aineias' method: usually he firmly sticks to his purpose of illustrating the theme he is discussing in each chapter. If he does provide more information than is strictly necessary in the immediate context, the additional details still contribute to the didactic purpose of the work. For example, his discussion of the Theban attack on Plataia in 431 BC combines points about assembling and organising troops in an emergency with details about organising the defence against an enemy who already controls the city.[10] Aineias' chapter on pre-arranged passwords includes an example which illustrates the danger of enemies sawing through the bar which secured the city-gates, a topic which is discussed in more detail later in the *Poliorketika*.[11]

Where Aineias offers any kind of chronological information, this is never very detailed, but at least serves the function to emphasise the notion that we are dealing with a specific event, perhaps even one the reader is expected to have heard of. There is no attempt to follow any of the dating conventions which were being developed in this period: if I say chronological, I really mean any kind of reference which allows us to connect a historical event to a specific time period. Some of these references are fairly easy to recognise, even today,

8 Ain. Tac. 31.28–29; cf. Hdt. 5.35.3.

9 Hdt. 5.52–53.

10 Ain. Tac. 2.3–6, cf. Thuc. 2.2–6.

11 Ain. Tac. 4.1–4; discussion of gates: Ain. Tac. 19.1–20.5.

when knowledge of events in Aineias' day is severely limited. For example, he describes the defence measures adopted in the city of Sparta against a Theban attack; although the Thebans invaded Lakonia twice, the fact that the city is shown under direct threat suggests that this refers to 362 BC.[12] A number of Aineias' vignettes come with similarly specific information, but often we, as modern readers, do not have enough alternative evidence to identify them with any certainty: for example, he describes Chalkedon under siege, and although we are not told who the aggressor is, we hear that the city is also under threat from a mercenary garrison supplied by a friendly but controlling Kyzikos.[13] The details seem specific enough, but the comparison between this passage and the Spartan example above illustrates that the reader requires additional information, preferably previous knowledge of the event, in order to identify it and to appreciate its full context—and this is true even for passages such as these, where Aineias is relatively generous with specific information.

Occasionally Aineias presents readers with historical details which presuppose quite detailed knowledge, for example when he refers to the 'second coup against democracy in Argos', probably in 370 BC.[14] Did he expect his readers to know the history of civil unrest in Argos well enough to identify the event correctly? How much local history of cities and regions other than their own could Greeks be expected to know in the late Classical period, especially if those events were not covered in the few historiographical works which were already widely known at the time? There is no clear answer to this question, since we have little evidence, beyond Aineias, of how current events and recent history were received in the Greek world outside Athens.

Even if readers were able to identify a specific event from Aineias' short descriptions, it is yet more doubtful whether they would have been able to fit these events into a more exact chronology and to synchronise them with historical developments elsewhere in the Greek world. Establishing the relative dates of different events was a difficult task. Great minds of the late fifth century were trying to tackle the problem—for example Hellanikos, who investigated local histories and chronological lists of officials,[15] Hippias of Elis, whose list of Olympic victors would eventually allow scholars to create a widely

12 Ain. Tac. 2.2.; cf. Xen. *Hell.* 7.5.9–10, Plut. *Ages.* 34, Polyain. 2.3.10; Buckler (1980) 210–211.

13 Ain. Tac. 12.3; possibly in the late 360s, see Hunter and Handford (1927) 142–143, but the evidence is not conclusive; cf. Whitehead (1990) 134.

14 Ain. Tac. 11.7.

15 Hellanikos *FGrHist* 323a (Attic History, organised by archons), *FGrHist* 4 F 74–84 (List of priestesses of Hera at Argos), *FGrHist* 4 F85–86 (*Karneonikai*); Jacoby (1913) 138–148.

understood system of chronology based on numbered Olympiads,[16] and Thucydides with his reckoning by winters and summers.[17] But even much later, if Pausanias can give us any indication of historical thinking outside expert circles, exact chronology probably remained the realm of specialised scholarship, while many continued to depend on historical parallels established through great events or relations between individuals and calculation by generations to establish rough chronological sequences, rather than insisting on an exact calendar of numbered years in order to synchronise events in different places more exactly.

Many of Aineias' historical vignettes do have a geographical setting: usually, Aineias merely gives the name of a city. A generous count produces 39 of such references, including examples which might be called broadly ethnographical, since they are dealing with specific institutions or customs of a *polis*, rather than one-off historical events. A simple toponym is often enough to give a narrative a more 'historical' feel, indicating that we are dealing with a specific situation. Aineias' selection of cities illustrates his own assumptions about his readership and their previous knowledge: most named locations are all large and well-known *poleis*. Only rarely does Aineias consider it necessary to offer extra information to define a city more closely: he tells us that Teos is 'an Ionian city of considerable size'; he also refers to 'Parion on the Hellespont', and 'Apollonia on the Pontus'.[18] The fact that all events he mentions should have occurred in well-known cities has to raise suspicion: did he just omit the name if a place seemed to obscure? One conspicuous example, where he refers to 'a city in the district of Achaia' without giving the individual name,[19] suggests that he was indeed happy to spare his readers the names of lesser known places. Educated Greeks at the time presumably had a sense of the location of at least the bigger cities in the Greek world, and Aineias' selection cannot have presented too much of a challenge.[20]

Just over half (25) of all the vignettes which are attached to a particular place also include some chronological information. In many cases, we are dealing with mere hints: the text included in the table of Aineias' historical vignettes represents this information in its entirety. There are twenty-two passages which

16 Plut. *Numa* 1.4; Mosshammer (1979) 87–88.
17 Thuc. 5.20.2–3.
18 Ain. Tac. 18.13, 28.6, 20.4.
19 Ain. Tac. 18.8.
20 This can also be seen in Xenophon's *Anabasis*, as Roy (1972) demonstrates specifically for Arkadians and individual cities in Arkadia.

provide a narrative of what seem to be unique events, but without a specific context, with neither time nor place indicated, although three of these examples include a reference to a famous individual.

Naming some of the protagonists also can help the reader with fitting a story into its historical context, and Aineias' text includes about twenty names in all. These names can be crucial for identifying an event correctly: for example, at 11.13, a mere reference to an impending oligarchic coup at Korkyra might have pointed at a number of similar incidents, especially the famous *stasis* described by Thucydides,[21] but the involvement of Chares suggests that the date is 361 BC, a reference to an incident much closer to the period when Aineias wrote his book.[22]

While historical precedents were evidently important to Aineias, it is striking that he hardly relies on the names of famous individuals to recommend his suggestions to his readers. A few of the famous generals who later gained some reputation for stratagems are mentioned: for example, Polyainos managed to collect whole lists of stratagems attached to Dionysios I, Dionysios II and particularly Iphikrates, while Aineias offers just one short example for Iphikrates, and one or two for Dionysios I.[23] This comparative absence of great leaders is in line with Aineias' attitude to the politics of warfare: he is firmly focused on the collective, namely the city as a community with its elected leaders. In Aineias' *polis*, anybody who makes a name for himself is more likely to earn suspicion as a potential threat to community cohesion rather than deserving more attention or admiration.[24] The largest group among Aineias' named historical characters are tyrants, followed by commanders of mercenary forces, and he also names a few Persians.[25] Many of these named individuals appear not as the problem-solvers, but as aggressors against whom the city has to be defended. This is very different from the attitudes of Aineias' close con-

21 Thuc. 3.70–83.

22 Cf. Diodoros 15.95.3; cf. Pritchett (1974) 57–58, Gehrke (1985) 96.

23 Ain. Tac. 24.16 (Iphikrates), 10.21–22 (Dionysios I), 40.2–3 (Dionysios I?), cf. 31.30–31 (probably Dionysios II, mentioned in a sample message).

24 See Pretzler below, p. 156–157.

25 Tyants: Ain. Tac. 4.8–12 (Peisistratos), 5.2 (Leukon of Bosporos) 10.21.2 (Dionysios I), 28.5 (Python of Klazomenai, probably a tyrant), 28.6–7, (Iphiades of Abydos), 31. 30–31 (Dionysios II), 31.33 (Astyanax of Lampsakos), 40.2–3 (Dionysios I?); mercenary commanders: Ain. Tac. 11.13–15 (Chares), 18.13–19 (Temenos the Rhodian, probably a mercenary commander), 24.3 (Charidemos of Oreos), 24.10 (Athenodoros of Imbros), 24.16 (Iphikrates); Persians: 31.25–27 (Artabazos), 31.35 (Glous the Persian Admiral), 40.4–5 (Datames).

temporaries Isokrates and particularly Xenophon, although both were grappling with the same problems of discord and political uncertainty in Greece which provide the poignant context for Aineias' efforts. Unlike Aineias, however, these writers express special admiration for larger-than-life characters and great leaders. It seems that Aineias deliberately avoided giving credit to some of the great men of his age: for example, he avoids any reference to Agesilaos, even when he relates an incident where the Spartan king acted as commander.[26] Aineias' approach to history, just like his work as a whole, is firmly rooted in the small-town politics which was increasingly under threat in this period, not least from larger-than-life characters with more power than most small *poleis* could usually bestow.

Looking at Aineias' attitude to chronological and geographical detail as well as his handling of historical characters, we see rather clearly that he did not consider it necessary to offer historical specifics with any consistency, and we have to assume that he did not expect his readers to identify all those rather vague episodes, either, let alone to understand more exactly where they fitted into a grand narrative of Greek history. At the same time, there are clearly historical vignettes which have links to better known events, and later on in this chapter, I shall return to the question what might be behind Aineias' decision to jog his readers' memories at least occasionally. But what would those readers actually remember? Most of Aineias' specific examples were more recent than the periods covered by established historical accounts, primarily Herodotos and Thucydides, and it is unlikely that by the 350s BC, when Aineias was writing, he was able to rely on a widely known literary narrative which could have served as an authoritative frame of reference for the period after the Peloponnesian War, let alone the more recent decades after the King's Peace of 386 BC which are the context for many of Aineias' examples. Even if parts or all of Xenophon's *Hellenika* or the *Hellenika of Oxyrhynchos* were available at the time, it is unlikely that such recent works were as widely known as the influential histories written two generations earlier.

This brings us to the question about Aineias' sources: again, this issue has been discussed before,[27] and I do not intend to pursue arguments about specific passages or sources in the *Poliorketika*. While *Quellenforschung* often involves hunting for literary sources to the exclusion of all else, experts have

26 Ain. Tac. 2.2., cf. Xen. *Hell.* 7.5.9–14.

27 Sources are discussed in detail in the various commentaries on Aineias' work, e.g. Hunter and Handford (1927), Dain and Bon (1967), Bettalli (1990), Whitehead (1990), Vela Tejada and García (1991); see also Vela Tejada (1991) 37–43, Burliga (2008). Studies of specific aspects: e.g. Brown (1981), Luraghi (1988) 166–167, David (1986) 120–122.

long agreed that Aineias relied on oral history, too, especially because many of the events which we can identify were relatively recent when he was writing.[28] These stories may have circulated among men with military interests as instructive examples of successful stratagems; in fact, it is possible that Aineias' vague sense of historical context was already a feature of some of his source material. The literary evidence suggests that military anecdotes could become attached to different individuals or contexts: for example, one of Aineias' suggestions for dealing with panic among the troops echoes an incident in Xenophon's *Anabasis* and then resurfaces in Polyainos' work as a stratagem of Iphikrates.[29] It is impossible to determine how these three accounts relate to each other, but some stratagems may have circulated as free-floating narratives remembered for a remarkable turn of events rather than for their specific context, and were ready to be attached to different personalities or to be deployed in different narrative settings.[30] We cannot rule out that Aineias himself contributed new material based on his own military experience, an approach which provides plenty of material for Lane Fox's chapter above—but it is unlikely that all 64 historical vignettes, or even a large proportion of this number, are based on Aineias' own career, and we have to ask in any case how such a collection of anecdotes may have been put together.

Was any of this material written down for military instruction before the 350s BC, and before Aineias wrote his work? A few names have been identified as possible writers on military theory and practice before Aineias, but, as so often on a quest for putative lost writers, it is highly doubtful whether they could have served as a major source for Aineias (e.g. Simon of Athens on horsemanship) or, in fact, whether they were indeed writers of the classical period (Demokritos, Daimachos).[31] In 11.2. Aineias himself apparently refers to a literary source, a book which contained stories of plots, presumably exactly the kind of historical vignettes under investigation here. It is not clear whether this was a collection by another author or one of Aineias' own works which also contained historical examples. The stories in this particular chapter are, in any case, qualitatively similar to what we see elsewhere in Aineias' work, and they also include events which probably happened in the 360s BC and were therefore still relatively recent. Although it is often impossible to trace Aineias'

28 Hunter and Handford (1927) xxxvi, Marinovic (1988) 201, Whitehead (1990) 38–39.

29 Ain. Tac. 27.11; cf. Xen. *Anab.* 2.2.19–21, Polyain. 3.9.4; cf. Schettino (1998) 124–126, 215–218.

30 Hunter and Handford (1927) 193; cf. Whitehead (1990) 176–177, Brown (1981) 388.

31 Burliga (2008) 94–95; cf Whitehead (1990) 36–38.

histories back to specific texts, especially because he does not name his sources, there are some passages which are clearly based on Herodotos[32] and one is taken from Thucydides.[33]

We do not have a lot of comparative material to work with, but where we have these parallels, it is possible to observe how the material was adapted to serve its new purpose: the language is close enough to suggest direct borrowing, but the narrative is abridged, and, where necessary, Aineias also inserted short remarks to allow better understanding of specific military details. Most intriguingly, he was also ready to add new details which are not in the original, and these examples are particularly instructive for our understanding of Aineias' approach to adapting historical material. For example, he adapted Thucydides' report of the Theban attack on Plataia in 431 BC: the passage is abridged, but echoes the original in many details, as Aineias reports how the Plataians secretly organised concerted action by breaking through walls between houses.[34] Thucydides' account does, however, not answer a crucial question: how did the Plataians find the time to carry out this plan? Aineias adds details of his own to tackle this problem, reporting that the magistrates were delaying the Theban intruders with negotiations. It seems unlikely that Aineias actually had access to additional material which would have added authentic historical information to Thucydides' report: he is less concerned with keeping his narrative as truthful as possible, than he is with explaining to his readers how his example might work in practice. In another passage, he paraphrases an episode from Herodotos' report of the Persian siege of Poteidaia. The attackers shoot arrows with messages for collaborators inside the city, but the plan goes wrong when one of the arrows hits a Poteidaian and the message is discovered.[35] Aineias adds his own speculation about why the arrow may have gone astray: after all, this is the kind of practical detail which he wants his readers to contemplate. Both these stories have a clear historical context, and well-educated men of the mid-fourth century may well have known the original texts. Aineias' treatment of these narratives, however, does not shrink from adding details which are not in the original, as long as these additions enhance

32 Ain. Tac. 31.9 ~ Hdt. 6.4; Ain. Tac. 31.14 ~ Hdt. 7.239; Ain. Tac. 31. 25–27 ~ Hdt. 8.128; Ain. Tac. 31.28–29 ~ Hdt. 5.35; Ain. Tac. 37. 6–7 ~ Hdt. 4.200; cf. Brown (1981).

33 Ain. Tac. 2.3–6 ~ Thuc. 2.2–6; Ain. Tac. 38.2. repeats an opinion on military matters expressed by Brasidas in Thuc. 5.9.8.

34 Ain. Tac. 2.3–6, based on Thuc. 2.2–6. Hunter and Handford (1927) 107–109, Celato (1967/8) 60, Losada (1972) 100, Lehmann (1980) 71, Brown (1981) 388, Gehrke (1985) 132, Whitehead (1990) 102–103.

35 Ain. Tac. 31.25–27; Hdt. 8.128.

the didactic value of the example, while any other details provided by the original source are ruthlessly shortened or omitted if they do not serve the primary purpose of the work.

Aineias usually adapts his historical vignettes very narrowly to illustrate the theme of the chapter in hand. The story is reduced to the bare bones necessary to get the point across. There is usually very little, if any, information about the consequences of the actions he is reporting: where mistakes lead to defeat a short comment might explain the consequences or highlight the lessons which should be drawn from the example,[36] but Aineias is not one to boast about the successes of stratagems he is suggesting: victory is usually merely implied or entirely left to the reader's imagination. Within the story, any information beyond what is strictly necessary in the immediate context, tends to contribute in some way to the general purpose of the work, for example by offering relevant material for another part of the book. For example, his discussion of the Theban attack on Plataia in 431 BC combines points about assembling troops in an emergency with details about organising the defence against an enemy who already controls the city.[37] Aineias' chapter on pre-arranged passwords includes an example which illustrates the danger of enemies sawing through the bar which secured the city-gates, a topic which is discussed in more detail later in the *Poliorketika*.[38]

Aineias' himself states in 28.7. that the didactic value of historical examples is uppermost in his mind:

> I have thought it best to collect these precepts to show the several precautions which should be taken at the various times, that no-one may be too ready to accept anything without due examination.[39]

A few of Aineias' historical vignettes explicitly include comments, usually at the end, which emphasise the essential lesson of the story.[40] The focus of all these narrative passages is strictly on technical details which can illustrate his point. We might dispute in some cases that he was entirely successful in communicating his views clearly, but on the whole, these examples do support his arguments, and at times they are indeed vital to illustrate what he is trying to

36 E.g. Ain. Tac. 2.6, 4.12, 23.5, 28.5.

37 Ain. Tac. 2.3–6, cf. Thuc. 2.2–6.

38 Ain. Tac. 4.1–4; discussion of gates: Ain. Tac. 19.1–20.5.

39 Ain. Tac. 28.7; "Δοκεῖ δέ μοι συναγαγόντι δηλωτέον τίνα δεῖ φυλάσσεσθαι καὶ ἐν οἷς καιροῖς ἕκαστα, ἵνα τις μηδὲν εὐήθως ἀποδέχηται."

40 Ain. Tac. 17.4, 18.21.

suggest. For example, the story of a coup accomplished by smuggling weapons into a city[41] does not just offer an exciting and entertaining account, it also invites the reader to think about the various dangers inherent in importing apparently harmless merchandise, and perhaps alert them to the potential dangers which might be lurking in every market place. Aineias' story and vivid description, down to the daggers pushed into melons and the spears wrapped up in wickerwork, helps him make this point much more effectively than a theoretical discussion of the same issues might have done. This unusual thoroughness has been used as evidence that this must be one of Aineias' own adventures[42]—but as we have already seen, he adds details to his material to enhance its didactic value, and this, too, may be a didactic embellishment of a fairly straightforward story, adding a range of important details (from experience, imagination, or both) to make the reader think more carefully about necessary security measures. For all we know, Aineias could simply have invented the whole incident, or extended a much simpler story, to illustrate the different methods of secretly providing weapons for a coup. If we want to grant our author some literary sophistication, we should not automatically conclude that more vivid details in a historical vignette can only come about as personal reminiscence, particularly if the narrative appears to be an effective vehicle for delivering a whole range of didactic details.

The discussion will now turn to those among Aineias' stories which do not include specific information about their historical context: they, too, can tell us something about Aineias' intentions, and particularly about his efforts to enhance his didactic writing. Twenty-two such passages have been identified for this study (see table): they were included because Aineias explicitly distinguishes them from generic advice by presenting them as references to actual events or specific circumstances. Due to a lack of contextual information, these passages might be considered of little use as historical source material, but they can provide more insight into Aineias' way of dealing with narrative. These 'a-historical' vignettes vary widely in their approach to narrative detail: from short summaries which only offer the most crucial details, to tantalisingly elaborate accounts which seem to call for more historical background. The most notable example is the story discussed above, where weapons and a commander are smuggled into a city to aid a coup, and the many details in the account seem to demand an explicit historical setting. A warning against ill-considered sorties is illustrated by a report of a complex plot in another anonymous

41 Ain. Tac. 29.3–10.

42 Hunter and Handford (1927) 198, Dain and Bon (1967) 64, n. 1.

city;[43] again, there are many specific details and the whole story seems to depend on a particular topographical setting, namely a location by the sea. Some of Aineias' examples are presented without a reference to a specific place, but nevertheless have close parallels in events known from other literary works. Thus, one story about a messenger who betrays a plot to the governor of an anonymous city essentially retells an episode of the Ionian revolt as recorded by Herodotos;[44] and when Aineias describes how secret messages can be concealed under the wax on a writing tablet, he was probably thinking of the letter in which the deposed king Damaratos warned the Spartans of a Persian attack—again, the source is probably Herodotos, although (yet again!) some details have been added to the story.[45] We should therefore note that in some cases, Aineias was probably able to provide more information about time, place and historical context, but chose not to do so.

While some of those 'free-floating' stories can be quite detailed and specific, there are others which lack any defining characteristics: these passages appear to be generic and are usually rather vague. In fact, some examples look like general advice recast as a narrative, for example 23.3, where Aineias relates in very general terms how the citizens of an unnamed city simulate civil unrest which allows them to stage a sortie and to overwhelm their enemies.

> The following device has been employed in making a sally. The citizens made a plausible pretence of sedition within the city, watched their opportunity for a sally, and made a surprise attack on the enemy with complete success.[46]

It is important not to impose artificial categories on Aineias' historical vignettes: just as some passages which come without a context can be very specific in narrative detail, some of the examples where he is at his most generic and vague actually have some contextual information attached to them, for example 31.6:

43 Ain. Tac. 23.7–11.

44 Ain. Tac. 31.9, cf. Hdt. 6.4.

45 Ain. Tac. 31.14, cf. Hdt. 7.239.

46 Ain. Tac. 23.3. "Ἐποιήσαντο δέ τινες καὶ τοιόνδε τεχνάσαντες·Στασιασμοῦ προσποιητοῦ μετὰ προφάσεως εὐλόγου γενομένου παρ' αὐτοῖς, καιρὸν τηρήσαντες καὶ ἐπεξελθόντες ἐπέθεντο παρ' ἐλπίδα τοῖς πολεμίοις καὶ κατώρθωσαν."

> Again, a message was brought to Ephesos in the following way. A man was sent with a letter written on leaves, the leaves being bound on a wound in his leg.[47]

This passage may be set in Ephesos, but how much historical information does it really offer? The story about this message has no specific feature which would link it to any particular event or period in history. Nevertheless, the place name seems to anchor it in real history.

In fact, the difference between generic advice and specific example can become rather blurred in the *Poliorketika*, and occasionally Aineias himself seems to be confused about this distinction, for example in 31.2:

> A message was once sent in the following manner. A book or some other document, of any size and age, was packed in a bundle or other baggage. In this book the message was written …[48]

Aineias is telling his story while also trying to explain that there is some flexibility in such a scheme for secret messages: this has an impact on the narrative, making it rather irregular. If this was a unique event, as the text seems to suggest, the book cannot have been 'of any age or size', because we are dealing with a specific object. It may be pedantic to point out this small discrepancy, but it seems very significant that the boundaries between specific example and general advice could become blurred in Aineias' own mind as he was striving for clarity and perhaps also variety.

We find more very generic narrative examples illustrating Aineias' advice on dealing with the bolt of a locked gate: he discusses a range of methods which are presented as short narratives: 'Once a gatekeeper cut a notch into the bolt … tied string around it and pulled it up again by the string; another prepared a fine net, pushed home the bolt enclosed by the net and afterwards drew it up …'—and so forth.[49] Were all the examples mentioned in this chapter real events? Some of them may have been, but it seems more likely that at least a part of these short descriptions of relevant techniques are the result of methodical analysis, trying to predict how the gate could be manipulated and

---

47 Ain. Tac. 31.6. "Εἰς Ἔφεσον δ' εἰσεκομίσθη γράμματα τρόπῳ τοιῷδε. Ἄνθρωπος ἐπέμφθη ἐπιστολὴν ἔχων φύλλοις ἐγγεγραμμένην, τὰ δὲ φύλλα ἐφ' ἕλκει καταδεδεμένα ἦν ἐπὶ κνήμην."

48 Ain. Tac. 31.2. "Εἰς φορτία ἢ ἄλλα σκεύη ἐνεβλήθη βυβλίον ἢ ἄλλο τι γράμμα τὸ τυχὸν καὶ μεγέθει καὶ παλαιότητι. Ἐν τούτῳ δὲ γέγραπται ἡ ἐπιστολὴ …"

49 Ain. Tac. 18.5–6 (extracts).

how one might prevent such activities, rather than relying on actual precedents for each example. The examples could just as well have been presented as general advice: 'Here are a few methods to remove the bolt if you do not have the key: cut a notch into the bolt, tie a string around it and use it to pull up the bolt ...'. It seems clear from the way in which these stories are presented that Aineias himself did not necessarily think of them as very specific historical examples: when he reports that 'bolts already in position are said to have been undone by pouring sand into the socket ...'[50] we are still given the notion that this did actually happen in some place, at some time, perhaps even more than once—but in this case Aineias does not give his readers the impression that he really minds whether this was indeed a real event. These borderline cases of 'historical' narrative suggest strongly that Aineias was willing to cross the boundary between didactic prose and narrative, and perhaps even found the distinction rather blurred already.

This part of the text strongly suggests that for Aineias, narrative was not just a straightforward means of reporting true events: he avoids a potentially monotonous list by presenting some of his technical advice as stories of specific events. This does not only allow him more variety of expression, it also helps the author to convey the seriousness of the threats which his instructions are meant to counter: 'these are not just theoretical possibilities', he seems to be saying, 'this can happen, and it has happened'. It is a common feature of traditional classical scholarship that writers (such as Aineias Tacticus) whose prose is considered inferior are also assumed to be rather naïve in their writing, combined with the assumption that little thought went into shaping their text effectively. The way in which Aineias handles the boundary between technical advice and illustrative examples would however suggest that he was by no means naïve about the use of narrative in his work: The *Poliorketika* is not a straightforward combination between advice and illustrative examples. Narrative is clearly used to shape the specifically didactic parts of the text as well, and we have to assume that the difference between the two categories is by no means clear-cut. As we have seen, Aineias probably introduced some narratives to enhance what might otherwise have been fairly dull lists, probably even just by adding that vital 'this once happened' to otherwise generic advice, and at the same time, he took existing historical episodes and added technical details to the story in order to emphasise the message he wanted to convey to his readers. This is not good news for historians desperate to extract otherwise unknown historical information from the *Poliorketika*.

50 Ain. Tac. 18.3.

Aineias was one of the earliest authors, if not the earliest, in prose at least, to grapple with historical examples in a didactic context, which means that he had to work out for himself how to combine advice and narrative. These were the early days of didactic literature, and historical examples evidently were not an indispensable ingredient of treatises offering practical advice, since Xenophon, Aineias' close contemporary, managed without them.[51] Aineias' pioneering role in developing historical vignettes in a new context is easily overlooked and underappreciated, since later developments in ancient literature make this mix between report and instruction seem so natural that it is difficult to pay attention to it. Thus we see Aineias experimenting with this new combination between narrative and didactic prose, and evidently more concerned with literary effect and didactic impact than with maintaining a strict separation between historical information and generic advice. Unlike in later periods with a stronger focus on a (by then) more substantial and revered literary tradition, there was perhaps also little to worry about being 'caught out' inventing or adapting a historical episode. Some of Aineias' 'historical examples', especially many of those which do not include any clear historical reference points, might have more to do with literary expediency than with an interest in reporting events which actually happened.

How did Aineias expect his work to be read, and particularly, how did he expect his readers to react to his historical vignettes? The readership, it is implied, are men who can expect to be in a leading position in their city at some point in their lives, and who might turn to the *Poliorketika* for practical instruction.[52] Aineias makes no assumptions about the exact nature of such readers' authority or their position in relation to the *polis* they might be defending (or perhaps, in fact, attacking). It is impossible to tell whether many, if any, of Aineias' readers would have matched even this very vague implicit profile, but we have to assume that, whatever their own practical needs, they would have been willing to engage with the overt didactic aims of the book. The book could be instructive even if such readers were not able to link Aineias' historical vignettes to specific events, but at the same time, they may have been exactly the kind of people who engaged with a contemporary tradition about clever stratagems. Although we cannot tell how much detailed knowledge of crucial events during the previous two or three decades could be expected of any reasonably educated reader in this period, Aineias' historical examples arguably have a different effect if the reader recognises at least some of them.

51 Burliga (2008) 95–97.

52 Whitehead (1990) 39–42.

One crucial effect of telling 'real' stories is that they could enhance Aineias' authority: such examples would give his advice more credibility as practical solutions which could be useful in the 'real world' of a Greek *polis* under threat. We can observe that, at least since the fifth century BC, historical examples and parallels played an increasingly important part in political persuasion, philosophy and historiography.[53] Aineias introduces this method of enhancing the credibility of an argument to a new genre. This was not an inevitable choice, as Xenophon shows when, in his own didactic works, he prefers to convince on the basis of his own authority as an expert rather than relying on historical examples. In the introduction to the *Peri Hippikēs*, Xenophon explicitly refers to his long experience of the subject.[54] *Poroi* deals with specifically Athenian issues, and the *Hipparchikos* was probably also aimed at an Athenian audience. These works were written late in Xenophon's life, when he was already well known and, despite his long exile, probably also respected in his native city.[55] We do not know whether Aineias, too, was widely known as an expert who could simply rely on his audience's respect. Since he does not state his credentials, at least not in the work as we have it, he seems to expect his readers to judge his advice on merit. This means that Aineias probably had to work a lot harder to establish his authority than Xenophon towards the end of his eventful life.

The selection of historical examples was important for this purpose: although I have spent the last few pages focusing on 'a-historical' vignettes, even suggesting that some may be entirely artificial constructs, I do think that the selection of those with an identifiable historical context does matter. It is striking just how many of Aineias' examples were so very recent: here he differs significantly from later authors for whom historical examples were enhanced by a respectable literary pedigree and a venerable age. Aineias relies on history in its raw form: things that happened, or were said to have happened, whether they were recorded in writing or not. Most of his historical vignettes were so recent—a few years, at most one or two decades before he wrote them down—that readers would judge their authenticity by first-hand memory, however hazy, rather than by the authority of a respected writer. Aineias appeals to readers' memories of actual events: the desired reaction would be 'I remember hearing about it when it happened', rather than 'oh yes, that's mentioned in Thucydides, too', which is the effect later authors would usually aim for. This approach also explains why geographical information and a preference

53 Pearson (1941), Perlman (1961), Harding (1987), Worthington (1994).

54 Xen. *Peri Hipp.* 1; cf. *Poroi* 1–2, where he emphasises his own perspective.

55 Diog. Laertius 2.54–55; Breitenbach (1967) 1576–1577, Anderson (1974) 193–194, Higgins (1977) 131–139.

for well-known cities was much more important to Aineias than chronological accuracy: he is using details that were most likely to jog his readers' memories.[56] Resonance with the audience's own personal memories would then also make the example, and the advice attached to it, seem more relevant and immediate. Aineias was not alone among his contemporaries in preferring recent historical examples to illustrate their point: Athenian rhetoric of the period demonstrates particularly well that recent history could have a strong emotional effect as orators drew on their audience's personal memories to enhance the impact of their arguments.[57] Aineias' recent but usually vague historical examples place his advice firmly in a reality which could appear consistent with his readers' own knowledge. Just as in Athenian assembly speeches, historical accuracy is not a priority: it is the persuasive effect that matters.

Aineias therefore plays an active part in a significant literary trend of his time which led to a rethinking of the meaning, shape and uses of history. Rhetoric probably led the way as writers experimented with new ways of presenting the past; Xenophon offers a number of innovative approaches, from the travelogue/adventure story that is the *Anabasis*, to biography (*Agesilaos*) and historical fiction (*Kyroupaideia*). Aineias' contribution to the diversification of historical writing seems a lot more modest, but we should appreciate his work particularly for its low key approach to history in a context where references to the past had not yet become an indispensable aspect of learned discourse. The *Poliorketika* demonstrates just how important history, especially recent history, was to Greeks in the fourth century and how it could be used in a new literary setting. In a political context, such as Demosthenes' assembly speeches or Isocrates' political pamphlets, references to previous events are essential, but Aineias is experimenting with historical examples to enhance his arguments in a context which was not explicitly historical or political.

To conclude, I am going to focus on three basic questions about historical vignettes in the *Poliorketika*—the answers have been emerging throughout the chapter, but they deserve to be summarised more explicitly.

First, we need to ask why Aineias decided to include these short narrative passages at all: as we have seen, didactic literature in this period was still being developed and could do without them. The answer emerging from the *Poliorketika* is not straightforward. On one hand, we see Aineias using narrative to enhance his prose, and we should not rule out the possibility that at times

56 Pearson (1941) 215–219, Harding (1987) 37–38; e.g. Demosth. 18.168–170.

57 Loraux (1981) 118–173.

he turned to narrative to convey some general advice, saying 'this happened at one time, and this was done to avoid it' where 'this might happen; you should do the following to avoid it' might have been a more straightforward way of giving instructions. This allows for more variation, especially where long lists of possible stratagems are involved. Even if the historical context was left vague, the weight of precedent would also enhance his argument: something which had happened before, a stratagem which had been used already, was surely more credible than a piece of advice which was potentially no more than the theoretical idea of the author. On the other hand, Aineias does refer to specific historical events as well. Just as contemporary orators were appealing to their audience's memories to enhance an argument, these historical references are used as a device to enhance Aineias' credibility as an advisor. If some of the episodes appear (quite possibly truthfully!) as if they might be based on personal experience, this would enhance his authority even further.

Second, it is worth considering how Aineias chose his examples. Here the main criterion is clearly their relevance to the advice he is trying to give, specifically relevance to the theme of the chapter is the main selection criterion. In later ancient literature there is often a separate motive to showcase the writer's acquaintance with the literary canon. There is little evidence in the *Poliorketika* that Aineias had similar concerns: his authority stems from military expertise, not scholarly erudition. Where specific historical details are involved, one crucial criterion for selection of a historical example, or for the decision to enhance a narrative passage with a specific historical context in the first place, may have been the question whether his readers would be able to recognise the story. Thus the places he chooses to name tend to be better known cities, and the events he highlights were relatively recent and more likely to resonate with readers' personal memories of current events in their own lifetime. Some examples are taken from earlier writers, Herodotos and Thucydides—again, these were probably widely read already in this period.

The third question is how Aineias adapted his examples to fit his purpose. There are few where we can compare his version with an original source, and in those cases, we see that he ruthlessly cut details which did not fit his specific purpose; but not only that, in his attempt to explain how exactly a certain outcome was achieved, he actually added technical details which cannot be found in the original. These additions are probably based on his own attempt to imagine how the reported result could be brought about. As with the selection of the examples, the main purpose is indeed didactic: Aineias shapes his historical vignettes to serve his immediate need to illustrate his points.

Despite this single-minded focus on didactic purpose, Aineias' selection of examples, particularly their geographical spread, can tell us more than his distinguished Athenian contemporaries about the shape of history and supraregional politics as they were perceived and discussed by people around the Greek world. This view of Greek history is more universal: it includes large cities from Sicily, Asia Minor and the Black Sea region, and refuses to give particularly large roles to Athens or Sparta. Aineias also offers a glimpse at the way in which events were remembered. We should observe (rather than merely criticise) the fact that many of Aineias' historical examples seem vague and perhaps quite moveable: these historical memories are hardly ever given a clear place within a general chronological framework, and there is no emphasis on the deeds of great men. Many of these stories could just as well be told about another place, time or political context: Aineias seems to be concerned with conveying a sense of history in order to exploit the value of precedents, but at the same time, he does not seem to be particularly worried about accuracy or an emphasis on specific privileged events.

Aineias' attitude to history also represents a calculated rejection of the conventions of historiography, as is quite appropriate for a writer pioneering a new and very different genre. The use of historical examples in deliberative or forensic rhetoric of the period is a better parallel: history becomes a means of persuasion, as the author tries to play on the memories of his audience to give his arguments and suggestions more credibility. While Attic orators usually expected to speak to the members of a well-defined community with collective memories, Aineias' audience was potentially a lot more diverse and had fewer predicable common reference points.

Because so much of classical literature is dominated by Athens, the *Poliorketika* almost looks like a deliberate distancing from this one city and its increasingly dominant cultural influence. But Aineias' conception of a Greek world with many places of interest is perhaps much more representative for anybody from the period who had no close links with Athens: at this point, it was still possible to use the whole Greek world as a reference point because Athenian literature was not yet dominating the memory of recent events, as it did later, when Attic Greek had become the defining characteristic for the Classical literary canon. This different viewpoint is exactly what makes Aineias' unorthodox handling of history so valuable. When we try to understand classical historiography, we should not just compare it with other examples of the genre: Aineias offers an important alternative perspective on how contemporary events were remembered and interpreted in the fourth century BC, and how historical narrative began to be employed in new literary contexts.

Aineias Tacticus: 'Historical Vignettes'

- List of passages which convey some sense of a specific setting and/or a specific event.
- The text summarises all information which defines the narrative as a specific event, or places it in a specific context. Passages which provide any form of historical specifics are presented in Italics, all personal names are in bold.
- Dates are based on Whitehead's commentary.

2.2.	Sparta	*When the Thebans broke into their city, the Lakedaimonians ...*		362
2.3–6.	Plataia	*The Plataians, when they discovered during the night that some Thebans had entered their city ...*	Thuc. 2.2.6	431
4.1–5	Chalkis	Chalkis on the Euripos was captured by an exile who started from Eretria ...		? 357/6
4.8–12	Athens	***Peisistratos**, when general at Athens ...*		560s
5.2	Bosporos	***Leukon**, tyrant of Bosporos ...*		c. 389–49
10.21–22	Syracuse	*It was thus that **Dionysios** (I) dealt with his brother **Leptines**, when he saw that he was in high favour with the people of Syracuse.*		c. 388–6
10.25		It has happened before now that when people's attempts to revolt and intrigue with the enemy were completely baulked ...		
11.3–6	Chios	When Chios was on the point of being betrayed, a magistrate who was in the plot ... it was peace time ...		?
11.7–10	Argos	*In Argos the following measures were taken against the revolutionary party ... the second attempt against the democracy ...*		370
11.10a–11.	Herakleia Pontica	At Herakleia Pontica, when under a democratic regime the wealthy party was plotting against the people and contemplating an attack ...		before 364
11.12	Sparta	A similar incident is said to have occurred <u>long ago</u> at Sparta ...		c8th?
11.13–15	Korkyra	*At Korkyra, when a revolt of the wealthy oligarchs against the people was impending (**Chares the Athenian**, who was stationed there with a guard, was in sympathy with this revolt) ...*		361/60

(cont.)

12.2–4	Chalkedon	*The people of Chalkedon during a siege received a garrison from Kyzikos, which was then their ally.*	late 360s?
12.5	Herakleia Pontica	*The inhabitants of Herakleia Pontica ... called in too strong a mercenary force ... the mercenary captain made himself tyrant.*	364
15.8–10	Abdera	*A raid made by the Triballians into the country of Abdera ... they lost more men, it is said, than any other city of the same size ever lost in so short a space of time.*	376/5
16.14–15	Cyrene, Barke	The people of Cyrene and Barke ... when they sent relief expeditions over their long carriage roads, used carts and chariots.	N/A
17.2–4	Argos	*At Argos a public festival took place outside the city, ... armed procession ... conspirators made ready and joined in the demand for arms to carry in the procession ...*	417
17.5–6.	Chios	When the people of Chios celebrate their Dionysiac festival ...	N/A
18.3		One of them poured sand into the socket in the day-time ...	
18.3–4		Bolts already in position are said to have been undone by pouring sand ...	
18.5		Once, too, a gatekeeper who had been deputed by his general to fasten the bolt ...	
18.6		Another prepared a fine net with a string attached ...	
18.6		The bolt has also been removed by being knocked upwards ...	
18.6		Again, it has been taken out with a small pair of pincers ...	
18.7		Another traitor succeeded in turning round the cross-bar without being noticed ...	
18.8–11	Achaia	At a city in the district of Achaia, where they were plotting secretly to let in mercenaries ...	360s?
18.12.		Once, too, the circumference of the bolt was measured ...	

18.13–19.	Teos	*An agreement was once made to betray Teos, a Ionian city of considerable size, to* ***Temenos the Rhodian****, with the complicity of the sentinel at the gate.*	?
18.20–21.		Another way in which a city was betrayed by a gatekeeper was this.	
20.4–5	Apollonia (Pontus)	At Apollonia on the Pontus ... the gates were so constructed as to be shut to the sound of a big hammer ...	N/A
	Aigina	And the same thing was done at Aigina.	N/A
22.20.	Naxos?	*After the battle of Naxos,* ***Nikokles****, the commander of the garrison ...*	376
23.3.		The citizens made a plausible pretence of sedition within the city ... and made a surprise attack on the enemy with complete success.	
23.4–5.		In another city the besieged surprised the enemy by sally ...	
23.7–11		I will mention here a scheme originated by certain magistrates ...	
24.3–14	Ilion	*I will give as an instance what happened in Aeolis to* ***Charidemos of Oreos****, after he had captured the town of Ilion by the following stratagem ...* ***Athenodoros of Imbros***	360
24.16		**Iphikrates** used even to say that rounds and sentries should not have the same password.	N/A
24.18.	Thebes	*Thebes during the capture of the Kadmeia ...*	379
27.7–10	Thrace?	***Euphratas****, the Spartan governor in Thrace, finding night alarms of very frequent occurrence in his army ...*	
27.11.		While the camp was in an uproar one night, the herald called for silence ...	
28.5.	Klazomenai	***Python of Klazomenai****, who had accomplices in the city ... seized Klazomenai ...*	c. 386?
28.6–7.	Parion (Hellespont)	***Iphiades of Abydos*** *was trying to take Parion on the Hellespont.*	362–59?
29.3–10		A trick once put into practice, which ... resulted in the capture of a city during public festival.	

(cont.)

29.11–12.		On similar occasions men in want of shields ... had recourse to importing osiers ...		
29.12.	Sikyon	*The Sikyonians for instance suffered a great disaster from neglect of these precautions.*		369?
31.2.		A message was once sent in the following manner ...		
31.6.	Ephesos	A message was brought to Ephesos in the following way ...		?
31.8		Again, a letter containing an offer of betrayal was once conveyed by a traitor ...		
31.9		Another man sent out a trooper with a note sewn up in his bridle rein.		
31.9		Here is another story about a letter. During a siege ...	Hdt. 6.4.?	
31.14		Again, a man has before now poured wax on a writing tablet ...	Hdt. 7.239	
31.23		Again, a note has been written on very thin papyrus ...		
31.24	Ilion	The men round Ilion, after all this time, and in spite of their efforts, are not yet able to prevent the Locrian maidens from coming into their city ... every year.		N/A
31.25–27.	Poteidaia	*<u>In earlier years</u> the following trick was once played. **Timoxenos** wished to betray Poteidaia to **Artabazos** ...*	Hdt. 8.128	479
31.28–29.	Miletos	*Again, when **Histiaios** wished to communicate with **Aristagoras** ... Miletos ...*	Hdt. 5.35	499
31.30–31		**DIONYSIOS (II) ... HERAKLEIDAS**		357
31.32.	Epiros, Thessaly	Dogs were often used in Epiros in the following way ... This method is used in Thessaly.		N/A
31.33	Lampsakos	*A letter was sent to **Astyanax, tyrant of Lampsakos,** containing information of the plot which proved fatal to him ...*		?
31.34	Thebes,	*The same delay caused the capture of the citadel in Thebes ...*		382
31.34	Mytilene	Something like this happened in Mytilene in Lesbos ...		?

31.35.	Persia	*When **Glous the Persian** admiral went up to see the king ...*		380?
37.6–7.	Barke	*There is an old story in this connexion, **Amasis**, while besieging Barke ...*	Hdt. 4.200	c. 512
40.2–3	Syracuse?	**Dionysios (1?)** once wished to occupy a city which he had conquered ...		?
40.4–5.	Sinope	*The men of Sinope, when at war with **Datames** ...*		c. 384–62

Bibliography

Anderson, G. (1993) *The Second Sophistic. A Cultural Phenomenon in the Roman Empire*. Routledge, London and New York.

Anderson, J.K. (1974) *Xenophon*. Duckworth, London.

Baragwanath, E. (2008) *Motivation and Narrative in Herodotus*. Oxford University Press, Oxford.

Bettalli, M. (1990) *Enea Tattico. La difesa di una città assediata (Poliorketika). Introduzione, traduzione e commento*. ETS editrice, Pisa.

Bowie, E.L. (1974) 'Greeks and their Past in the Second Sophistic', in M.I. Finley (ed.), *Studies in Ancient Society*. Routledge and K. Paul, London, Henley and Boston. 166–209.

Breitenbach, H.R. (1967) 'Xenophon', *Paulys Realencyclopädie der classischen Altertumswissenschaft* 9A: 1567–2052.

Brown, T.S. (1981) 'Aeneas Tacticus, Herodotus and the Ionian revolt', *Historia* 30: 385–393.

Buckler, J. (1980) *The Theban Hegemony. 371–362 BC*. Harvard University Press, Cambridge, MA.

Burliga, B. (2008) 'Aeneas Tacticus Between History and Sophistry: The Emergence of the Military Handbook', in J. Pigoń (ed.), *The Children of Herodotus: Greek and Roman Historiography and Related Genres*. Cambridge Scholars Publishing, Newcastle upon Tyne. 92–101.

Celato, S. (1967–1968) 'Enea Tattico. Il problema dell'autore e il valore dell'opera dal punto di vista militare', *Atti e memorie dell'Accademia patavina di scienze lettere ed arti* 80: 53–67.

David, E. (1986) 'Aeneas Tacticus, 11.7–10 and the Argive revolution of 370 BC', *American Journal of Philology* 107: 343–349.

Dain, A. and Bon, A.M. (1967) Énée le Tacticien. Poliorcétique. Les Belles Lettres, Paris.

Gehrke, H.-J. (1985) *Stasis. Untersuchungen zu den inneren Kriegen in den griechischen Staaten des 5. und 4. Jahrhunderts v. Chr.* Beck, Munich.

Gould, J. (1989) *Herodotus*. Weidenfeld & Nicolson, London.

Harding, P. (1987) 'Rhetoric and Politics in Fourth-Century Athens', *Phoenix* 41: 25–39.

Higgins, W.E. (1977) *Xenophon the Athenian*. State University of New York Press, Albany.

Hunter, L.W. and Handford, S.A. (1927), *Αἰνείου Πολιορκητικά. Aineias on Siegecraft*. Clarendon Press, Oxford.

Jacoby, F. (1913) 'Hellanikos (7)', *Paulys Realencyclopädie der classischen Altertumswissenschaft* 8: 104–153.

Krentz, P., Wheeler, E.L. (1994) *Polyaenus: Stratagems of War*. Ares Publishers, Chicago.

Lehmann, G.A. (1980) 'Krise und innere Bedrohung der hellenischen Polis bei Aeneas Tacticus', in Eck, W. et al. (eds), *Studien zur antiken Sozialgeschichte*, Festschrift Friedrich Vittinghoff. Böhlau, Cologne. 71–86.

Loraux, N. (1981), *L'invention d'Athènes*. Mouton, Paris and La Haye.

Losada, L.A. (1972), *The Fifth Column in the Peloponnesian War* (*Mnemosyne* Supplement 21). Brill, Leiden.

Luraghi, N. (1988) 'Polieno come fonte per la storia di Dionisio il Vecchio', *Prometheus*: 164–180.

Marinovic, L.P. (1988) *Le mercenariat grec au IVe siècle avant notre ère et la crise de la polis*, translated by J. Garlan and Y. Garlan. Annales Littéraires de l'Université de Besançon, Paris.

Mosshammer, A.A. (1979) *The Chronicle of Eusebius and Greek Chronographic Tradition*. Bucknell University Press, Lewisburg.

Pearson, L. (1941) 'Historical Allusions in the Attic Orators', *Classical Philology* 36: 209–229.

Perlman, S. (1961) 'The Historical Example, its Use and Importance as Political Propaganda in the Attic Orators', *Studies in History: Scripta Hierosolymitana* 7: 150–166.

Pretzler, M. (2007) *Pausanias. Travel Writing in the Ancient World*. Duckworth, London.

Pretzler, M. (2010) 'Polyaenus the Historian? Stratagems and the Use of the Past in the Second Sophistic', in Brodersen, K. (ed.) *Polyaenus. New Studies*. Verlag Antike, Berlin. 85–107.

Pritchett, W.K. (1974) *The Greek State at War. Part II*. University of California Press, Berkeley, Los Angeles, London.

Roy, J. (1972) 'Arcadian Nationality as seen in Xenophon's *Anabasis*', *Mnemosyne* 25: 129–136.

Schettino, M.T. (1988) *Introduzione a Polieno* (*Studi e testi di storia antica* 10). Edizioni ETS, Pisa.

Swain, S. (1996) *Hellenism and Empire. Language, Classicism, and Power in the Greek World, AD 50–250*. Oxford University Press, Oxford.

Vela Tejada, J., (1991) *Estudio sobre la lengua de la poliorcética de Eneas el Táctico*. Departamento de Ciencias de la Antigüedad de la Universidad Zaragoza, Zaragoza.

Vela Tejada, J., Martín García, F. (1991) *Eneas el Táctico, Poliorcética. Polieno, Estratagemas*. Gedros, Madrid.

Whitehead, D. (1990) *Aineias the Tactician. How to Survive Under Siege*. Clarendon Press, Oxford.

Whitmarsh, T. (2001) *Greek Literature and the Roman Empire. The Politics of Imitation*. Oxford University Press, Oxford.

Worthington, I. (1994) 'History and Oratorial Exploitation', in Worthington, I. (ed.), *Persuasion: Greek Rhetoric in Action*. Routledge, London and New York. 109–129.

CHAPTER 6

Creating *Koine*: Aineias Tacticus in the History of the Greek Language*

José Vela Tejada

A collective work devoted to Aineias Tacticus needs to include a linguistic approach, because Aineias also plays a leading role in the history of the Greek language. Apart from its contribution as a source for the study of the historical and literary context of the fourth century BC, the *Poliorketika* of Aineias Tacticus provides first-rate evidence from a linguistic point of view. As a matter of fact, this is the first extant work written in a form of Attic Greek by a non-Attic author, which might suggest that speakers of other Greek dialects which lacked a widely read literary expression began to adopt Attic as the literary language which conveyed authority. Writers began to compose prose works not using a pure Attic dialect, but a literary variety consisting of an Attic base enriched with features of Ionic origin.[1] This is precisely the context for a linguistic study of the *Poliorketika*: the beginning and development of Greek *koine*.

The Attic Origins of *Koine*

Broadly speaking, we can observe the beginnings of this process in Athens, in the second half of fifth century BC, when a blending between Ionic and Attic dialects occured.[2] The Ionic dialect had prestige, because it was the language of epic, elegy, iambic poetry, philosophy, historiography and medicine. Even the earliest Attic authors, such as Solon, wrote their poems in Ionic. Therefore, when early Attic literature emerges in tragedy or in the prose of Gorgias, Antiphon, Thucydides and the Old Oligarch, we are faced with a new liter-

* This study has been carried out under the general framework of the Team *Byblíon* H 52 (supported by the Gobierno de Aragón, Spain, and the European Social Fund).

1 According to López Eire (1981–1982) 24, Athenians were familiar with Ionian traits present in tragedy and scientific and literary prose. See also Horrocks (2010) 67.

2 Horrocks (2010) 81, dealing with traditional opinions of a creolized version of Attic, stresses for *koine* "the inherent implausibility of such a pidgin in the Attic-Ionic context."

 | DOI: 10.1163/9789004352858_007

ary language (the latest variant after the Ionic, Lesbian, Boeotian and Dorian dialects) characterized by an Attic base enhanced with terms from the literary Ionic tradition.

In sociolinguistic terms,[3] every language has, as a rule, two levels of 'cultural identification',[4] namely of High Culture and of Low Culture. Due to the prestige acquired by writers or intellectuals who cultivated Attic prose, this literary language turned into the high variety (H) of the Attic dialect. This cultured register then gained prominence over epichoric Attic, so that the high variety (H) was eventually integrated as 'standard' into the low variety (L), creating a *koiné diálektos*[5] (an idiom Athenians would have encountered particularly during the dramatic festivals). The Old Oligarch (*Ath.Pol.* 2.8) provides the earliest evidence for the sociolinguistic conditions in Athens during the second half of the fifth century BC:

> ἔπειτα φωνὴν πᾶσαν ἀκούοντες ἐξελέξαντο τοῦτο μὲν ἐκ τῆς, τοῦτο δὲ ἐκ τῆς· καὶ οἱ μὲν Ἕλληνες ἰδίᾳ μᾶλλον καὶ φωνῇ καὶ διαίτῃ καὶ σχήματι χρῶνται, Ἀθηναῖοι δὲ κεκραμένῃ ἐξ ἁπάντων τῶν Ἑλλήνων καὶ βαρβάρων.
>
> And then, since they heard every kind of dialect, they have taken something from each; and while Greeks mostly maintain their own dialect, way of life and mode of dress, the Athenians have adopted a mixture from all Greeks and Barbarians.

Thus, in the fifth century BC, Athens[6] not only became the 'heart of an Empire' (the Delian League was founded in 478 BC after the victory in the Persian Wars),

3 Brixhe (1979) 237 notes: "l'usage quotidien de la langue nous montre faite de variant et d'invariant, traversée par une foule de variations, qui sont certes quelquefois aléatoires: à la région, à l'ethnie, au sexe, à l'âge, à la classe sociale, au contexte du discours." Along the same lines, López Eire (1981–1982) 22, skilfully applied the ideas of Joshua Fishman (see the next footnote). With regard to Aineias see our 'Introducción' in Vela (1991) 64–76.

4 Fishman (1966) 428 introduced the concept of *diglossia* referred to a situation in which two dialects or languages are used by a single language community. He also introduced the pattern of sociolinguistic levels in the expected use of language.

5 As Meillet (1913) 264 remarked, "les classes cultivées parlent partout une même langue; les parles locaux, réduits au rang de patois, ont dû persister longtemps chez les gens de culture inférieur et dans les petites localités rurales [...] le dialecte est remplacé par une sorte de κοινή patoisée." See also López Eire (1981–1982) 21–22.

6 Hunter and Handford (1927) XLIV, were right in observing that *Κοινή* arose in Athens with the help of a strong literary tradition, jealousy guarded and kept pure to the last: "But the language they spoke is no longer the language of the people. It is already a language with a conscious

but also turned into the centre of Greek culture (Thucydides 2.41 calls Athens 'the School of Hellas'—'λέγω τήν τε πᾶσαν πόλιν τῆς Ἑλλάδος παίδευσιν εἶναι'). From this period, the Ionic-Attic *koine* began to develop into a lingua franca among the Greeks.

The decisive development towards *koine* coincided with the appearance of the Macedonian Kingdom on the Greek scene. In 413 BC, King Archelaos I, who admired Athens and its culture, hosted poets including Agathon and Euripides in Pella, where he had moved the capital from Aigai, and there they wrote new works for him, for example the tragedies *Archelaos* and *The Bacchae*. There were other artists, too, for example the musician Timotheus, Euripides' friend, and painters including Zeuxis, who was particularly celebrated at that time. Later, when Macedonia imposed hegemony over Greece, Attic *koine* was the official language at the court of Philip II, and Aristotle taught this linguistic variety to his son Alexander the Great. Subsequently, while Alexander conquered the Persian Empire, he spread *koine* as the universal language for communication.[7]

The Language of the Poliorketika

Any study of the language of the *Poliorketika* must bear in mind this process through which this originally Ionic-Attic *koine* eventually displaced the other dialects. Although the language of the treatise has been defined by some scholars as 'pure Attic', a synchronic study[8] allowed me to show that Aineias' language differs from the Attic standard with regard to phonetics, morphology, syntax and lexicon.

It includes linguistic features which do not have parallels in Attic inscriptions of the fifth and fourth century BC, while they agree with epigraphic documents of the Hellenistic period (third century BC), when *koine* replaced

mission, that of preserving the speech of Hellas in its purest and subtlest form: carefully chosen, carefully expurgated; no longer possessing or desiring to possess that unbounded capacity of reception and expansion which marks the speech of the street and the marketplace." See also Meillet (1913) 264, and López Eire (1981–1982) 36.

7 According to Horrocks (2010) 84, the expansion of *koine* naturally took place at the expense of the ancient dialects, written and spoken, while in the new Hellenistic kingdoms *koine* was the only written standard from the beginning, and the spoken language of the Greco-Macedonian aristocracy. See also Crespo (2010) 119–136.

8 This study took shape as my PhD later published as a monograph; see critical discussion in Vela (1991) 64–76.

the common Attic dialect. But if we compare this language with Attic literary sources of the classical period, we find meaningful similarities. These mainly consist of Ionic traits incorporated to improve literary Attic in the fifth century BC, as we have noted. In this context we can only provide a summary overview of these features, but this allows us to show that the *Poliorketika* is an example of the continued use of this variant of Attic, but deployed, at this point in the first half of the fourth century BC, outside its original geographical and dialectal limits.

Aineias certainly does not limit himself to a predefined bookish language. On the contrary, in comparison with earlier and contemporary examples, he uses an innovative idiom. This shows that, at the same time as the development of a literary *koine*, the coexistence of traces of both high and low registers leads to a restructuring of the language: after the high variety enters into conversational language, we see a development of regularization and simplification of this new combination. This can be observed in the *Poliorketika* by monitoring features in a number of linguistic categories.

1 *Phonetics*

In discussing phonetic features we start with consonants, since this is where we find some of the most meaningful characteristics to distinguish the phenomenon of *koine*. Thus, the Ionic groups -σσ- and -ρσ- entered the Attic dialect, replacing epichoric -ττ- and -ρρ-. In fact, when the Atticists of the first and second centuries AD tried to imitate 'pure Attic', one of the most characteristic traits recovered were -ττ- and -ρρ-.[9]

Attic inscriptions[10] adopted these features late: -σσ- from 350 BC and -ρσ- in the second century BC respectively. Literary sources, however, are more innovative, although different authors represent various stages of this development. Thus, the most Attic features of speech are found in the Old Oligarch, while Antiphon and Thucydides only use -σσ-. Aineias employs both groups, but includes a greater number of forms with -σσ-.[11]

9 For further information see Schmid IV (1896) 137 and 579.

10 Cf. Meisterhans and Schwyzer (1900) 101.

11 The statistical data from Antiphon, Thucydides and the Old Oligarch have been obtained from the TLG database; on Aineias' statistics see Vela (1991) 95.

	ΣΣ	ΤΤ
Antiphon	5	0
Thucydides	θαλασσ- 188	θαλαττ- 0
Ps. Xenophon	2	16
Aineias	78	24

Looking at the other consonant group, we always find -ρσ- in Aineias, with the only exception of πόρρω and πόρρωθεν,[12] and there our author coincides with late *koine* (this group is also regular in the New Testament,[13] even in Modern Greek). Thucydides[14] registers 63 occurrences of the root θαρσ- versus none (0) of θαρρ-. In contrast, Antiphon[15] consistently chooses the Attic option -ρρ- (14) instead of -ρσ- (0).

	ΡΣ	ΡΡ
Antiphon	0	14
Thucydides	63	0
Aineias	always	only πόρρ-

The next distinguishing features are the present tense forms of γίγνομαι and γιγνώσκω which also appear in their Ionic versions[16] γίνομαι and γινώσκω in some readings of manuscript M, the archetype of the *Poliorketika*. However, most of the editions usually emend them to forms with γιγν-, with the only exception of Hunter and Handford,[17] who sometimes even correct Aorist forms with γεν- to γιν-. In my view, only seven instances of the γιν- form in the archetype can be accepted (14.2; 20.4; 22.5, 6, 24; 31.20, 33), in contrast with 39

12 See Barends (1955) s.v. πόρρω and πόρρωθεν, and Vela (1991) 98.

13 Blass, Debrunner and Rehkopf (1976) 27.

14 Schrader (1998) s.v. θαρσ-.

15 Van Cleef (1964) s.v. θαρρ- and πορρ-.

16 In Attic inscriptions only γιγν- is attested until 306/5 BC. Inscriptions after 250 BC provide the latest evidences and forms as γιν- was by then generally used in *koine*. See Threatte (1980) 561–564.

17 Hunter and Handford (1927) LXVIII.

Attic forms with γιγν- (Pref. 1; 3.1; 4.3; 7.2, 4; 10.4, 23; 11.11; 12.2, 4; 13.1; 15.3, 4; 16.4, 5; 18.1 bis, 2, 4, 10, 11; 22.7, 10, 19, 25; 23.4; 24.14; 27.4, 6, 7, 8; 31.19 bis, 20, 27; 32.7, 10; 39.5 bis).

The *Poliorketika*[18] also displays some hiatus of Ionic provenance, mainly in contracted thematic declension: ἀθρόος (1.6), ἀθρόους (2.8 and 16.7), and ἀθρόα (30.2), facing to νοῦν (11.1, 22.15), and εὔνους (1.6); but also τείχεος (18.14, 22.14, 33.3, 38.6, 7), πάθεα (22.18), ἔτεα (31.24), and εἴδεος (24.1) from -σ- stems. Even though *koine* finally lost these instances of hiatus, these occurrences bear witness to a literary origin[19] before a definitive selection between Attic or Ionic traits had taken place.

Due to the weakness of intervocalic -ι-, the adverb ἀεί is attested in Ionic from Herodotos and the Corpus Hippocraticum onwards. After its integration into literary Attic—not yet present in Thucydides' work,[20] but anticipated in the Old Oligarch and Aineias—,[21] it was to become standard in *koine*:

	αἰεί	ἀεί
Thucydides	128	0
Ps. Xenophon	0	2
Aineias	0	15

Contrary to previous instances, in the reflexive pronoun Aineias[22] prefers the Attic contracted form αὑτοῦ to the Ionic ἑαυτοῦ, even though *koine* would later replace the contracted variant in order to avoid confusion with forms

18 Vela (1991) 122–123, 125–126.

19 In Attic inscriptions hiatus is regularly resolved by contraction: see Meisterhans and Schwyzer (1900) 126, and 131–137. We must also remember a literary Ionic influence documented in the *New Testament*: see Blass, Debrunner and Rehkopf (1976) 39.

20 See Schrader (1998) s.v. αἰεί.

21 Cf. Schrader, Vela and Ramón (2008) s.v. αἰεί and ἀεί, and Vela (1991) 79–80, Barends (1955) s.v. αἰεί and ἀεί, respectively. According to Meisterhans and Schwyzer (1900) 30–31, the fall of intervocalic -ι- in common Attic is attested in inscriptions from 361 BC—see also Threatte (1980) 322.

22 Cf. Barends (1955) s.v. ἑαυτοῦ and αὑτοῦ, and Vela (1991) 132–134. For Thucydides Schrader (1998) s.v. ἑαυτοῦ; for the Old Oligarch see Schrader, Vela and Ramón (2008) s.v. ἑαυτοῦ.

of the anaphoric pronoun αὐτός, which is also employed as reflexive. In Attic inscriptions[23] we can observe the progress of this restitution:

	ἑαυτοῦ	αὑτοῦ
Thucydides	250	0
Ps. Xenophon	3	0
Aineias	5	24
Attic inscriptions		
403 to 300 BC	31	23
300 to 30 BC	100	7

2 *Morphology*

In Morphology we can note some innovations aimed at simplifying the inflectional system:

'Attic' declension in -εως (evolved from *-η[ϝ]ος), in spite of not being regular in thematic -o- stems, subsists in Attic inscriptions until 250 BC.[24] Nonetheless in literary Attic[25] ναός appears from the fifth century: a more precise form without metathesis, spreading most probably from literary Doric to Attic tragedy. For instance, in Euripides we count 63 instances of ναός in contrast to 28 examples of νεώς as nominative singular or accusative plural. The new form became more common in literary *koine*, as can be seen even in Xenophon's *Anabasis* (although the old variant is more common in the *Hellenica*). Our treatise[26] again represents the most innovative option:

	ναός	νεώς
Euripides	63	28
Xenophon	5 *Anabasis*	1 *Anabasis*
	4 *Hellenica*	7 *Hellenica*
Aineias	2	0

23 See Meisterhans and Schwyzer (1900) 153–154. See also Threatte (1996) 324.

24 Cf. Meisterhans and Schwyzer (1900) 127.

25 Data from Euripides and Xenophon's *Anabasis* and *Hellenica* were obtained from the TLG database.

26 Cf. Barends (1955) s. v. ναός, and Vela (1991) 124–125.

We should assign a literary extraction to the Ionic forms ἄστεος (1.1), instead of ἄστεως, and νηῶν (16.21) alongside νεῶν (11.3), and equally to the two occurrences of the 'Aeolic' optative:[27] ἀποδείξειεν (3.1) and πεισθεῖεν (5.1).

Θέλω and ἐθέλω are attested in Greek from Homer, even if ἐθέλω is the most regular. Nonetheless, because of pronunciation within a sentence, aphaeresis of the initial ἐ- was common, and in *koine*[28] θέλω finally became the standard form—as it is in modern Greek—, but still keeping the long augment η- in its past indicative forms. In contrast with Thucydides,[29] our author[30]—like Antiphon—[31] is closer to *koine* in using ἐθέλω only five times, four of which are past tense forms beginning with ἠθελ-:

	Θέλω	ἐθέλω
Antiphon	5	1
Thucydides	0	70 (ἠθελ- 24)
Aineias	17	5 (4 ἠθελ-)

In verbal morphology one of the most important developments is the creation in Ionic of the -σαν ending which eventually replaced -ν endings. One of these instances is the third person plural imperative, in which -ντων and -σθων become -τωσαν and -σθωσαν. As far as *koine*[32] is characterized by an analogical simplification of inflections these new endings turn up early in literary Attic,[33]

27 The 'Aeolic' optative is a Homeric creation extended to all Greek literary languages including the latest Attic. It occurs even in the *New Testament*: see Blass, Debrunner and Rehkopf (1976) 54. Despite the disappearance of the Optative mode in Greek, the Atticist Movement recovered it as an imitation of classical models: see Schmid IV (1964) 588.

28 In Attic inscriptions from 250 BC flexion from θέλω is regular and long augment ἤθελον and ἤθελα are strengthened: see Meisterhans and Schwyzer (1900) 178, and Threatte (1980) 426.

29 Cf. Schrader (1998) s.v. ἐθέλω and θέλω.

30 See Barends (1955) s.v. ἐθέλω and θέλω, and Vela (1991) 85–87.

31 Cf. Van Cleef (1964) s.v. ἐθέλω and θέλω.

32 As a matter of fact, these features must be understood within a general process of simplification and regularization of the verbal system marked by the elimination of the athematic flexion in -μι, the radical aorist and misleading doublets: see Vela (1991) 144–151. In Attic inscriptions thematized forms appear first from 300 BC; cf. Threatte (1996) 619. In the *New Testament* athematic verbs in -νυμι (e.g. μείγνυμι) are usually replaced by thematic verbs (e.g. μίσγω): see Blass, Debrunner and Rehkopf (1976) 63–69.

33 Data from Euripides and Xenophon have been taken from the TLG database.

namely in Euripides and Antiphon. In the fourth century BC Xenophon provides 23 examples and Aineias[34] two:

Euripides	ἴτωσαν (*IT* 1480); ἔστωσαν (*Ion* 1131)
Antiphon	γενέσθωσαν (Fr 49, 32)
Xenophon	ἴτωσαν (*Anab* 1.4.8); θεραπευσάτωσαν (*Hiero* 8.4); ἐννοησάτωσαν (*Poroi.* 4.41, 5.5); ἔστωσαν (*Peri Hipp.* 10.8, 12.5; *Cyn.* 2.4, 4.6, 7, 8, 10.2, 3); ἰχνευέτωσαν and προΐτωσαν (*Cyn.* 4.3, 4); ποιείτωσαν (*Cyn.* 4.4); διωκέτωσαν (*Cyn.* 4.5); μεταθείτωσαν (*Cyn.* 4.5, 6); ἐπανίτωσαν (*Cyn.* 4.5, 6); ἐχέτωσαν (*Cyn.* 10.2)
Aineias	ἔστωσαν (10.25), διαιτάσθωσαν (13.3)

3 *Indeclinable Forms*

Some indeclinable forms, halfway between phonetics and morphology, also have to be considered.

The evolution of the preposition ενς in Ionic-Attic resulted in a double form ἐς/ εἰς. Although poetry[35] uses both variants for metrical reasons, from Homer onwards ἐς predominates in Ionic, while in Attic inscriptions[36] εἰς is the most attested. The prestige of Ionian literature probably explains the widespread use of ἐς in Thucydides, whereas Aineias as well as the Old Oligarch again represent[37] a more advanced stage:

	εἰς	ἐς
Thucydides	0	1804
Ps. Xenophon	10	0
Aineias	171	3

34 Vela (1991) 143–144.

35 Because of that, some Atticist authors such as Dio Chrysostom wrongly considered ἐς as an Attic form: see Schmid IV (1896) 579.

36 Thus, the long variant εἰς began to prevail in the 4th century BC: see Meisterhans and Schwyzer (1900) 213–214, and Threatte (1980) 178. In this respect, Xenophon provides crucial evidence, since he only used εἰς in the *Hellenica*, whereas ἐς also appears in the *Cyropaedia* and *Anabasis* (see TLG database).

37 See Schrader (1998) s.v. εἰς and ἐς; Barends (1955) s.v. εἰς and ἐς; Vela (1991) 106–107; Schrader, Vela and Ramón (2008) s.v. εἰς and ἐς, respectively.

The preposition ξύν was simplified to σύν early on, except in the Attic dialect;[38] thus in the genesis of *koine*, Ionic σύν took the place of ξύν.[39] As a consequence, our author together with Antiphon yet again appears as more innovative than Thucydides:[40]

Antiphon	σύν/ συν- (91)	only preverb ξυν- (8)
Thucydides	σύν (1)	always ξύν (also in preverbs)
Aineias	σύν/ συν- (139)	only preverb ξυμ- (2)

The combination of the numeral ἕν with the root *ϝεκ- to compose a new preposition of cause evolved differently in Ionic and Attic: with lengthening in the Ionian εἵνεκεν, compared to Attic ἕνεκα without lenghtening. Nevertheless, from the fourth century BC a hybrid form ἕνεκεν (actually an Ionicized form of the Attic ἕνεκα) begins to appear and is generally used in Attic inscriptions from the second century BC.[41] We can find it in literary sources of the fifth and fourth century BC, in which ἕνεκα is still the most common; but the exception is again Aineias who provides statistics typical of *koine*, followed by the Old Oligarch who shows a similar tendency:[42]

	ἕνεκεν	ἕνεκα
Antiphon	2	40
Thucydides	1	63
Ps. Xenophon	1	1
Aineias	9	2
Plato	6	458

38 From 410 BC σύν is the most attested variant and finally ξύν disappears in Attic inscriptions after 378 BC: see Meisterhans and Schwyzer (1900) 220–221, and Threatte (1996) 768.

39 In spite of its Attic colouring, ξύν is an unusual variant among most of the Atticists, save for Dio Chrysostom and Arrian: cf. Schmid IV (1896) 580.

40 Cf. Barends (1955) s.v. σύν; Vela (1991) 104–105; Van Cleef (1964) s.v. σύν and ξύν; Schrader (1998) s.v. σύν and ξύν, respectively.

41 Cf. Meisterhans and Schwyzer (1900) 215, and Threatte (1996) 660–668.

42 Statistics for Antiphon come from Van Cleef (1964) s.v. ἕνεκα and ἕνεκεν, and for Thucydides see Schrader (1998), s.v. ἕνεκα and ἕνεκεν; for the Old Oligarch and Aineias, respec-

Eventually hypothesis after crasis -ε(ἰ)ἄν- gave rise to a new phonetic form ἐάν. The resulting hiatus is solved by contraction: Attic ἄν, Ionic ἤν, which also appears in literary Attic, where it is most frequently used in Thucydides.[43] However, both contracted forms can be confused with the modal particle ἄν and the exclamation ἤν, so that non-contracted ἐάν beame accepted in *koine*,[44] to the extent of shaping the conditional particle in Modern Greek (ἐ)άν:

	ἐάν	ἄν	ἤν
Andokides	23	8	0
Thucydides	12	0	258
Ps. Xenophon	13	1	2
Aineias	63	23	0

In conclusion we can emphasise the main features of Aineias' language:

- Traces common to Ionic and literary Attic tend to replace those of the epichoric Attic: -σσ- instead of -ττ-, -ρσ- instead of -ρρ-, ἀεί from αἰεί, imperative endings -έτωσαν and -σθωσαν in place of -ντων and—σθων.
- Local forms are avoided, with the exception of those supported by common Greek (e.g. the Attic retroversion). Attic forms are preferred to those only attested in Ionic (ἕνεκεν vs. εἵνεκεν).
- Ionic forms replace epichoric peculiarities and incorporate into literary Attic innovations which had not yet developed locally: σύν from ξύν, γίνομαι and γινώσκω from γίγνομαι and γιγνώσκω; some cases of hiatus come from Ionian too.

tively, Schrader, Vela and Ramón (2008), s.v. ἕνεκα and ἕνεκεν and Vela (1991) 107–109; data from Plato obtained from the TLG database.

43 Cf. Schrader (1998) s.v. ἐάν and ἤν. With regard to data of Antiphon, the Old Oligarch and Aineias, see, respectively, Van Cleef (1964) s.v. ἄν and ἐάν, Schrader, Vela and Ramón (2008) s.v. ἄν and ἐάν, Barends (1955) s.v. ἐάν and ἄν; Vela (1991) 81–83.

44 In Attic inscriptions from 5th to 3rd century BC we only find 6 occurrences of ἄν or ἤν: see Meisterhans and Schwyzer (1900) 255–256, and Threatte (1996) 672–674. In the *New Testament* only 5 instances of ἄν are documented; see Blass, Debrunner and Rehkopf (1976) 84: "Eine Anzahl alter Partikeln fehlt ganz, andere sind nur in beschränktem, zB ἄν (= ἐάν), oder werden anders als im Klassischen gebraucht, zB ἐάν."

- With the exception of αὑτοῦ (instead of ἑαυτοῦ) variants common in Hellenistic Greek are more frequent in Aineias than in other classical sources: ἐάν/ ἄν/ ἤν, θέλω-ἤθελον, εἰς/ ἐς restitution of τελείως from τελέως, and ναός instead of Attic inflection νεώς.

4 *Syntax*

In addition to these mainly phonetic and morphological examples, we can also examine Aineias' use of syntax.[45] Among the features of *koine* we note especially a change in the use of prepositions: standard syntactic structures changed significantly as prepositional phrases replaced the use of the simple case, to the extent that we can speak of a deep restructuring of the prepositional system[46] which went hand in hand with a shift in the internal balance of the case system. As the case was weakening and the use of prepositional phrases was increasing, the whole system of prepositions also changed, and some prepositions with rather vague meanings began to disappear. The resulting system is defined by economy and functionality: a smaller number of prepositions convey meaning with more accuracy. Then the use of prepositions—actually a subsidiary inflectional issue—replaced simple cases, to such an extent that the Dative case eventually disappeared altogether in Greek *koine*.

For this purpose, we shall now present a detailed description of the functional status of prepositions, gathering comparative references from Aineias' work and from other sources representing several stages of Greek *koine*. To this end a detailed survey of statistical percentages will verify the degree of evolution of the prepositional system in our author.[47]

---

45 Other syntactic features were widely studied in our monograph—Vela (1991) 162–234: the use of verbal periphrasis instead of simple future and perfect; the increasing to employ the passive voice; the decline and disappearance of the Optative mode. They all deal with the general process of linguistic simplification, which characterizes the Greek *koine*.

46 For a detailed study see Vela (1993) 235–239.

47 For further information and a wide analysis of prepositions see Vela (1993) 240–245.

PREPOSITIONS + 1 CASE

AINEIAS TACTICUS	ATTIC AND KOINE
ἀνά (+ACC): 2 20.2; 31.24	Antiph(o Sophista) o/ Ps. Xen(ophon) o/ Thuc(ydides) 2/ Polyb(ius) 24/ N(ovum) T(estamentum) 13.
εἰς/ἐς (+ACC): 174 Pr. 2, 4 (bis); 1.5 (bis), 9 (bis); 2.1, 2; 3.1, 2, 5 (ter), 6 (bis); 4.10, 11; 6.1, 4 (bis); 7.2 (ter), 3; 8.1, 2 (bis), 3; 9.1 (bis), 2; 10.1 (bis), 12, 13, 22, 25 (bis), 26; 11.3, 8, 10, 10a (bis), 14, 15; 12.1, 4; 13.1, 4 (bis); 14.1 (bis); 15.7, 8, 9 (bis); 16.2, 7, 12, 13, 14, 15 (bis), 18, 19 (bis); 17.2, 4, 5; 18.1, 3 (bis), 4, 7, 9, 14, 18 (bis), 20; 20.1; 22.4, 9 (bis), 12, 13 (bis), 17, 19, 26, 29; 23.7 (bis), 10 (bis), 11 (ter); 24.5, 8, 12; 25.4; 26.3, 6; 27.2, 5, 7, 14; 29.1, 7, 11; 30.1, 2; 31.2, 4 (ter), 6, 8 (bis), 9, 11, 13, 14 (ter), 15 (bis), 16 (ter), 17, 18 (bis), 19 (bis), 20 (bis), 21, 22, 23, 26 (bis), 27, 29 (ter), 31 (bis), 35 (bis); 32.1, 3, 8 (bis); 33.2 (ter); 37.1, 2, 3, 4, 9 (bis); 38.7; 39.1 (bis), 2; 40.4, 6. [ἐς: 3.6; 33.2; 40.4]	Thuc. (1804), Ps. Xen. (10) and Polyb. (7919) use this prepositional shift widely, including the locative sense.
ἀντί (+GEN): 6 1.6; 22.17; 31.7, 21, 31; 40.4.	Thuc. 51/ Antiph. 6/ Ps. Xen. 1/ Polyb. 27.
ἀπό (+GEN): 38 2.5; 3.2; 4.1, 9; 6.4; 10.11; 15.6; 16.22; 18.15; 22.2, 9 (ter), 11, 12, 16, 19 (bis), 24; 24.1, 17; 25.3; 26.2, 14 (bis); 27.13, 15; 28.1; 29.9; 31.8, 18, 33; 32.5; 33.1; 34.2; 39.3; 40.1 (bis).	Thuc. 637/ Antiph. 15/ Ps. Xen. 11/ Polyb. 620/ NT 20.
πρό (+GEN): 5 6.1; 15.5; 16.2; 24.13; 26.3.	Thuc. 80/ Antiph. 2/ Ps. Xen. 0/ Ain. Tac. 5/ Polyb. 226/NT 48.

AINEIAS TACTICUS	ATTIC AND KOINE
ἐκ/ἐξ (+GEN): 83 Pr.1; 1.2; 2.2 (ter), 4; 3.1; 4.1, 5, 6 (bis), 8, 9; 6.1, 2, 5 (bis), 6; 10.3, 23; 11.2 (bis), 3, 7, 13, 15; 13.3; 14.2; 15.5, 10; 16.2, 6, 11, 21, 18, 19; 17.6; 18.19; 22.3, 9 (bis), 10, 14 (bis), 19, 29; 23.7 (bis); 24.18; 26.1, 9, 11, 13, 14; 27.10, 12, 15; 28.5; 29.8; 31.2, 8, 18, 19, 22, 24, 32; 32.1, 2 (ter), 8 (quat); 33.4 (bis); 36.2 (bis); 37.8; 38.7; 39.2; 40.1 (bis).	Thuc. 899/ Antiph. 102/ Ps. Xen. 12/ Polyb. 2130/ NT 100.
ἐν (+DAT): 235 1.2, 3, 6, 9; 2.1, 3, 4, 6, 7, 8; 3.3, 4; 4.1 (bis), 3, 8; 5.1; 6.7; 7.1, 3, 4; 8.2, 3, 5; 9.1; 10.3, 4 (ter), 5, 8, 11, 15 (bis), 18, 20 (bis), 21, 22, 23, 24; 11.3, 6, 7 (bis), 8 (quater), 10 (bis), 10a (ter), 11 (bis), 12, 13, 14; 12.5; 13.1, 3; 14.1 (ter), 2; 15.3, 4 (bis), 10; 16.4 (bis), 7 (bis), 9, 11, 12 (bis), 14, 20, 21; 17.1 (bis), 2, 3, 4 (bis); 18.1 (bis), 6, 8, 9 (bis), 12, 13 (bis), 19 (bis); 20.1, 4, 5; 21.1 (bis), 2; 22.1 (bis), 4, 5, 5a (bis), 7, 12, 13, 14, 16 (bis), 18, 19, 20, 21 (quater), 26; 23.1, 5 (bis), 6 (bis), 11; 24.3 (bis), 4, 5 (ter), 7 (bis), 11, 12, 14, 18; 25.2 (ter), 4; 26. 1 (bis), 4 (ter), 6, 7, 8 (bis), 9, 10; 27.2 (bis), 7, 8 (bis), 12; 28.1, 2, 3, 5 (bis), 7; 29.1, 2, 3, 4, 6 (quinquies), 7 (bis), 8 (bis), 9, 12 (bis); 30.2; 31.2, 3 (bis), 4a, 7, 8, 12 (bis), 14 (bis), 18 (quater), 21, 22, 27, 30 (bis), 31, 32, 33, 34 (bis), 35; 32.8; 35.1; 37.3, 4, 5; 38.1 (bis), 5 (ter), 6 (bis), 7; 39.1, 2 (bis); 40.1 (ter), 2, 3, 4, 8 (bis).	Thuc. 1793/ Antiph. 139/ Ps. Xen. 48/ Polyb. 2441.
σύν (+DAT): 6 10.26; 11.8; 17.1, 2, 4; 23.8.	Thuc. 38/ Antiph. 2/ Ps. Xen. 0/ Polyb. 120.

PREPOSITIONS + 2 CASES

AINEIAS TACTICUS	ATTIC AND KOINE
διά	
ACC (36): Pr.3; 2.6; 4.3; 5.1 (bis), 2 (bis); 6.1; 10.17; 11.1;15.2, 6, 9; 16.8, 12, 20; 18.1; 22.6, 13, 14; 23.6, 7; 24.7, 18; 26.6, 7 (bis), 8, 12; 27.2, 9, 11; 31.27, 34; 36.1; 38.3.	Thuc.(book I) 61/ Antiph. 39/ Ps. Xen. 26.
GEN (18): 6.5, 6; 10.2, 3, 10, 26; 18.20; 20.1, 2; 22.4, 5 (bis), 19, 24; 23.4; 31.3 (bis), 16.	Thuc.(I) 28/ Antiph. 11/ Ps. Xen. 5.
κατά	
ACC (84): Pr.1 (bis); 1.1, 5; 2.2; 2.4; 3.1, 3, 6; 4.2, 4, 5 (bis); 8.4 (bis); 10.4, 5, 8, 10 (bis), 12, 13, 19, 21; 11.2, 6, 10; 13.1; 15.2; 16.4 (bis), 8, 11, 13; 18.4, 16, 20, 21; 20.2, 4; 22.2, 3, 5, 5a, 9, 16 (bis), 21, 23; 23.4, 7, 11 (bis); 24.2, 5; 26.7, 12, 14; 27.3 (bis), 4 (bis), 5, 12, 15; 28.2 (bis), 3, 4, 5, 6, 7; 29.7, 8, 10, 12; 31.13, 31; 32.8; 37.5, 9; 38.3; 40.7 (bis).	Thuc. (I) 120/ Ps. Xen. 13/NT 398.
GEN (3): 2.8; 16.21; 38.7.	Thuc. (I) 1/ Ps. Xen. 0/NT 73.
ὑπέρ	
ACC (0)	Thuc. 6/ Antiph. 0/ Ps. Xen. 0.
GEN (4): Pr.2, 3; 22.29; 23.1.	Thuc. 58/ Antiph. 30/ Ps. Xen. 0.

PREPOSITIONS + 3 CASES

AINEIAS TACTICUS	ATTIC AND KOINE
ἀμφί (0)	Thuc. 3 (+AC.)/ Antiph. 0/ Ps. Xen. 0.
ἐπί 19	
ACC (47): 1.8; 4.5, 6; 2.5; 3.6; 6.6; 7.2; 10.20, 23; 13.1; 15.1, 2, 3; 16.1, 4, 5, 17; 17.1; 18.2, 20 (bis); 19.1; 20.5; 22.2, 9, 19, 23; 24.4, 5, 8, 10 (bis), 15 (ter); 25.3; 26.3; 28.1; 29.4, 10 (bis); 31.6, 27; 32.6; 33.1; 39.7; 40.6.	Thuc. (I) 102/ Antiph. 16/ Ps. Xen. 6/NT (predominant).
GEN (21): 4.7; 9.3; 10.15, 24; 16.14; 17.3; 18.14, 20; 22.4, 19; 26.5, 13; 27.2, 7; 28.4 (bis), 6; 30.1; 31.10; 36.1; 38.1.	Thuc. (I) 34/ Antiph. 9/ Ps. Xen. 0.
DAT (33): 2.6; 3.1; 5.1; 6.1 (bis), 4; 10.16, 24; 12.2, 4; 16.15, 16; 22.4, 9, 17, 20, 21; 26.13; 27.12, 15; 29.3; 30.1; 31.6, 23 (bis); 32.8; 38.4; 39.4; 40.6 (ter), 7 (bis).	Thuc. (I) 56/ Antiph. 30/ Ps. Xen. 2.
μετά	
ACC (10): 8.1; 10.20; 18.5, 16; 22.20; 24.9; 27.1; 31.18, 32; 39.6.	Thuc. (I) 32/ Antiph. 5/ Ps. Xen. 0.
GEN (40): 3.5; 4.5, 11, 12; 10.11, 15, 20, 21, 22, 24 (bis); 11.10 (bis); 12.2; 16.4, 11; 17.1, 4, 6; 18.13, 18; 22.2, 8, 20, 23; 23.3, 6; 24.5, 7, 9, 10; 26.10; 27.9, 14; 31.24; 38.2, 3; 39.5 (bis); 40.3.	Thuc. (I) 57/ Antiph. 17/ Ps. Xen. 0/ NT (predominant).
DAT (0)	Thuc. (I) 0/ Antiph. 0/ Ps. Xen. 0/NT 0.
παρά	
ACC (21): 1.2; 2.4; 3.5; 10.2; 16.21, 22; 22.27; 23.3; 24.2, 14; 31.2, 4a, 9a, 13, 14 (bis), 33, 35 (bis); 39.2; 40.1.	Thuc. (I) 36/ Antiph. 11/ Ps. Xen. 1.
GEN (13): 6.2; 10.6; 11.7; 13.2 (bis), 4; 18.5; 22.7, 27, 28; 24.6; 31.9a, 9b.	Thuc. (I) 11/ Antiph. 12/ Ps. Xen. 5.
DAT (7): 10.9; 18.11; 22.10 (bis); 23.3; 27.13; 31.25.	Thuc. (I) 8/ Antiph. 5/ Ps. Xen. 0.
περί	
ACC (36): 1.4; 7.1; 10.21; 15.6 (bis), 7; 16.11; 17.1; 18.1, 8, 12; 20.1; 21.1; 22.3, 7, 11, 17, 18, 19, 23; 24.3, 13; 27.1, 5; 28.4, 6; 31.9, 19, 20, 24, 26, 32, 33, 35; 33.2; 34.2.	Thuc. (I) 17/ Antiph. 5/ Ps. Xen. 0.
GEN (19): 7.4; 8.5 (bis); 14.2; 18.1; 21.1 (bis), 2; 22.25; 24.1; 29.1; 31.1, 3, 4, 8, 35 (bis); 40.8 (bis).	Thuc. (I) 50/ Antiph. 71/ Ps. Xen. 10.
DAT (0)	Thuc. (I) 5/ Antiph. 1/ Ps. Xen. 0.

(cont.)

AINEIAS TACTICUS	ATTIC AND KOINE
πρός	
ACC (90): Pr.2; 1.1 (bis), 2, 3 (bis), 6, 8; 2.8; 3.3, 4, 6; 4.3 (bis), 5, 6, 12; 8.2 (bis), 3; 9.1; 10.6, 25, 26; 11.5, 7, 15; 16.7, 12, 16, 18 (ter), 20, 21; 17.1, 3, 5; 18.9, 11 (bis), 12, 17, 18 (bis), 19 (bis), 21; 22.5, 6, 15, 22, 24; 23.8, 9; 24.11 (bis), 17, 18; 25.1; 26.8, 13, 14; 27.2, 8; 28.6 (bis), 7; 29.8; 31.4 (bis), 9, 10, 29, 32; 32.1, 7, 8 (bis), 12 (bis); 33.3 (bis); 36.2; 37.6, 7; 38.1; 39.7 (bis); 40.4.	Thuc. (I) 98/ Antiph. 21/ Ps. Xen. 4.
GEN (0)	Thuc. (I) 3/ Antiph. 7/ Ps. Xen. 0/.
DAT (5): 17.3; 27.9; 31.33; 39.2, 4.	Thuc. (I) 8/ Antiph. 6/ Ps. Xen. 10.
ὑπό	
ACC (3): 20.3; 26.1; 31.8.	Thuc. (I) 3/ Antiph. 8/ Ps. Xen. 0.
GEN (36): 2.6 (bis); 3.3; 4.1; 5.1; 6.7; 7.2; 10.10; 12.5; 13.3; 15.2; 16.13, 21 (bis); 18.2, 13, 16, 17, 20; 20.3, 4; 22.11, 12; 24.14, 16; 27.1, 6; 31.4, 8, 9b; 32.1, 5; 33.3; 37.3; 39.2; 40.2.	Thuc. (I) 56/ Antiph. 124/ Ps. Xen. 6.
DAT (1): 37.9.	Thuc. (I) 1/ Antiph. 0/ Ps. Xen. 0.

In fact, the reorganisation of the prepositional system is not exclusive to *koine*: even earlier, DAT(ive) uses with prepositions are reduced to 7, in contrast with those 13 + GEN(itive) and 15 + ACC(usative). Nevertheless, this survey of data highlights that in the early stages of *koine* a reorganisation of the prepositional system did indeed take place, following a tendency to simplify language. Accordingly, given the unclear distinction between functional roles of prepositions linked to more than one prepositional phrase, opposition is neutralized and there is a tendency to generalise just one case, at the same time as the number of prepositions is reduced. Statistical patterns[48] found in Aineias' treatise are consistent with these developments:

48 The aftermath of this process throughout the Hellenistic and Roman *koine* and the Byzantine period will be the definitive disappearance of ἀνά, ἐν, ἐπί, σύν, ὑπό and ὑπέρ, while ἐκ, περί and πρό survived a little longer. Consequently, from 18 prepositions in Classical Greek, only six are remaining unchanged in Modern Greek. The preposition ἀπό is the only one

- Disappearance of ἀμφί and prepositional phrases of μετά, περί + DAT and πρός + GEN.
- Decline due to limited functionality of ἀνά, ἀντί, πρό, σύν + GEN, and μετά, ὑπέρ + ACC.
- Confusion of ἀπό and ἐκ with continuing preeminence of ἐκ.
- Preservation of ἐν + DAT, and διά, εἰς, κατά + ACC.
- Prepositions with three cases do not disappear but tend to be limited to a single prepositional phrase: ἐπί, παρά, περί, πρός + ACC, and ὑπό, μετά + GEN.
- In general, prepositional phrase + ACC is the best maintained in contrast with prepositional phrase + DAT, consistent with the earlier weakening of this case.

In conclusion, the deep relationship between the reduction of random functions and the simplification of the prepositional system is attested early in Aineias' work; the information gained from close analysis of his text is therefore crucial for our understanding, from a diachronic perspective, the nominal syntactic restructuring which began to take shape even in the earliest versions of *koine*.

5 *Vocabulary*

After dealing with grammatical details we shall approach the lexical field, which can contribute more relevant data.[49] If we take into account the short length of the work, Hunter and Handford[50] were correct when they stated that Aineias' vocabulary is one of the most extensive and peculiar in Greek literature. He uses literary terms from the most distinguished literary traditions[51]

to keep its old semantic identity; πρό, παρά, ἀντί, μετά, κατά express at times the same meaning as Classical Greek by metaphorical extension, while another four prepositions are direct descendants of Classical Greek, but with phonetic changes: με (<μετά); γιά (<διά) σε (<εἰς); μαζί (<ἅμα + σύν). See Vela (1993) 246–247.

49 Hunter and Handford (1927) LXII, emphasize that we can count a total of over hundred words "which he is either the first extant Greek authors to use, or the first to use in a peculiar sense."

50 Hunter and Handford (1927) LXII.

51 Van Groningen (1938) 334, perceived in Aineias a bookish vocabulary: "le vocabulaire d' Énée le Tacticien n' est pas celui d' un précurseur de la, mais celui d' un auteur technique qui a enrichi et embelli son fonds personnel par des emprunts faits à quelques auteurs qui lui étaient familiers." However it seems incoherent not to consider as innovative, from a lexical point of view, a language recognized as so innovative in other grammatical fields. Furthermore, Aineias uses a significant number of words not attested in previous sources: see Vela (1991) 324.

alongside vocabulary not attested before. Hunter and Handford[52] established four categories of words:

- words attested in epic or lyric poetry (documented or not in Ionic prose).
- words located for the first time in the recited portions of tragedy.
- words found first in Ionian prose and not in poetry.
- words witnessed for the first time in the later prose (e.g. Xenophon, Aristotle, Theophrastus, etc.).

In addition, there are words which Aineias has in common with Thucydides or Plato, but which are not found in pure Attic (these are generally technical terms and nouns with suffixes -μα and -σις), words attested only in writers of the post-Aristotelian period, and finally 41 *hapax legomena* or words not found in any extant Greek writer except Aineias.

Thus, we can test the same process as in the previous linguistic fields: at first, vocabulary of Ionic origin[53] is integrated into early Attic literature, but not in the orators, who used pure Attic. Therefore these words fit in with the development of literary Attic, which in turn provided the basis for the genesis of *koine*.

Authors who contributed to this expansion of vocabulary in tragedy and Attic prose were innovators and also created new words; these terms enriched the vocabulary of *koine*, as can be seen in literary works of the Hellenistic and Roman periods. In the case of our treatise, verbal and nominal compounds,[54] addition of prefixes δυσ-, εὐ-,[55] double preverb[56] before the verbal stem, and, above all, suffixes play a significant role in lexical configuration:

---

52 Hunter and Handford (1927) LIII–LXII.

53 One of the most distinctive features of our treatise is the Ionic origin of much of its lexicon. On this aspect Hunter and Handford (1927) LIII, point out that "it is in his diction above all that he betrays himself, and the more closely we study it, the more do we recognize the prevalence of the Ionic element so strong in the diction of Xenophon."

54 In our treatise the most extensive compounds are the verbal dependent compounds: denominative verbs in -εω (e.g. πολιορκέω: 12.3, 31.9, 37.6, and 39.1) well attested from Homer in the Ionian literary tradition. In some cases we see substantives derived from these verbs (e.g. πολιορκία: 10.20 and 23, 40.8): cf. Vela (1991) 314–315.

55 These prefixes are inherent to the development of *koine*—Vela (1991) 309. As a matter of fact, a large number of occurrences are recorded in the *New Testament*: see Blass, Debrunner and Rehkopf (1976) 94.

56 Verbs prefixed with two preverbs (τριπλᾶ) are typical in Herodotos' prose (e.g. συνεκβάλλω: 3.108), from which they enter in the Attic prose of 5th century (e.g. ξυνδιαβαίνω: Thuc.

- *nomina actionis*[57] -μα, -σις, -σία, and -μος.
- *nomina agentis*[58] with -τηρ, -τωρ -της, the most productive.
- adjectives[59] with suffixes -ικός and -τος.
- new thematic verbs[60] suffixed with -έω, -εύω, -όω, -άζω, -ίζω, -νω.

As a result, we encounter a language very suitable for creating new technical terminology and specialized lexica. Apart from these lexical procedures, the systematic nominalization[61] of infinitives, participles, adjectives, adverbs and so forth facilitates abstraction of thought and adds expression to Greek vocabulary[62] alongside the restructuring process we have observed in the grammatical field.

6 *Style*

In order to complete a linguistic analysis of the language used in the *Poliorketika* and to achieve a more accurate knowledge of our author, we need to pay special attention to style. Since this treatise is a literary work, stylistic affinities are key elements to set the work in a context. We have to consider the fact that we are dealing with an author who was hypothetically born in Stymphalus, Arkadia, or at least came from the Peloponnese. Nevertheless, he wrote in Attic *koine* and, as can be inferred from stylistic analysis, he followed the Attic prose style which was the prevalent literary template in this period by using λέξις εἰρομένη (the parataxis as characteristic of the earliest stage of Greek prose) in combination with traces of λέξις κατεστραμμένη (in the *Preface*), a wide range of figures of repetition, antithesis—following the Gorgian scheme—, and μεταβολαί (*Variatio*).[63]

4.62), to the extent that simple verbs like βαίνω appear only once in Thucydides (5.7.1) and Xenophon (*Anab.* 3.2.19). See Schrader (1998) s.v. βαίνω, and Vela (1991) 312–313.

57 Cf. Vela (1991) 299–302.

58 Cf. Vela (1991) 303–305.

59 Cf. Vela (1991) 306–308.

60 The creation of new thematic verbs—mostly with suffix -έω—is essential in the process of simplification of the Greek language in *koine*, at the same time eliminating athematic verbs: see Blass, Debrunner and Rehkopf (1976) 86, and Vela (1991) 309–312.

61 Cf. Vela (1991) 318–319.

62 Thesleff (1966) 89, underlines the exactness of expression—based upon a consistent terminology—and the abstractness of expression—founded upon the wide use of abstract nouns—as 'typical tendencies' of a scientific style.

63 See Hunter and Handford (1927) LXXVII: although they often suggest Thucydides' influence on our author, they also assert that use of μεταβολή by Aineias is not so much due to imitation of Thucydides' style as to the carelessness of an unskilful writer.

Otherwise, in contrast to other prose works, Aineias' language shows a style consisting of an amalgamation of argumentation, and regression of emotion and stylistic ornamentation more typical of technical prose, the τέχναι,[64] characterized by a scientific style.[65] This kind of style began to appear in Ionian prose from 440 BC, having its roots in ὑπομνήματα and practical catalogues from the Archaic Age. From these early texts arose handbooks, τέχναι, whose subject deals with practical knowledge, and finally became general in scientific-technical learning.[66]

There is no doubt that in this period, the need for specialist knowledge[67] was steadily increasing, and a writer who wanted to disseminate this knowledge was not forced to discuss it and present his view in front of an audience. On

64 As Zhmud (2006) 45 stresses, in the second half of the fifth century most activities involving skills based on knowledge and experience were subsumed under the notion of τέχνη. In this respect, according to Thesleff (1966) 107, "Xenophon's Socratic Οἰκονομικὸς λόγος [...] and the rest of the technical works of Xenophon and the Tactics of Aineias, represent for us the 4th century τέχναι with moderately consistent and wide application of the so-called scientific style." See also Van der Eijk (1997) 99–120.

65 Thesleff (1966) 89, defines the scientific style as "a continuous, systematic, and discursive, though non-rhetorical and non-emotional prose."

66 In the first period of Greek science, which developed in general terms until the fourth century BC, the basic principles are established and τεχνίται begin to create a scientific language. Then we see the emergence of handbooks (τέχναι, according to the name that they received in the fifth century BC), most probably the type of older organised scientific writing: they can adopt the form of collections, as happened in the first manuals of rhetoric or treatises. The characteristic that defines them is practical utility and the fact that they are addressed to an audience with experience. At this stage, Zhmud (2006) 45 gives a prominent role to the Sophists in teaching sciences: "with the Sophists, too, who appeared on the intellectual scene at this period, knowledge was, as a rule, looked upon as purely utilitarian." Even Socrates shows a utilitarian attitude towards science, as seen in Xenophon, *Mem.* 4.7.1–8.

67 We can speak of a technical revolution in fourth-century BC Greece; and a 'military revolution', as Cuomo (2007) 67 puts it, was a part of this, reaching its peak in the Hellenistic period: "not just because it saw the rise of new machines, but also because it gave wider currency and respectability to some features essential to technical knowledge," but there was a general belief that "one becomes a good soldier and general through training and learning." At this point, regardless of a hypothetical personal and literary relationship with Xenophon, we believe that the most significant difference between both authors lies precisely in this point: whereas Aineias' τέχνη is focused on the practical preparations for a commander, for the historian it is part of an education; the τέχνη στρατηγική is a component of the *paideia* of the individual—see. Xen., *Cyr.* 1.5.1. For further information see Wilms (1995) 110–115.

the contrary, he could write his treatise in the isolation of his library or study. The hegemony of the written word had started and from then on authors begin to corroborate their works with references to a written record.[68]

However, Aineias' life is a mystery and he does not provide personal information. It is evident that he read the eminent historians of the fifth century, but we cannot find a definite link to Athens. Hunter and Handford[69] often insist in the influence of the Sophists, whereas Thesleff[70] undervalues their contribution. The fourth century also saw an increase in systematic scholarship and Aineias displays a wide knowledge of military and poliorcetic innovations, above all from Western Greece (Aineias refers twice to Dionysios I, the tyrant of Syracuse, in 10.21 and 40.2–3, whose military advances were inspired by the

68 Van der Eijk (1997) 93–99, thinks that the development of written prose must be understood as a consequence of "a new attitude towards knowledge," resulting from an aim to store data obtained from practical experience and to systematize them in order to make them accessible to other experts. In fact in Greece, towards the end of the fifth century, and at Athens in particular, there was a 'literate revolution' and the transformation into a society of written culture in which writing began to prevail as a transmitter over the ancient oral transmission. For a study in depth, see Havelock (1982) 3–38, and Thomas (1989)—above all 15–94—, who in n. 56, 7, remarks: "for though they used oral tradition, they may have rearranged their material considerably (as indeed a study of oral traditions suggests) and engaged in extensive research."

69 For instance, Hunter and Handford (1927) XLV: "As Athens became conscious, under the influence of the sophistic teaching, of her own paramount capacity for literary expression, the tendency in prose is more and more to purity, select, and expurgate." Thus Xenophon (*Mem.* 3.1) refers to Dionysodoros—most probably the Sophist of Chios visiting Athens to teach tactics—, and to Simon of Athens (*Peri Hipp.* 1.1 and 3; 11.6) as a master of *hippiké téchne*. See Van der Eijk (1997) 81.

70 Thesleff (1966) 106 only allows for an indirect contribution of the sophists to the intellectual background of prose writing: "The influence of the Sophists on technical writing hardly was of any importance. They were popularizers and propagators and are more likely to have made use of existing τέχναι than to have inspired the composition of such texts." We can consider, however, the influence of the Sophists on promoting the book as a means of preserving knowledge, which was already an underlying goal in historiography after Herodotos—even in the *Preface* when he says: "[...] so that what has been done by mankind be not forgotten in time and that great and wonderful deeds achieved by Greeks as well as Barbarians may not lose their glory" '*μήτε τὰ γενόμενα ἐξ ἀνθρώπων τῷ χρόνῳ ἐξίτηλα γένηται, μήτε ἔργα μεγάλα τε καὶ* θωμαστά, τὰ μὲν Ἕλλησι τὰ δὲ βαρβάροισι ἀποδεχθέντα, ἀκλεᾶ γένηται'—and in Thucydides' famous declaration that his work is written as "[...] a possession for all time rather than for an immediate performance"—'κτῆμα ἐς αἰεὶ μᾶλλον ἢ ἀγώνισμα ἐς τὸ παραχρῆμα' (1.22). Cf. Van der Eijk (1997) 90.

Carthaginians). Archytas[71] of Tarentum, a scientist of the Pythagorean School as well as a good friend of Plato, played an important role in these developments. Philosopher, mathematician, astronomer, statesman and strategist, the Tarentines elected him *strategos*,[72] and for seven years he remained undefeated in Tarentine campaigns against their southern Italian neighbours.

Nonetheless, what is most remarkable is that, despite being a professional soldier and not a proper literary author, Aineias knew literary tradition[73] and cultivated prose which contributed to the consolidation of the Attic dialect as the language of prestige even beyond its original borders. According to Hunter and Handford,[74] despite being Arkadian by birth, coming from a region most remote from the urbane influence of Athens, Aineias wrote 'in a dialect which

---

71 Thanks to some of the fragments of his work (DK 47), we know Archytas (born about 435/439 BC) used the Sophistic theory of τέχνη, which was initially oriented toward practical knowledge, in theoretical descriptions. Diogenes Laertius' biography informs us that he was the first to turn mechanics into a system by applying mathematical principles—'Οὗτος πρῶτος τὰ μηχανικὰ ταῖς μαθηματικαῖς προσχρησάμενος ἀρχαῖς μεθώδευσε' (8.83). His invention of a mechanical dove is well known. The Aristotelian treatise on *Mechanical Problems* probably derives its main features from Archytas' mechanics. For further information see Zhmud (2006), 60–70.

72 Cuomo (2007) 36 notes that "Athens in the fifth or fourth century BC saw the rise of a new breed of political men," whom she depicts as technicians. In this regard, we have suggested—Vela (2004) 146—that Aineias' work, in which his professional knowledge and personal experiences are collected, could relate to the specialization that the art of war had reached: "could have been addressed to any *polis* in a critical situation and to anybody with special responsibility: officers with civil and military responsibilities; magistrates temporarily invested with military functions; a general given unrestricted powers (στρατηγὸς αὐτοκράτωρ); or even a military expert hired from outside the community."

73 Within the framework of Greek prose, Aineias seems to have been mainly influenced by reading historians. In fact, the only literary sources attested in the *Poliorketika* are Herodotos and Thucydides. From Herodotos (4.200) he takes the reference to the siege of Barke by Amasis (37.6–7); from 5.35.3, the notice of the correspondence between Histiaios and Aristagoras (31.28–29); from 7.239.3, the delivery of Demaratus' secret letter to the Lacedaemonians (31.14); from 8.128, sending of correspondence between Timoxenus and Aratabazus. From Thucydides 2.4 he takes the account of the Theban entry into Plataea in 431 BC (2.3–6). The reference in 11.12 to the revolt of the *Parthenioi* resembles more that of Ephorus quoted by Strabo (278–280). And finally, the account of the defeat of the Megarians by Pisistratus (4.8–11) follows an unknown tradition confirmed by Justin (2.8) and Frontinus (2.9.9), but divergent from that of Plutarch, *Sol.* 8, who attributes victory to Solon. Cf. Hunter and Handford (1927) XXXIV–XXXVII, and Vela (1991) 37–43.

74 Hunter and Handford (1927) LII.

is not his own. It was, to be sure, the recognized international speech of his day; still, it looks as if he is not quite at ease while expressing his ideas on paper, and we must not forget that he is, as far as we know, the first Peloponnesian who used Attic Greek as his medium of expression. He had probably some sort of schooling, if not at Athens, in some other literary centre, such as Sikyon was in his day.'

A Language for Universal Communication

With regard to language, the aim of Aineias' work is stated in his own words, when in chapter 24.1, *Of Passwords*,[75] he points out: 'When giving passwords to armed forces drawn from a mixture of states and nationalities [τὸ στράτευμα μιγάδες ὄντες ἀπὸ πόλεων ἢ ἐθνῶν], take care to avoid the uncertainty which can arise if two names are applicable to a single entity [τὸ ἓν εἶδος δύο ὀνόματα, ἀμφιβόλως παραδοθήσεται].' Then Aineias gives pairs of passwords and adds in § 2: '[...] Ares and Enyalios, Athena and Pallas, sword and dagger, torch and light, and suchlike—all of them hard to remember [δυσμνημόνευτά] if contrary to the normal usage of the different nationalities [ἑκάστῳ ἔθνει τῶν ἀνθρώπων], and dangerous [βλάβην φέρει] to use if given out in dialect form ⟨rather⟩ than in universally intelligible language [κατὰ γλῶσσάν τις παραγγέλλῃ ⟨μᾶλλον⟩ ἢ κοινόν τι ἅπασιν].'[76]

Therefore, in § 3 Aineias advises not to issue such passwords to a mixed force of mercenaries or to allies of different nationalities ['Εν μιγάσι δ' οὖν ξένοις οὐ δεῖ τὰ τοιαῦτα παραγγέλλειν, οὐδὲ ἐν ἔθνεσι συμμάχοις]. Aineias goes on telling an incident where troops sent from Imbros by Athenodorus in order to relieve Charidemus in the city of Ilion were discovered because they did not remember the password properly: they said 'Tyndaridae' while Charidemus' password was 'Dioscuri'.

---

75 24.1. Παραδιδόντα δὲ συνθήματα δεῖ προνοεῖν, ἐὰν τύχῃ τὸ στράτευμα μιγάδες ὄντες ἀπὸ πόλεων ἢ ἐθνῶν, ὅπως μή, ἂν παρέχῃ τὸ ἓν εἶδος δύο ὀνόματα, ἀμφιβόλως παραδοθήσεται, [...] 24.2. καὶ ἄλλοτε δὲ Ἄρης Ἐνυάλιος, Ἀθηνᾶ Παλλάς, ξίφος ἐγχειρίδιον, λαμπὰς φῶς, καὶ ἄλλα ὁμότροπα τούτοις, ἅπερ δυσμνημόνευτά ἐστιν παρὰ τὰ νομιζόμενα ἑκάστῳ ἔθνει τῶν ἀνθρώπων καὶ βλάβην φέρει, ἐὰν κατὰ γλῶσσάν τις παραγγέλλῃ ⟨μᾶλλον⟩ ἢ κοινόν τι ἅπασιν. 'Εν μιγάσι δ' οὖν ξένοις οὐ δεῖ τὰ τοιαῦτα παραγγέλλειν, οὐδὲ ἐν ἔθνεσι συμμάχοις. We follow the translation of Whitehead (1990) 75. In the commentary to this chapter (p. 164), Whitehead points out that on passwords Aineias is drawing on material earmarked for even fuller exposition in another treatise, *Encampment*, quoted in 21.2.

76 See Whitehead (1990) 75, and commentary 165.

As we can see, this anecdote does not present a dialectal misunderstanding but the mistake is due to cultural differences, and Aineias says later on (24.14): 'The passwords you give should therefore be both easy to remember and related as far as possible to the business in hand,'[77] and shortly after (24.15) he recommends to use, as far as possible, words that will be intelligible to all the troops [καὶ ὡς μάλιστα ὁμότροπα τούτοις καὶ κοινότατα πᾶσιν].

Our author probably was a commander[78] experienced in leading fourth-century mercenary armies—seeing that he distinguishes accurately between ἐθνῶν and ξένοις—and we can note his awareness of the need not to make use of a local language [κατὰ γλῶσσάν τις παραγγέλλῃ ⟨μᾶλλον⟩ ἢ κοινόν τι ἅπασιν]. And this is precisely the aim that would help to expand *koine* all over Greece: not to serve as a new epichoric language, but as a common idiom supported by cultural prestige. This would indeed become the principle underlying the idea of oikoumene expanded by Alexander the Great in the last decades of the same century throughout the East and Egypt: a language for a wider universal communication.[79]

In conclusion, the language of the Poliorketika constitutes valuable evidence for the initial stage of the evolution of *koine*. It supports the hypothesis which locates the beginnings of this phenomenon in mid-fifth century Athens and some of its features help us track the further development of *koine* in a later period. We should note that the *Poliorketika* was written by a man who did not come from Attica: if we accept that Aineias came from Arkadia or at least from the Peloponnese, we must assume that he could not draw on a vernacular literary tradition native to his own place of origin. Given this context, the language of the treatise exhibits a high degree of complexity and it provides the earliest evidence that, as Meillet noted,[80] the prestige of literary Attic began to transcend its natural boundaries and to make an impact on Hellenic culture as a whole.

77 Παραγγέλλειν οὖν χρὴ τὰ συνθήματα εὐμνημόνευτά τε καὶ ὡς μάλιστα ἀδελφὰ ταῖς μελλούσαις πράξεσι γίγνεσθαι. See Whitehead (1990) 76.

78 According to Cuomo (2007) 65, "Aineias emphasizes experience, preparedness and foresight over anything else." The possession of military experience clearly determines his own qualification to narrate military events: see Vela (2004) 144.

79 As stated by Horrocks (2010) 4, the so-called Hellenistic *koine* "eventually came to dominate the Greek-speaking world, having first been endorsed and adopted by the all-conquering Macedonians and then carried throughout the East as an administrative and cultural language by the campaigns of Alexander the Great." Crespo (2010) 122–123.

80 See Meillet (1913) 260–264.

Bibliography

Barends, D. (1955) *Lexicon Aeneium*. Assen Vangorcum, Utrecht.

Blass, F., Debrunner, A. and Rehkopf, F. (1979) *Grammatik des neuestestamentlichen Griechisch*. Vandenhoeck-Ruprecht, Göttingen.

Brixhe, C. (1979) 'Sociolinguistique et langues anciennes. A propos de quelques traitements phonétiques en grec', *Bulletin de la Société de linguistique de Paris* 74: 237–259.

Crespo, E. (2010) 'The Significance of Attic for the Continued Evolution of Greek', in Caragounis Ch.C. (ed.), *Greek, a Language in Evolution: essays in honour of Antonios N. Jannaris*, Olms, Hildesheim. 119–136.

Cuomo, S. (2007) *Technology and Culture in Greek and Roman Antiquity*. Cambridge University Press, Cambridge.

Fishman, J. (1966) *Language Loyalty in the United States*. Mouton, The Hague.

Havelock, E.A. (1982) *The Literate Revolution in Greece and its Cultural Consequences*. Princeton University Press, Princeton.

Horrocks, G. (2010) *Greek: a History of the Language and its Speakers*. Wiley-Blackwell, Chichester.

Hunter, L.W. and Handford, S.A. (1927), *Αἰνείου Πολιορκητικά. Aineias on Siegecraft*. Clarendon Press, Oxford.

López Eire, A. (1981–1982) 'Fundamentos sociolingüísticos del origen de la koiné', *Cuadernos de Filología Clásica* 17: 21–53.

Meillet, A. (1913) *Aperçu d'une histoire de la langue grecque*. (eighth edition: 1975) Klincksieck, Paris.

Meisterhans, K. and Schwyzer, E. (1900) *Grammatik der Attischen Inschriften*. Wiedmann, Berlin.

Schmid, W. (1887–1897) *Der Atticismus in seinem Hauptvertretern. Von Dionysius von Halikarnass bis auf den zweiten Philostratus, I–V*. Stuttgart; new imprint 1964: Olms, Hildesheim.

Schrader, C. (1998) *Concordantia Thucydidea*. Olms-Weidmann, Hildesheim; Zürich; New York.

Schrader, C., Vela, J. and Ramón, V. (2008) *Xenophontis operum Concordantiae: volumen quintum: Opuscula*. Olms-Weidmann, Hildesheim; Zürich; New York.

Thomas, R. (1989) *Oral Tradition and Written Record in Classical Athens*. Cambridge University Press, Cambridge.

Threatte, L. (1980) *The Grammar of Attic Inscriptions. I: Phonology*. Walter de Gruyter, Berlin.

Threatte, L. (1996) *The Grammar of Attic Inscriptions. II: Morphology*. Walter de Gruyter, Berlin.

van Cleef, F.L. (1964) *Index Antiphonteus*. Olms, Hildesheim.

van der Eijk, P.J. (1997) 'Towards a Rhetoric of Ancient Scientific discourse', in Bakker,

E.J. (ed.) *Grammar as Interpretation: Greek Literature in its Linguistic Contexts*. Brill, Leiden. 77–129.

van Groningen, B.A. (1938) 'Le vocabulaire d'Énée le Tacticien', *Mnemosyne* 6: 329–334.

Vela Tejada, J. (1991) *Estudio sobre la lengua de la Poliorcética de Eneas el Táctico*. Departamento de Ciencias de la Antigüedad de la Universidad Zaragoza, Zaragoza.

Vela Tejada, J. (1993) 'La reestructuración funcional del sistema preposicional griego en la koiné', *Habis* 24, 235–247.

Vela Tejada, J. (2004) 'Warfare, History and Literature in the Archaic and Classical Periods: the Development of Greek Military Treatises', *Historia* 53.2: 129–146.

Whitehead, D. (1990) *Aineias The Tactician: How To Survive Under Siege*. Clarendon Press, Oxford.

Wilms, H. (1995) *Techne und Paideia bei Xenophon und Isokrates*. Teubner, Stuttgart.

Zhmud, L. (2006) *The Origin of the History of Science in Classical Antiquity*. Walter de Gruyter, Berlin.

CHAPTER 7

Writing and Other Forms of Communication in Aineias' *Poliorketika*

Peter Liddel

1 Introduction

Fundamental to understanding the text of Aineias' *Poliorketika* is a statement made in its preface: that this advice is for those who whose *megista*—that is, their shrines, country, parents and children—are endangered (Pref. 2). Aineias' work offers guidance on organising resistance to a military threat or a siege; this resistance will, at the same time, inspire fear among enemies. Aineias envisages, therefore, an emergency situation, but while his resistance aims at physically saving the community, it has the purpose also of bolstering its reputation and making an impact upon the psychology of the enemy. Communication—by any means possible—is central to Aineias' aims.

At a time of crisis, it is likely that pragmatism will dictate the form of communication deployed. But it would be wrong to think of Aineias' advice as objectively pragmatic: the recommendations of any treatise-writer are necessarily culturally specific, and so we can reasonably expect to find ancient Greek prejudices and ideologies on communication embedded into Aineias' view of the city under siege. Furthermore, Aineias does more than reflect socio-cultural realities: the *Poliorketika* is, after all, a work of literature, and one which offers a quirky form of entertainment; chapter 31 in particular, with its historical allusions and curiosities,[1] makes a contribution to Greek literary perspectives on written forms of communication. There are, therefore, three related issues at stake in an analysis of Aineias' ideas about communication: the practical uses of writing recommended by Aineias for the city under siege, the question of what Aineias' work suggests about ancient ideologies of writing and the written word and their relationship to other forms of communication, and Aineias' literary perspectives on the meaning of writing. As a historian with interests in epigraphy I aim primarily to deal with the first two areas.

---

1 Burliga (2008).

 | DOI: 10.1163/9789004352858_008

Ancient Greek views on forms of communication sometimes formulate hierarchical perspectives: in Plato's *Phaedrus*, writing was deemed as subordinate to oral communication: it provided its *hypomnemata* (*Phaedr.* 275a, 276a).[2] Modern scholarship, on the other hand, has identified many more roles played by writing in classical Greece: as well as supporting oral communication, it expressed community-sponsored statements on benefaction, policy and administration; it offered opportunities for the accrual of political capital; in fourth-century Athens, and perhaps many other cities, writing was deployed so widely in public, private, monumental, commemorative, and administrative contexts that it appears to take on social, political and religious significance well beyond anything that Plato outlines.[3] But we cannot assume that this was the case in every fourth-century Greek city: the epigraphical habit, while far from being a merely Athenian phenomenon, was unevenly dispersed across the Greek world.[4] It seems likely that Aineias has in mind the defence of a medium-sized Greek *polis* with only limited resources:[5] this may have been the kind of state in which there was far less community-organised use of public writing than there was in Athens. Like other military and tactical writers, Aineias does not quote state inscriptions or documents,[6] nor does he deploy demonstrably epigraphic language,[7] nor does he explicitly conceive of his work as possessing any sense of epigraphic-style permanency.[8]

Nevertheless, Aineias has much to say about writing, and his most valuable contribution is to display how writing is intertwined with other forms of communication in the besieged Greek city. I will suggest that Aineias' treatise offers

2 On this argument, and Alkidamas' critique of the written format of speeches, see Rösler (2009) 436–437.

3 For a sophisticated view of the way in which inscriptions introduced a performative context for the decrees they bore, see Osborne (1999). Generally, on the roles of writing Thomas (1989); Hedrick (1999); Bodel (2001) 23; Pébarthe (2006); Gagarin (2008); Liddel and Low (2013); Ceccarelli (2013).

4 On the diffusion of inscriptional habits, see Lewis (1997); Liddel (2009), (2010).

5 For Aeneias as a man of the *polis*, see Shipley, this volume, p. 52.

6 The bibliography on the deployment of documents and inscriptions is large and expanding. The most important volumes are Biraschi (et al. 2003) and Moggi (2010); most recently, see Smarczyk (2006); Davies (2010); Lane Fox (2010); Liddel and Low (2013).

7 However, this is not to say that there are not epigraphical parallels for Aineias' language: see Pippidi (1968); Vela Tejada (1991) 278; Charneux (1983); Mitsos (1983).

8 See Moles (1999) for this view of Thucydides' work. As Bettalli (1986) 82–85 suggests, he transferred to a written medium—possibly for the first time—pieces of advice that previously had existed only in oral form.

a valuable perspective on the role of writing in the Greek community. The perspective he offers is refreshingly different from that offered in recent assessments of writing and the epigraphical habit, which tend to focus upon the relatively grandiose evidence of inscribed decrees and which are predictably Athenocentric.[9] As Bettalli observes, Aineias emphasises the significance of obtaining knowledge but places even greater importance on the means of transmitting knowledge:[10] writing (and reading) is an extremely important medium of transmission.

Aineias' views on writing may be best understood in the context of the various forms of communication that would be deployed in a Greek city. I therefore set out to analyse the relationship between writing and other forms of communication, considering the possibilities of a hierarchical relationship, but also the ways in which different forms of communication complement and reinforce each other. In this paper, my exploration will start out with signs and signals (section 2), proceeding into oral communication (section 3), considering documentary and epigraphical parallels for the kinds of advice offered by Aineias (section 4), and ending with some consideration of secret messages, with particular reference to chapter 31 (section 5).

2 Signs and Signals

Aineias asserts from early on that establishing secure and mutually-comprehensible means of communication of any kind is vital to defending a city facing a military emergency. Extended discussion of long-distance signalling occurs in several passages.[11] Chapter 4 asserts the importance to the besieged of establishing σύσσημα (pre-arranged signals) 'as soon as possible' for distinguishing friends from foes (4.1). This is a principle substantiated with two case-studies in which the absence of pre-arranged signals led to military disasters: these consist of descriptions of the Chalcidian failure to defend their city against the Eretrians (4.1–5) and the botched Megarian raid on the Athenian women celebrating the Thesmophoria (4.8–11); the conclusion is offered that unam-

9 I aim not to take Hedrick (1999) to task, but rather to build on his important conclusions. On the relationship between inscriptions and the epigraphical habit, see now Meyer (2013). Ceccarelli's study of letters takes the whole Greek world as its focus: Ceccarelli (2013).

10 Bettalli, this volume, p. 173.

11 As Bettalli also notes in this volume, p. 174, citing 6.4–7, 7.2–4, 22.21–22, 26.12–14.

biguous signals are essential for the organisation of musters and expeditions (4.12).[12]

The importance of pre-arrangement of signals—recognised, according to Herodotos' sources, by the Alkmaionidai (Hdt. 6.115)[13]—is re-iterated in chapter 9, where a context for their dissemination is offered: the people may be forewarned, at an assembly, that a trumpet signal will indicate the urgency of muster (9.1–2).[14] In this passage, Aineias draws out the potential consequences in an unexpected direction, suggesting that reporting this news to the enemy might divert them from what they are planning to do, and will both arouse fear among an enemy force and inspire one's friends.[15] Not only, then, does the sounding of the trumpet serve as a convenient signal upon which the besieged might organise themselves, but there is a psychological element: it offers possibilities of producing disarray among the enemy.

There is occasional insight into other non-verbal forms of communication, their shortcomings and also their subversion: when guards get separated, whistling is a possibility, but there is always a risk that it might over-excite guard dogs (24.17–18); passwords should be devised in such a form as to avoid regional confusion and so that they are easy to remember (24.1–3,[16] 14), and they may be accompanied by signs and gestures (25.1–4). Other forms of signing include the cap-raising of rebellious Spartans at 11.12, which, we are told, was undermined by the authorities who outlawed the gesture. More often, however, Aineias offers a general reference to signing rather than details of the medium: we are not told, for instance, at 10.13, about how a call to arms is to be signalled or how the order that all shops or stores are to be closed is to be disseminated (10.14).[17] Aineias urges the reader to consider his circumstances: a signal can be given at sunset for those outside the walls to return to the city, but if the people are scattered across a wide area of land, relay-stations are required (7.2).

12 4.12: "Δῆλον οὖν ὅτι μετὰ *συσσήμων καὶ μὴ ἀγνοουμένων πρὸς ἀλλήλους*, τὰς ἀθροίσεις καὶ τὰς πέμψεις δεῖ ποιεῖσθαι".

13 Cf. Hodge (1991).

14 For use of the trumpet in calling assemblies, see Dem. 18.169; D.S. 16.84.3.

15 Cf. also 27.4 for the trumpet being understood as a call to arms; for its military uses, see Krentz (1991).

16 On dialect in Aineias, see Vela Tejada, this volume, p. 119–120.

17 Further potential deployments of signals are offered in chapter 7: they are to be used to sign the evacuation of the countryside (7.2), that dinner should be made, that guards should take up their posts (7.4) or as a way of calling for help (15.1).

The trumpet is particularly useful, for, as Whitehead observes, it will attract the attention even of those not expecting a sign;[18] furthermore, it was less prone to interference than the voice, as Asklepiodotos and other later military writers noted.[19] Generally, however, Aineias' exposition of non-verbal signals is restricted and non-comprehensive: even though, as Brown has argued, he is familiar with another story in Herodotos book 6, he fails to mention the possibility of shield-raising as that writer alleges took place after the battle of Marathon (Hdt. 6.115).[20]

Fire-signals, which were of course widely known in classical Greece (Aesch. *Ag.* 281–316),[21] emerge as the other principal form of non-verbal signal: the end of chapter 6 introduces us to the importance of burning torches, as we are told (6.7) that day scouts should be told to raise their signals now and then just as the night scouts raise their torches. At the close of chapter 7, it is revealed that he is talking about fire-signals, but with characteristic economy, Aineias refers us to his lost *Paraskeuastike* (7.4).

By far the most sophisticated means of signalling known to Aineias, and perhaps his own innovation, is the telegraphing of verbal messages by fire-signals: these were the subject of brave attempts at reconstruction by Hunter.[22] This mode of communication is detailed in Polybius' critique of Aineias' treatise on stratagems (*Peri ton strategikon*).[23] But whereas Aineias' method, reliant upon the signalling of previously agreed possibilities, can be used to reveal one of a number of 'the most obvious and universal possible events in war' (Polyb. 10.44.4),[24] Polybius offers a more sophisticated technique which allows

---

18 Whitehead (2002) 111, with 210.

19 Asklep Takt. 12.10; cf. Aelian Takt. 35.2; Arr Takt. 27: "σαφέστατα μὲν γάρ ἐστι τὰ διὰ φωνῆς δηλούμενα—οὐ μὴν πάντοτε δυνατὸν διὰ κτύπον τῶν ὅπλων ἢ διὰ πνευμάτων σφοδρῶν ἐμβολάς, ἀθορυβώτερα δὲ τὰ διὰ τῶν σημείων· ἀλλ' ἐνίοτε καὶ τούτοις ἐπιπροσθοίη ἢ ἡλίου ἀνταύγεια ἢ παχύτης ἀέρος καὶ κονιορτοῦ ἢ καὶ ὄμβρου πλῆθος" ("The most distinct commands are those given by the voice, but they may not carry at all times because of the clash of arms or heavy gusts of wind; less affected by uproar are the commands given by signals; but even these may be interfered with now and then by the sun's glare, thick fog and dust, or heavy rain." tr. Loeb).

20 Brown (1981). But Hodge (1991) argues against the idea that the shields were raised with the intent of reflecting sunlight; esp. 238–240 for other Greek testimonies of raising shields, but read *Hell.* 2.1.27 for Hodge's 3.1.27.

21 See also Bettalli, this volume, p. 175, Hershbell (1978), and Ceccarelli (2013) 86–88.

22 Hunter and Handford (1927) 123. On ancient telegraphy, see Foucault (1934) and Hershbell (1978).

23 Suda, s.v. *Aineias*; Polyb. 10.43–47; on Aineias at 10.44 = Loeb fr. 3.

24 Polyb. 10.44.4: "τὰ προφανέστατα καὶ καθολικώτατα τῶν ἐν τοῖς πολεμικοῖς συμβαινόντων".

the construction of words or even sentences from scratch;[25] he claims that this system, developed by Kleoxenos and Demokleitos, which he himself perfected, 'is quite definite and capable of dispatching with accuracy every kind of urgent messages' (Polyb. 10.44.6).[26]

Generally, Aineias recognises the importance of a capacity for flexibility and responsiveness among the bearers of messages: we are told that the general should ensure that trumpeters and dispatch-bearers should encamp and remain close to the general's headquarters, so that if a signal or dispatch is needed, they are ready at hand (22.3). For the city under attack, there are to be three day-scouts, chosen according to their experience in war, lest they make a distracting announcement out of ignorance (6.1). These scouts communicate both by signals and by word of mouth, and if the topography of the city offers no places from which signals can be issued to its territory, they must work in relay (6.4–5 cf. 22.22–23). Horsemen may be used if horses are available (6.6). It is at this point that Aineias' thoughts on the shortcomings of long-distance non-verbal signals emerge: he says that there are some matters (presumably, for instance, news of unexpected developments) that must be reported as they cannot be signalled (6.5).[27] This is an indication of the primacy of verbal (and written) communication over signs and signals. At this point, it is appropriate to consider in more detail the profile of oral communication. Asklepiodotos viewed communication by voice as 'most distinct' (σαφέστατα: Asklep. *Tact.* 12.10); what did Aineias make of it?

3 Oral communication

There are clear references in the treatise to specifically oral communication particularly for the dissemination of messages within communities. The clear-

---

25 For references to diagrams and other notes on fire signals, see Whitehead (2002) 113 and Hunter and Handford (1927) 123; for detailed discussion of Aineias' fire-signals in the context of their development, see Rihll, this volume, p. 281–287.

26 Ὁ δὲ τελευταῖος (τρόπος), ἐπινοηθεὶς διὰ Κλεοξένου καὶ Δημοκλείτου, τυχὼν δὲ τῆς ἐξεργασίας δι' ἡμῶν, (πάντη πάντως) μέν ἐστιν ὡρισμένος καὶ πᾶν τὸ κατεπεῖγον δυνάμενος ἀκριβῶς διασαφεῖν, κατὰ δὲ τὸν χειρισμὸν ἐπιμελείας δεῖ καὶ παρατηρήσεως ἀκριβεστέρας.

27 6.5: "Εἶναι δὲ τοὺς ἡμεροσκόπους καὶ ποδώκεις, οἳ ὅσα μὴ οἷά τε ⟨διὰ⟩ τῶν σημείων δηλοῦν, ἀλλ' αὐτῶν τινα δέῃ ἐξ ἀνάγκης ἀγγέλλειν, δύνωνται ταχὺ ἀφικνεῖσθαι, καὶ ὡς ἐκ πλείστου ἀγγέλλωσιν". ("Also, the daytime scouts must be fleet of foot, able, in circumstances where signalling is impossible and information has to be conveyed by word of mouth, to return quickly with the news even over the longest distances." tr. Whitehead).

est example concerns the conveyance of information to *polis* authorities: at 4.8 Aineias refers to a report made to Peisistratos that the Megarians were about to attempt an attack in the night; we are told that Peisistratos planned an ambush 'upon hearing this' (ἀκούσας).

There are other points at which we might think of messages being whispered from one ear to another: in a reference to an episode at Plataia, when there were Thebans in the town, some of the Plataian magistrates passed secretly a message (παρήγγελλον κρύφα) that the citizens ought not go out of their houses singly, but should gather in each others' houses by breaking through partition walls (2.4): the use of παραγγέλλω makes it likely, although not certain, that this refers to a message sent round orally (cf. LSJ, s. v., II), although this is one of those occasions upon which Aineias does not attempt to distinguish between written and oral communication.

Highly significant to the subject of communication is chapter 10, in which Aineias accounts for *kerugmata* (proclamations). It opens with a statement that cattle- or slave-owning citizens are to be instructed to place them in safety among neighbours (10.1–2). Again, a little later, there is revealed a psychological element: at 10.3 Aineias says that *kerugmata* are issued 'to intimidate and deter conspirators'; these include the evacuation of citizens into the city, the celebration of festivals in the city, restrictions on private gatherings, sacrifices, and public dining, and announcements on the fate of those who abscond with, or make contact with, exiles (10.4–6).

As we shall see later, such emergency measures were enacted in late-fourth century Athens through *psephismata* (decrees) of the popular assembly. But what is the significance of the term *kerugma*? When we look at the work of the roughly-contemporary Xenophon, *kerugma* can be used of an emergency announcement: in his description of the Spartan occupation of the Theban Kadmeia in 379, he says that the Theban polemarchs issued a proclamation to their people ordering them to come out into the streets and that when the Spartan governor heard about this 'night-time *kerugma*', he sent to Plataia and Thespiai for help (Xen. *Hell.* 5.4.9–10; cf. also. Xen. *Agesil.* 1.33.4). Moreover, the inscribed evidence for *kerugmata* is dominated by Hellenistic documents from Delphi which refer to the proclamation of honours at a particular venue or on a particular occasion. One example is the Chian decree of 246/5 recognising the reorganized Delphic Sotereia, which appears to contain provisions for a *kerugma*—a proclamation—which announced a key element of a broader *psephisma*.[28] We can therefore be certain that the *kerugma* is a proclamation,

28 FD III 3:215 lines 21–27 (246/5 BC):

presumably one given authority by the instruments of state institutions, to be read out loud, rather than a synonym for a *psephisma*, and it seems to be the case that Aineias is trying to impress upon his readers the orality of the transactions in chapter 10.

Indeed, chapter 10.18–19 offers us an example of a *kerugma* that is explicitly to be made orally: this consists of an announcement made to mercenaries—discharging them if they are discontented with conditions in the city—and a warning against undermining morale. The announcement is introduced with the provision that it must be made after a 'call for silence in camp, and when everyone is listening' (10.18).[29]

While it is unclear by what means passwords are to be 'given out' (παραδιδόντα: 24.1), there are other clearly oral transactions to be deployed in emergencies: at the onset of panic, soldiers are told to stand in their places and to sing a Paian (27.3);[30] we read also of the way in which Euphratas, the Laconian governor in Thrace, deterred the spread of panic among his men by ordering that anyone who stood up during a panic should be treated as an enemy (27.7–8); a third way of addressing such scenarios was for a herald to command silence (27.11).[31]

"[ἀ]νειπεῖν τὸν ἱεροκήρυκα Διονυσίοις ἐν τῶι θεάτρωι, ὅταν οἱ τῶν παί[δων χόροι μέλλωσι ἀγω]νίζεσθαι τόδε ⟨τὸ⟩ κήρυγμα· ὁ δῆμο[ς ὁ] Χίων στεφανοῖ τὸ κοινὸν τῶν Αἰτω[λῶν χρυσῶι σρεφάνωι ἀρετῆς ἕνεκεν καὶ εὐσεβείας τῆς εἰς τοὺς θεούς· ἐψήφισται δὲ κ[αὶ ἀποδέχεσθαι τὸν ἀγῶνα] ὃν συντε[λ]οῦσιν Αἰτωλοὶ τὸμ μὲμ μουσικὸν ἰσοπύθιον, τὸν δὲ γυμν[ικὸν καὶ ἱππικὸν ἰσο]νέμεον ταῖς τε ἡλικίαις καὶ ταῖς τιμαῖς· τοῦ δε κηρύγματος ἐπιμ[εληθῆναι τὸν ἀγωνοθέ]την· ἀναγράψαι δὲ καὶ τὸ ψήφισμα τόδε εἰς στήλην λιθίνην καὶ στῆ[σαι ἐν τῶι ἱερῶι τοῦ Ἀπόλ]λωνος τοῦ ἐν Δελφοῖς τοὺς πρῴτους πλευσουμένους θεωρούς." ("The sacred herald is to announce this *kerugma* at the Dionysia in the theatre, when the *choroi* of the boys are about to compete: "The demos of the Chians crowns the *koinon* of the Aetolians with a gold crown because of its virtue and reverence towards the gods; and it has voted also to recognise the contests which the Aetolians are holding, that in music as equivalent to the Pythian and that in athletics and in equestrian contests as equivalent to the Nemean both as regards the ages and the honours." The *agonothetes* shall be responsible for the *kerugma*; and this *psephisma* shall be inscribed on a stone *stele* and erected in the sanctuary of Apollo at Delphi by the first sacred envoys to sail." tr. Burstein (1985) no. 62, adapted).

29 10.18: "Ἐν δὲ ξενικῷ στρατοπέδῳ τοιάδ', ἀναγγείλαντα σιγήν, πάντων ἀκουόντων κηρῦξαι".

30 On the singing of the Paian in emergency circumstances, see Käppel (1992) 45–46 on the Paian in war.

31 Soph. *Ant.* 454–455 contrasts Creon's *kerugma* with the *agrapta nomima* of the Gods. However, there is no firm indication that the *kerugma* is a written one. Hunter and Handford (1927) xxxv suggest Xen. *Anab.* 2.2.20 as a source or parallel.

Any discussion of oral announcements in this work has to also recognise the fact that Aineias, at 38.5, refers to a lost section of his treatise, or perhaps a separate work, entitled *Akousmata* (literally, 'things heard'); this may have consisted of addresses to be made by a general or commander: the context offered is the general's advice to those who are fighting on the wall. His words should consist of advice, praise, appeal, but not *orge*, though he might reprimand the wealthy for their indiscipline. Again, Aineias communicates a sense of the psychological significance of oral communication and exhortation.

We have not yet encountered much indication that the use of the written medium is significant in organising the defence of the city under siege. Before I discuss references to written forms of communication and in particular the secret messages of section 31, it is necessary to explore in more detail, by reference to Attic parallels, the possibility that the orally-transmitted *kerugmata* of chapter 10 could plausibly have had some basis in written documentation.

4 Decrees and the Regulatory/Exhortative Use of Writing

Aineias' advice that citizens should be notified on occasions when they are required to remove animals or slaves to the property of neighbours (εἰς τοὺς προσοίκους: 10.1; cf. 10.3) suggests the kind of arrangement that would have to be organised by regular or (more likely, given Aineias' near-silence on the role of demotic, ekklesiastic, or bouleutic institutions in decision-making) emergency state institutions. Indeed, there are plenty of good examples of state-organised evacuations, and provisions for them are sometimes mentioned on inscriptions.[32] But such provisions are mentioned in epigraphical contexts only in inscribed versions of prior agreements between communities, such as the alliance between Erythrai and Hermias, which states that there is to be mutual immunity from taxation of items deposited on account of war (RO 68 lines 1–5);[33] there are also extant honours for those who enabled an evacuation at Rhamnous in the third century BC (IRhamnous 3 lines 6–19; cf. OGIS 748). Unsurprisingly, given the fact that most inscribed decrees concerned non-ephemeral matters, there are no extant inscribed versions of decrees actually ordering an evacuation. So, while Aineias is thinking of the kind of action that would be set in motion by decree of an assembly, there is no strong reason to

32 Mueller (1975).

33 Cf. *IG* IX 1. 98; *SIG*^{3} 588, 633; *IC* III iii 4 lines 21–27.

believe that he is thinking of a written notice. Yet it is relevant to note that there are three Athenian parallels are available for the measures recommended at 10.3–5:

(a) The evacuation of citizens residing outside the city walls to within the city and the celebration of normally extramural festivals inside the city (10.3–4) have a significant parallel in the decree of Kallisthenes of Skirophorion 346, which forbade anyone from spending the night outside the city. Portions of the decree are preserved in Demosthenes *On the False Embassy* (Dem. 19.86, 125), while a version of dubious authenticity appears in some versions of the speech *On the Crown* (Dem. 18.37).[34] Evacuation is also known to have taken place in the fifth century: Thuc. 2.14.1–2.
(b) The restrictions on the travel of citizens or metics away from the city at 10.8 also has a parallel: in autumn 338, after the battle of Chaironeia, it was proposed that anyone leaving the city would be classified as a traitor (Lycurg. 1.53).[35]
(c) Thirdly, the suggestion that, 'when sudden terror strikes a state which has not been put on a war footing, the quickest way to organise citizens for its defence would be to assign by lot to each tribe the task of proceeding immediately to a section of the wall and mounting guard over it' (3.1) is almost mirrored in the evidence for what the Athenians did after Chaironeia: according to Aischines 3.27, on the penultimate day of Thargelion, in the archonship of Charondas, Demosthenes made a decree that on the 2nd and 3rd days of Skirophorion assemblies of the tribes should be held and men should be chosen from each tribe as overseers and treasurers for work upon the walls.

None of these three examples is epigraphically attested, and so while, as decrees of the Athenian state, they may have been written down and stored in an archive at Athens, it is unlikely that their primary mode of dissemination was by written notice. Moreover, Aineias gives no indication that he envisages a state within which there is an organised system of recording or disseminating the enactments in written form.

On occasions, however, when Aineias' advice touches on pieces of legislation that have more long-term significance, there do exist inscribed overlaps.

34 On the authenticity of this decree, see Canevaro (2013) 243–248.

35 Cf. Hyp. *Athenog.* 29, suggesting that the legislation applied only to metics: see Whitehead (2000) 335–336.

The first concerns rewards for informers against conspirators against the state: at 10.15 we are told that 'For whoever points out anyone conspiring against the city, or reports anyone as doing any of the things above-mentioned, a reward in money shall be announced, and the reward shall be displayed openly in the agora or on an altar or in a shrine.'[36] Whitehead offers a late fifth-century epigraphical parallel in the shape of ML 83 (I line 1; II line 1), which outlines financial reward for whoever informs against an insurrection (*epanastasis*).[37] It seems more likely that Aineias envisages that there is, as at Thasos, to be a written notice of the reward rather than the actual money being put on display at the agora.

Other overlaps may be found between Aineias' advice and later epigraphical evidence: the designation of enemies of the state with cognates of ἐπιβουλεύω (10.15) and the encouragement of informing against traitors is attested both in a civic oath of the Chersonese and in a treaty of *sympoliteia* between Smyrna and Magnesia from the third century BC (*IOSPE* I^2 401 lines 18–22; *Smyrna* 14 lines 67–69). The inscribed evidence for anti-tyranny legislation offers also parallels for Aineias' rewards for actions against exiled rulers. Aineias says at 10.16 that it is necessary to announce in advance (προκηρύττεσθαι) a reward concerning information about them, and that if the tyrannicide himself is killed, then the reward will be paid to his children. A fourth-century law from Eretria outlines rewards for tyrannicides, and, as does Aineias, introduces the principle that rewards are to be bestowed upon the offspring of the tyrannicide who dies in his attempt (*SEG* li 1105 lines 10–15). Moreover, Aineias at 10.17 adds that if an assassin emerges from among the associates of the exile, he too is to be rewarded. Comparable provisions are alluded to in a law from Ilion dating to the third century which states that that those associates of the insurgent who join in against him shall he rewarded (*IK Ilion* 25 lines 43–53).[38] But one key difference to observe here is that the Greek epigraphic evidence often frames anti-tyranny measures within the context of an oath (see for instance *IOSPE* IV 79 13–23; *IC* III iv 8),[39] whereas there is a lack of divine element and a sidelining of religious matters in Aineias' treatise.[40]

36 Ain. Tac. 10.15. "ὃς ἂν καταμηνύσῃ τινὰ ἐπιβουλεύοντα τῇ πόλει, ἢ ὅ τι ἂν τῶν προγεγραμμένων τις πραττόμενον ἐξαγγείλῃ, ἀνηγγέλθαι τε αὐτῷ ἀργύριον καὶ τὸ ἀγγελθὲν ἐμφανῶς προκεῖσθαι ἐν ἀγορᾷ ἢ ἐπὶ βωμοῦ ἢ ἐν ἱερῷ, ἵνα προχειρότερόν τις τολμήσῃ μηνύειν τι τῶν προγεγραμμένων".

37 Whitehead (2002) 125. There is also a parallel from Abdera (*BCH* 66–67 (1942–1943) 189, no. 3) offering a talent for information.

38 On anti-tyranny legislation, see Koch (2001) and Teegarden (2014).

39 See also *RFIC* 70 (1942) 15 of Kos, *Labraunda* 51 of Mylasa, and *Teos* 48.

40 Cf. Whitehead (2002) 98, commenting on the 'distinctly businesslike' attitude to religion.

Another area where later epigraphical evidence offers parallels to Aineias' advice concerns rewards to those who import goods to the city (10.12). We can compare the rewards made at Rhamnous to Epichares (*IRhamnous* 3), and there are plenty of other parallels too.[41]

Crucially, with the exception of the Thasian example, all of these epigraphical parallels are later, Hellenistic, ones; they suggest that Aineias was envisaging the kind of ordinances that were often written down by Greek cities mostly after the end of the fourth century BC; however, what we lack is decisive evidence to say that he was thinking of deploying the written medium in the dissemination of these provisions: there is nothing to suggest he conceived of a role of stone inscriptions in the security of the community.

There are, however, a few firmer indications that Aineias envisaged restrictions on citizens' behaviour that would have required the deployment of written registers: at 10.7, we are told that men who own more than one set of weapons are to register (ἀπογράφεσθαι) them. Residents from neighbouring states were to be registered too (10.10).[42] There are no certain epigraphical parallels for these provisions, though it is plausible that in Athens there were lists of metics kept by the Polemarch.[43] At another point, Aineias talks about conscripting (τετάχθαι) half of the men enrolled in the army (τοὺς ἡμίσεας τῶν προγεγραμμένων: 22.26). The verb προγράφω could plausibly refer to the use of a written list (see LSJ, s. v.). While it is often impossible to be certain about the function of inscribed lists of individuals,[44] we should remember that in Athens, until the early fourth century, conscription was carried out by the creation of tribal *katalogoi* of those liable for service; moreover, even after the introduction of conscription by age class in the early fourth century there existed written lists of age-classes on bronze *stelai*.[45] At Helisson in Arkadia, too, lists of citizens were inscribed after it became a *kome* of Mantinea (RO 14 lines 16–20). It is reasonable to suggest, therefore, that Aineias had in mind a role for written records of citizens in conscription.

41 Other parallels of rewards for importers: Bettalli (1990) 241–242, pointing to Migeotte (1984) 42, 60, 85, 117. For further epigraphical parallels, see Engelmann (1983) 126–129. Migeotte collects examples of honours for those who have made loans, but his epigraphical testimonia start in the 330s and are concentrated in the Hellenistic period.

42 On the treatment of foreigners, see Boëldieu-Trevet and Mataranga (2003).

43 Whitehead (1977) 83.

44 Such as those from early Hellenistic Eretria (*IG* XII 9.245–249), described by Hansen (2006) 95 as 'rosters of citizens'.

45 *Ath Pol* 53.4–5; Christ 2001; Liddel (2007) 185.

5 Secret messages

But the most extensive discussion, to a degree that it stands out from the rest of the work, of the use of written communication, is of course section 31,[46] where Aineias offers an account of modes by which secret messages might be sent. The chapter starts with the expression Περὶ δὲ ἐπιστολῶν κρυφαίων ('On Secret Messages') and resembles a sub-treatise in its own right:[47] it is an account of secret means of written communication, but one which does not appear to attempt any internal consistency or flow. Debidour (2006) has already assembled a list of the modes of communication and comparable citations in Greek literature and other writers; I want to pick out a few aspects of the chapter which are revealing of what he has to say about the written word.

Noteworthy in the first place, perhaps, is the variety of (a) means of physical transference of written objects, (b) means of concealment[48] and (c) of the materials used for writing. While not as outlandish as the inscribed mushroom brought to Trajan (Cass. Dio 68.8.1),[49] the written materials range from the ingenious adaptation of objects commonly associated with writing—the marking of books or seemingly innocuous documents (2–3); the writing on a wooden part of a writing tablet, and covering it with wax or whitewash (11); the use of a dedication (12), concealed papyri (23, 20, 8, 9)—to objects less normally associated with writing: famously, tattooed, shaved heads (28–29).[50] Indeed, there are a few examples of inscriptions on durable objects attested in the material record: section 7 offers up lead sheets which are to be rolled up and placed inside earrings—the use of lead of course is well known from its use in cursing and letter-writing, and the habit of rolling them up may have made augmented

46 For a comprehensive treatment of secret messages in Aineias section 31, see Debidour (2006).

47 As also Whitehead (2002) 183.

48 Moreover, it has been widely recognised (cf. Whitehead (2002) 183) that most of Aineias' devices fall under the heading of steganography (in which the message is concealed rather than encoded) rather than cryptography proper, of which there are a few examples, such as the drilled and threaded astragal or length of wood (17–22). The omission of vowels also constitutes an encoded message, and one that relies upon prior agreement on signifying vowels by dots (31.30).

49 This told him that the Buri and other allies urged him to turn back and make peace; it was depicted on Trajan's column: Austin and Rankov (1995) 65 and pl. 3.

50 The example of leaves which were bound to a wound (6), as Whitehead (2002) 184 points out, reminds us of the use of leaves for petalism at Syracuse, mentioned by Diodorus Siculus 11.87.

their secrecy:[51] we could compare, for instance, Lesis' letter to Xenokles from the Athenian agora asking for help in a desperate situation.[52] Aineias also suggests the use of a concealed tin plate in a herald's sandal at 4. The use of tin for writing is also known from the fourth-century Athenian decree on the sacred *orgas*, where its thinness is exploited for its use in the drawing of a lot (RO 58 lines 23–24).

The second point of interest is that chapter 31 represents a mixture of historical (or pseudo-historical) examples and speculative advice on the methods of sending secret messages: some of the hypothetical examples are so intricate as to sound clumsy or impracticable, such as the concealment of lead letters inside earrings (7),[53] or the 'most secret and most troublesome' embroidered knucklebones (16–19), which, as he acknowledges, are even more laborious to decipher. Section 2, on the other hand, begins with an apparently real example (ἐπέμφθη ἐπιστολὴ ὧδε … 'a message was once sent in the following manner …') of picking out and marking letters in a book in order to compose a sentence; moreover, the use of message-carrying dogs is substantiated by the claim that it was an Epirote and Thessalian custom (31–32). Famously, Aineias also draws on examples that very closely resemble episodes in Herodotos: Herodotos' Damaratos wrote on the wooden part of a tablet (31.14; cf. Hdt. 7.239.3–4); he mentions Histiaios' agitating message to Aristagoras using a tattooed slave (31.28; cf. Hdt. 5.35.3[54]), and Timoxenos' treachery of Poteidaia to Artabazos (31.25; cf. Hdt. 8.128.1). Even though Aineias presents them as real examples, these stories may well have fictional elements.[55] It is surely the case that Aineias used historical examples in an aim to enhance his authority as a writer and

---

51 Eidinow and Taylor (2010) 35; Thomas (2009) 25–26 for curses on other materials, Eidinow (2007) 285.

52 Jordan (2003) no. IX.
"Λῆσις {ις} ἐπιστέλλει Ξενοκλεῖ καὶ τῆι μητρὶ μηδαμῶς περιιδεν αὐτὸν ἀπολόμενον ἐν τῶι χαλκείωι, ἀλλὰ πρὸς τὸς δεσπότας αὐτο ἐλθεν καὶ ἐνευρέσθαι το βέλτιον αὐτῶι. Ἀνθρώπωι γὰρ παραδέδομαι πάνυ πονηρῶι· μαστιγόμενος ἀπόλλυμαι· δεδέμαι· προπηλακίζομαι· μᾶλλον μᾶλλον". ("Lesis is sending a letter to Xenokles and to his mother by no means to overlook that he is perishing in the foundry but to come to his masters and find something better for him. For I have been handed over to a man thoroughly wicked. I am perishing from being whipped; I am tied up; I am treated like dirt—more and more!" tr. Jordan). The Berezan letter is another call for help: see Bravo (1974).

53 An example introduced with the potential optative + ἄν.

54 For other literary instances, see Debidour (2006) 226–227.

55 Brown (1981) 385–393 argued that 31.9 alludes to the betrayal of Histiaios of Miletos' letters by Hermippos at Hdt. 6.4.1–2. Chapter 31 is the chapter that includes most references to material in Herodotos; but note also 37.6–7 with Hdt. 4.200.1–3. Brown notes that Aineias

also with a dicactic purpose in mind.[56] But one also wonders, in its variety of allusion, material and method, whether chapter 31 was composed with more than just practical advice in mind.[57] The deployment of secret written messages was of great interest to military writers,[58] but they were also deployed as entertaining props elsewhere:[59] in Aristophanes' *Thesmophoriazusai*, Mnesilochos, parodying Euripides' lost *Palamedes*, sent a message written on votive wooden slabs (769–776).

Third, it must not be forgotten that chapter 31 has a practical value too: some examples are offered in order to demonstrate methods of secure communication, although the relationship to reality or pragmatism are not always clear or indeed stable. The levels of secrecy employed vary: some examples offer the extra security of the ignorance of the message-bearer whether human (4) or canine (31–32),[60] while others actually require the messenger to knowingly hand it over (9). On the other hand, depositing the message in an agreed place keeps the identity of the recipient private (31).

Fourth, it is significant that whereas some of Aineias' devices aim at the simple transference of news to outsiders or insiders, some of them offer the possibility of sending a return message: this is the case with the messenger who conceals a tin plate in his sandal (4–5) or the inflated bladder (10); and he suggests that a reply can be sent on the wooden tablet waxed over (14). This is vital: Aineias' devices offer the chance of opening up lines of communication either side of the siege.

Fifth, it is worth noting one significant condition attached to some of Aineias' devices, already commented upon in this paper (above, section 2 on σύσσημα): the existence of a prior private arrangement between the sender and

---

may well not have read book 1 of Herodotos, as he omitted mention of a letter concealed inside a hare (1.123–124). On overlaps with Herodotos, see Burliga (2008).

56 See Pretzler, this volume, pp. 79–85.

57 Burliga (2008), noting that Aineias cited Herodotos' episodes both to illustrate his arguments but for the sake of entertainment.

58 Indeed, about a century after Aineias, Philon of Byzantium, referred to a work on sending written messages secretly (102 lines 47–50 at Garlan (1974) 324), and shared an interest in the use of writing in coded communication (e.g. 90 lines 28–40 at Garlan (1974) 306); the subject was of interest to Polybius (8.17; 10.44), Frontinius (*Stratag*. 3.13) and Aulus Gellius (17.9). For a putative treatise *De secretis epistolis* and the reference of other later writers to secret messages, see Rance, this volume, pp. 347–354.

59 Lewis (1996) 143–146.

60 Another connection between written messages and secrecy is found in Antiphon, *Herodes* 5.53, where he says that written messages are necessary for both secret or excessively lengthy messages: see Thomas (2009) 25–26.

receiver (31.1). Aineias suggests that a dedicated tablet concealing a message should be decorated by some prearranged sign (12); when Timoxenos wished to hand over Poteidaia to Artabanos he is said to have prearranged spots at which they would shoot arrows carrying important information (31.25); codes too have to be pre-arranged (31.30); later on, Aineias observes that the deposit of a message at a certain place could be indicated by a commercial transaction (31): what we should note here is that the efficacy of these secret messages is reliant on prior agreement through oral transactions.

Finally, Aineias shows us his awareness of the dangers of subversion: he offers warnings about the shortcomings of written communication. There is a danger that a message carrier, for instance, might go not to the traitor but to the commanding officer of the city thereby double-crossing the informers (9–9b). At 16 he offers a methodological statement on secrecy: that it is the most secret thing of all (πασῶν ἀδηλοτάτη) to send messages without using letters; elsewhere he warns us that letters must be opened as soon as they are sent (33), and adds also that gatekeepers should be painstaking in their attention to the possibility of secret message (35). He knows about the potential risk that letters pose towards the security of a city: he warns, in his discussion of the treatment of exiles, that outgoing and incoming letters are to be censored (10.6), and he insists that there has to be a definite risk or a penalty awaiting the person who exchanges letters with an exile. Written communication, therefore, also has its shortcomings.

Moreover, at the close of this chapter, Aineias himself reflects on the reliance of communicating messages upon the written word when he gives a short account of Glous, the Great King's nauarch who, since it was forbidden to bring ὑπομνήματα ἐν βιβλίῳ (notes on papyrus or a tablet) before the king, made notes by writing in the spaces between his fingers (31.35).[61] There is a place, after all, for Plato's view of writing as the maidservant of oral communication, in Aineias' treatise.

6 Conclusions

Communication in Aineias can be divided into three overlapping, mutually reliant, and sometimes indistinguishable, types: (a) visible non-verbal signs and sounds; (b) messages disseminated by telegraph, by word of mouth, or

61 Debidour (2006) 228 compares Xenophon's account of the 'pushing their hands through the sleeve' at *Hellenika* 2.1.8.

proclamation; (c) concealed or encrypted written messages. Aineias asserts the superior accuracy of announcements over non-verbal means of communication by demonstrating their deployment as exhortations and deterrents, warnings and regulations, devices to confuse the enemy, and, in written form, as a means of sending secret messages. Communication by proclamation or audible sound offers also the advantage of introducing a considerable psychological element, which aims not just at the telegraphing of precise details, but also at the raising of morale.

As we have seen, Aineias reflects on the shortcomings of both non-verbal and verbal communication. Non-verbal communication is inexact; on the other hand, letters, despite their reputation in Greek literature for secrecy,[62] if they are not encrypted or hidden from their carriers, run the risk of compromising security. He offers solutions to the question of security in chapter 31. It is, therefore, impossible to identify any simplistic or ideological hierarchy to modes of communication in the situations that Aineias describes: this is no surprise, given that, in the ancient Greek world, written forms of communication often amounted to transcriptions of oral messages.[63] Moreover, in this sense his work resembles the flexibility of Asklepiodotos' position, which maintains that 'one cannot find signals to which the phalanx has been accustomed suitable for every circumstance that arises, but now and then new signals must be found to meet the situation but it is hardly likely that all the difficulties appear at the same time, so that a command will be indistinguishable both by bugle, voice and signal' (Asklep. 12.10).[64]

What of public writing? There is little in Aineias' text to suggest that he envisages the extensive use of openly written notices in the city under siege, but the inscribed evidence for anti-revolutionary legislation of the later fourth and the third centuries BC suggest that some of his advice was epigraphically paralleled at least a little later in Greek history, and on one occasion (Thasos) a little earlier. However, it is clear from chapter 31, as it is from other writers, that the main deployment of writing in military emergencies was for secret messages. Like other forms of sophisticated communication, they functioned

62 On the secretive connotations of letters in Greek literature, see Lewis (1996) 143–146; Ceccarelli (2013) 129, 195–197, 203–205, 232–234, 283–284, 331; on accuracy, Lewis (1996) 144. See also above, note 60 on Antiphon 5.

63 Ceccarelli (2013) 331.

64 Asklep. 12.10: "οὐ ῥᾴδιον πρὸς πάσας τὰς ἀνακυπτούσας χρείας εὐπορῆσαι σημείων, οἷς προσήθισται ἡ φάλαγξ, ἀλλ' ἐνίοτε πρὸς τοὺς καιροὺς ἀνάγκη καινὰ προσευρίσκειν, πλὴν ἀδύνατον ἅπαντα συμπεσεῖν, ὥστ' ἄδηλον εἶναι καὶ σάλπιγγι καὶ φωνῇ καὶ σημείῳ τὸ παράγγελμα".

most effectively within a 'knowledge community' of their own,[65] among those who have a prior expectation of the message that they are expecting to receive: this is a synecdoche for the broader subject of defending the city from attacks, which, at 16.19–20, Aineias tells us, is heavily reliant on acquaintance (συνηθεία) with the local terrain.

Aineias contributes, then, to our understanding of the cultural history of writing: in addition to offering us a view of communication in the city under siege, Aineias offers us a view of the place of writing within the broader sociology of the city.[66] This is a clear alternative to the uses of reading and writing offered by Aristotle: money-making, household management, learning, political affairs (Politics 1338a15–19). Moreover, there gaping differences between Aineias' use of writing and the view that we get from the inscribed record: not only are Aineias' deployments of writing mostly ephemeral, but we note the absence from Aineias' world of a divine element, which is so important to Greek monumental epigraphy; we should consider also that there is a complete lack of reference to long-term pledges in the form of curses or oaths or formal alliances.[67] Another key difference is the fact that writing, in Aineias' world, brings with it the opportunity of fostering two-way communication:[68] he therefore offers us, as does the evidence for letter writing,[69] an alternative to the univocal, monological form of communication offered by inscribed state decrees, laws, calendars, dedications or even graffiti. In Aineias, the written medium is reserved for communicating, or opening up lines of communication, with allies who are distant or are located on the other side of enemy lines, or for state-registration purposes. Writing is deployed in pursuit of short-term and clearly-defined goals; it is rational and does not concern itself with implications for man's relationship with the gods. Aineias' Poliorketika therefore offers us a view of writing in action, in a state of crisis, in which it is deployed usually for secretive communication the implications of which are ephemeral. It

65 On different groups of readers and different kinds of literacy, see Eidinow and Taylor (2010); Thomas (2009).

66 On what Aineias offers us on the sociology of the city, see Bengston (1962) 461–468; Whitehead (2002) 22–33; Dain and Bon (1967) xxvii–xxx; Winterling (1991).

67 It is clear that Aineias is deeply suspicious of the potential of deploying external aid in a city under siege, as he makes very clear in his warnings at chapter 12; cf. 26.7. But on the uses of foreigners, see Boëldieu-Tevet and Mataranga (2003) 25–28; Roy, this volume, p. 206–213.

68 Bettalli, this volume, p. 173–174, observing that the problem is not so much obtaining information but transmitting it.

69 Eidinow and Taylor (2010); Ceccarelli (2013).

stands in contrast to the legislative, commemorative, monumental and even performative deployment of public writing that is commonly associated with ancient Greek epigraphical habits of the fourth century and later.[70]

We might end with a reflection on a literary perspective to this subject: chapter 31 suggests that Aineias viewed written communication as worthy of writing about for its own sake, and it is just possible that later readers, among them perhaps Ovid (Ars Amatoria 3.623–624 cf. Aineias Tacticus 31.4, 6),[71] thought so too.[72]

Abbreviations

References to inscriptions follow the conventions of *Guide de l'épigraphiste*, 3rd ed., (Paris, 2000), 17–18, unless indicated below:

IKIlion: Frisch, P. (1975), *Die Inschriften von Ilion*. IGSK 3: Köln.

IOSPE: Latyshev, V. (1885–1901), *Inscriptiones antiquae orae septentrionalis Ponti Euxini graecae et latinae*: St. Petersburg.

IRhamnous: Petrakos, V. (1999), *Ο δήμος του Ραμνούντος II, Οι επιγραφές*. Greek Epigraphical Society, Athens.

Labraunda: McCabe, D.F. (1991), *Labraunda Inscriptions. Texts and List*. The Princeton Project on the Inscriptions of Anatolia, The Institute for Advanced Study, Princeton.

ML: Meiggs, R., and Lewis, D., (eds) (1988), *A Selection of Greek Historical Inscriptions to the End of the Fifth Century BC*. Rev. ed. Oxford University Press, Oxford.

RO: Rhodes, P.J., and Osborne, R. (eds) (2003), *Greek Historical Inscriptions 404–323 BC*. Oxford University Press, Oxford.

Smyrna: McCabe, D.F. (1988), *Smyrna Inscriptions. Texts and List*. The Princeton Project on the Inscriptions of Anatolia, The Institute for Advanced Study: Princeton.

Teos: McCabe, D.F. (1985), *Teos Inscriptions. Texts and List*. The Princeton Project on the Inscriptions of Anatolia, The Institute for Advanced Study: Princeton.

70 For a related attempt to use a body of evidence to rethink the context of writing using a body of evidence pertaining to non-elite social groups, see Eidinow and Taylor 2010, esp at 46.

71 Gibson (2003) 155, 179, 337, 339.

72 I would like to thank Maria Pretzler for her organisation of a rich and dynamic conference on a well-chosen topic at a wonderful venue; I would also like to thank the editors of the volume for their efforts in its publication.

Bibliography

Austin, N. and Rankov, B. (1995) *Exploratio: Military and Political Intelligence in the Roman World from the Second Punic War to the Battle of Adrianople*. Routledge, London.

Bengtson, H. (1962) 'Die griechische Polis bei Aeneas Tacticus', *Historia* 11: 458–468.

Biraschi, A., Desideri, P., Roda, S. and Zecchini, G. (eds) (2003) *L'uso dei documenti nella storiografia antica. Incontri Perugini* XII, Napoli.

Bettalli, M. (1986) 'Enea Tattico e l'insegnamento dell'arte militare', *Annali della Facoltà di Lettere e Filosofia dell'Università di Siena* 7: 73–89.

Bettalli, M. (1990) *Enea Tattico. La difesa di una città assediata (Poliorketika). Introduzione, traduzione e commento*. ETS editrice, Pisa.

Bodel, J. (2001) 'Epigraphy and the Ancient Historian', in J. Bodel (ed.) *Epigraphic Evidence: Ancient History from Inscriptions*. Routledge, London. 1–56.

Boëldieu-Trevet, J. and Mataranga, K. (2003) 'Étrangers et citoyens: le maintien de l'ordre dans une cité assiégée selon Énée le Tacticien', in M. Molin (ed.), *Les régulations sociales dans l'Antiquité: actes du colloque d'Angers*, 23 et 24 mai 2003. Pr. Universitaires de Rennes, Rennes. 21–41.

Bravo, B. (1974) 'Une lettre sur plomb de Berezan: colonisation et modes de contact dans le Pont', *Vestnik Drevnej Istorii* 1: 110–187

Brown, T.S. (1981) 'Aeneas Tacticus, Herodotus and the Ionian revolt', *Historia* 30: 385–393.

Burliga, B. (2008) 'Aeneas Tacticus between History and Sophistry: the Emergence of the Military Handbook', in Pigoń J. (ed.) *The Children of Herodotus. Greek and Roman Historiography and Related Genres*. Cambridge Scholars Publishing, Newcastle. 92–101.

Burstein, S. (1985) *The Hellenistic Age from the Battle of Ipsos to the Death of Kleopatra*. Cambridge University Press, Cambridge.

Canevaro, M. (2013) *The Documents in the Attic Orators: Laws and Decrees in the Public Speeches of the Demosthenic Corpus*. With a chapter by E.M. Harris. Oxford University Press, Oxford.

Ceccarelli, P. (2013) *Ancient Letter Writing. A Cultural History (600 BC–150 BC)*. Oxford University Press, Oxford.

Charneux, P. (1983) 'Sur quelques inscriptions d'Argos', *Bulletin de Correspondance Hellénique* 107: 251–267.

Christ, M.R. (2001) 'Conscription of Hoplites in Classical Athens', *Classical Quarterly* 51: 398–422.

Dain, A. and Bon, A.M. (1967) *Énée le Tacticien. Poliorcétique*. Les Belles Lettres: Paris.

Davies, J.K. (2010) 'Data-Base or Check List? How best to track a document', in Moggi M.

(ed.) *Documenti nella Storiografia Antica. Prospettive Informatiche*. Mediterraneo Antico 10.1–2: 1–18.

Debidour, M. (2006) 'Le secret et les messages secrets dans la Poliorcétique d'Énée le Tacticien', in Olivier, H., Giovannelli-Jouanna, P. and Bérard, F. (eds), *Ruses, secrets et mensonges chez les historiens grecs et latins*. Université Jean-Moulin Lyon 3, Lyon. 213–241.

Eidinow, E. (2007) *Oracles, Curses, and Risk Among the Ancient Greeks*. Oxford University Press, Oxford.

Eidinow, E. and Taylor, C. (2010) 'Lead-Letter Days: Writing, Communication and Crisis in the Ancient Greek World', *Classical Quarterly* 60: 30–62.

Engelmann, H. (1983) 'Inschrift und Literatur', *Zeitschrift für Papyrologie und Epigraphik* 51: 123–130.

Foucault, J. de (1934) 'La télégraphie aérienne dans l'antiquité,' *Hum(RES)* 16, 111–114.

Gagarin, M. (2008) *Writing Greek Law*. Cambridge University Press, Cambridge.

Garlan, Y. (1974) *Recherches de poliocétique grecque*. Bibliothèques de l'Ecole française d'Athènes et de Rome—Série Athènes 223, Paris.

Gibson, R. (2003) *Ovid. Ars Amatoria Book 3*, Cambridge University Press, Cambridge.

Hansen, M. (2006) *The Shotgun Method. The Demography of the Ancient Greek City-State Culture*. University of Missouri Press, Columbia and London.

Hedrick, C. (1999) 'Democracy and the Athenian Epigraphical Habit', *Hesperia* 68: 387–439.

Hershbell, J. (1978) 'The Ancient Telegraph: War and Literacy', in Havelock E. and J. Hershbell, *Communication Arts in the Ancient World*. Hastings House, New York. 81–92.

Hodge, A.T. (1991) 'Reflections on the Shield at Marathon', *Annual of the British School at Athens* 96: 237–259.

Hunter, L.W. and Handford, S.A. (1927), *Αἰνείου Πολιορκητικά. Aineias on Siegecraft*. Clarendon Press, Oxford.

Jordan, D. (2003) 'A Letter from the Banker Pasion', in Jordan D. and Traill J. (eds) *Lettered Attica. A Day of Attic Epigraphy*. Canadian Archaeological Institute, Athens.

Käppel L. (1992) *Paian. Studien zur Geschichte einer Gattung*. de Gruyter, Berlin and New York.

Koch, C. (2001) 'Prozesse gegen die Tyrannis', *Dike* 4: 169–217.

Krentz, P.' (1991) 'The salpinx in Greek warfare', in Hanson V. (ed.) *Hoplites. The Classical Greek Battle Experience*. Routledge, London.

Lane Fox, R. (2010) 'Thucydides and Documentary History', *Classical Quarterly* 60: 11–29.

Lewis, D. (1997) 'Democratic institutions and their diffusion', in Rhodes P.J. (ed.) *Selected Papers in Greek and Near Eastern History*. Oxford University Press, Oxford: 51–59.

Lewis, S. (1996) *News and Society in the Greek Polis*. Duckworth, London.

Liddel, P. (2007) *Civic Obligation and Individual Liberty in Ancient Athens*. Oxford University Press, Oxford.

Liddel, P. (2009) 'The Decree Cultures of the Ancient Megarid', *Classical Quarterly* 59: 411–436.

Liddel, P. (2010) 'Epigraphy, Legislation, and Power within the Athenian Empire' *Bulletin of the Institute of Classical Studies of the University of London* 53: 99–128.

Liddel, P. and Low, P. (eds) (2013), *Inscriptions and their Uses in Greek and Latin Literature*. Oxford University Press, Oxford.

Mataranga, K. (1993) 'Un étrange proxène chez Énée le Tacticien', *Cahiers du Centre Gustave Glotz* 4: 223–227.

Meyer, E. (2013) 'Inscriptions as Honours and the Athenian Epigraphical Habit', *Historia* 62: 453–505.

Migeotte, L. (1984) *L'emprunt public dans les cités grecques. Recueil des documents et analyse critique*. Sphinx and Belles Lettres, Quebec and Paris.

Mitsos, M.T. (1983) 'Une inscription d' Argos', *Bulletin de Correspondance Hellénique* 107: 243–249.

Moggi, M. (ed.) (2010) *Documenti nella Storiografia Antica. Prospettive Informatiche*, (Atti del Seminario, Istituto Italiano per la Storia Antica, Roma: 30 novembre 2007), Mediterraneo Antico 10.

Moles, J. (1999) 'Anathema kai Ktema: the Inscriptional Inheritance of Ancient Historiography', *Histos* 3, 27–69 (electronic publication: http://research.ncl.ac.uk/histos/documents/1999.03MolesAnathemakaiKtema2769.pdf, accessed 03/07/2017).

Mueller, H. (1975) 'φυγῆς ἕνεκεν', *Chiron* 5: 129–156.

Osborne, R. (1999). 'Inscribing Democracy', in Osborne R. and Goldhill S. (eds) *Performance Culture in Athenian Democracy*. Cambridge University Press, Cambridge. 341–358.

Pébarthe, C. (2006) *Cité, démocratie et écriture: histoire de l'alphabétisation d'Athènes à l'époque classique*. De Boccard, Paris.

Pippidi, D.M. (1968) 'Notes de lecture, XX: Pour un commentaire épigraphique des Poliorcétiques d' Énée le Tacticien', Studii clasice 10: 240–243.

Rösler, W. (2009) 'Books and Literacy' in G. Boys-Stones, B. Graziosi, and Vasunia P. (eds) *The Oxford Handbook of Hellenic Studies*. Oxford University Press, Oxford. 432–441.

Smarczyk, B. (2006) 'Thucydides and Epigraphy' in Rengakos A. and Tsakmakis A. (eds) *Brill's Companion to Thucydides*. Brill, Leiden. 495–522.

Teegarden, D.A. (2014) *Death to Tyrants! Ancient Greek Democracy and the Struggle Against Tyranny*. Princeton University Press, Princeton.

Thomas, R. (1989) *Oral Tradition and Written Record in Classical Athens*. Cambridge University Press, Cambridge.

Thomas, R. (2009) 'Writing, Reading, Public and Private "Literacies": Functional Literacy and Democratic Literacy in Greece', in Johnson W. and Parker H. (eds) *Ancient Literacies. The Culture of Reading in Greece and Rome*. Oxford University Press, Oxford.

Vela Tejada, J. (1991) *Estudio sobre la Lengua de la Poliorcetica de Eneas el Tactico*. University of Zaragoza, Zaragoza.

Whitehead, D. (1977) *The Ideology of the Athenian Metic*. Cambridge Philological Society: Supplement 4.

Whitehead, D. (2000) *Hypereides. The Forensic Speeches*. Cambridge University Press, Cambridge.

Whitehead, D. (2002) *Aineias the Tactician. How to Survive Under Siege*. A Historical Commentary, with Translation and Introduction. 2nd edn. Classical Press, Bristol.

Winterling, A. (1991) 'Polisbegriff und Stasistheorie des Aeneas Tacticus: zur Frage der Grenzen der griechischen Polisgesellschaften im 4. Jahrhundert v. Chr', *Historia* 40: 193–229.

CHAPTER 8

The *Polis* Falling Apart: Aineias Tacticus and *Stasis*

Maria Pretzler

"*Stasis*—civil strife—is worse than war to the same degree as war is worse than peace."[1] Aineias may have agreed with Herodotos' general sentiment about the terrifying nature of conflict within a community, particularly because it had such a potential to increase the risks of war, but in the *Poliorketika*, there are no neat distinctions between conflict inside the *polis* and outside aggression. In fact, the term *stasis* is never used in the work, but it will be used in this chapter as a shorthand for various symptoms of internal disunity which Aineias clearly identifies as a serious threat to his city under siege. Traitors inside the city can inspire or invite outside attack, sabotage defence efforts, and finally open the gates to bring about utter defeat; at the same time, the pressures of war are also likely to cause new tensions among the besieged, compromising loyalty to the community. *Stasis* and war fuel each other, make each other worse, and any competent commander in charge of a besieged city keeps the attackers in check while also watching his own back. Aineias knows that the most devastating blow may well come from the enemy within, not from the army outside: it is necessary to watch out for traitors, plotters, collaborators and people simply losing the trust and hope necessary for an effective defence. The *Poliorketika* therefore covers situations where it is wise to have the watchmen on the wall face outside as well as inside.[2]

The Greek term *stasis* is often translated as 'civil war' or 'civil strife', modern terms which capture many of the more extreme phenomena which characterised *stasis* in the Greek city. The term 'revolution' is also commonly used, but is also more problematic, because its history, especially since the late eighteenth century, places it firmly within a concept of statehood and governance which is very different from an ancient *polis* community, and its applicability is in question.[3] Losada employed the term 'Fifth Column' for cases of collaboration between inside plotters and an enemy outside, drawing on terminology which developed in the context of the Spanish Civil War.[4] The collaboration

1 Hdt 8.31.

2 Ain. Tac. 9.12–13.

3 Eckstein 1965.

4 Losada 1972, 4–14.

 | DOI: 10.1163/9789004352858_009

between an outside aggressor and part of the besieged population, or perhaps just a few determined plotters, is only part of the ancient *stasis* phenomenon, but this aspect is particularly relevant to Aineias' concerns. Ancient and modern commentators have also analysed these civic conflicts in terms of social class, as a contest between the interests of the wealthy few and the poor masses, a concern which is to an extent reflected in the *Poliorketika*.[5]

Gehrke, in his comprehensive monograph on *Stasis*, prefers to adopt Eckstein's term 'Internal War' ('Innerer Krieg'). This, too, conjures up ideas of physical fighting. For my purposes in this chapter, I would like to draw the net a lot wider: *stasis* at its worst did indeed involve horrific violence, but that is just the most excessive expression of a wider phenomenon where community cohesion and basic solidarity with the *polis* disintegrate. Eckstein focuses on the deviation from generally agreed social norms; Gehrke develops this further, emphasising specifically the suspension of normally accepted rules of political disagreement, stepping outside the framework of constraints imposed by laws, social norms and customs.[6] In the end, *stasis* primarily requires a number of people to decide that their own individual interests, or those of a specific interest group, are more important than—and no longer coincide with—those of their *polis* and its community as a whole. In the ancient Greek world, this step constitutes a monumental transgression, the ultimate desperate and bold act, because one's *polis* is not just the prime source of individual rights and status, but also the focus of a free man's identity, based on history, relationship with the divine and a general sense of belonging. For the purpose of this chapter, I use the term *stasis* in this deliberately vague fashion, as a catch-all term for any action by citizens or inhabitants of Aineias' besieged *polis* which suggest that they no longer see the defence of the city as the ultimate priority.

Aineias is often mentioned in passing as an important source for tensions inside Greek cities in the fourth century, but there has in fact been little detailed analysis of his attitude to *stasis* as part of special studies of the phenomenon.[7] The earliest scholar to discover Aineias as a source for conflict within ancient societies was Pöhlmann, who researched ancient 'communism' and 'socialism'

5 Aristotle *Politics*, e.g. 1307a20–33, 1308b25–1309a33, cf. 1279b18–27, with Winterling 1993, Weed 2007, 118–123, cf. Fuks 1979–1980, 53–56; for more general discussion see Fuks 1974, esp. 59–61, Lehmann 1989; cf. Ste Croix 1981, 285–289.

6 Gehrke 1985, 6; cf. Eckstein 1964, 8–12, Eckstein 1965.

7 Gehrke 1985, 2, Lonis 1996; see also van Wees 2008 on *Stasis* in Archaic Greece; Losada 1972 focuses on the Peloponnesian War and Thucydides; Lintott 1982 does not discuss Aineias at all.

mainly to present an argument against such movements in his own time.[8] Few ancient historians showed interest in tackling the question of *stasis* in a less politicised manner until Lehmann took up the question, culminating in a 1989 article specifically dealing with Aineias and the socio-political crisis of the *polis*.[9] A year later, Whitehead's introduction to his translation of the *Poliorketika* emphasised the importance of the topic, provided an overview of Aineias' approach, and tried to defuse the question about Aineias' own political leanings.[10] For the purpose of this volume, it was thought that the topic warrants another detailed examination.[11]

The two most comprehensive discussions of *stasis* dating from the Classical period are Thucydides' account of the events in Korkyra in 427BC, and Aristotle's treatise on the subject in book V of the *Politics*. Both accounts deserve attention in their own right, but here they are introduced briefly to provide some context for Aineias' approach. Thucydides' account of *stasis* in Korkyra[12] remains the ultimate model of the disintegration of a Greek *polis* into horrific carnage: it describes a slippery slope from robust disputes between politicians of different factions through the law courts to the point where random killing and sacrilege is the order of the day, and part of the city centre is burned down. While this is going on, ships of interested outside powers are circling the island like vultures, occasionally prodding the two parties into further action, even when their action is ostensibly designed to calm things down. Later chapters demonstrate that this conflict dragged on for years and caused further trouble for the beleaguered Korkyraians. This is an extreme case, yet Thucydides deliberately sets it up as an exemplar for developments in his own time,[13] evoking images of extreme destruction in the reader's mind each time the narrative reaches another case of *stasis*. Thucydides' account is a classic in historical analysis as well as dramatic historiography, carried out with the diagnostic eye of a medic analysing the symptoms of a deadly disease.[14] By comparison, Aineias is much more focused on the root causes of the phenomenon and on early prevention wherever possible: he tries to get to potential plotters before their actions become material for a dramatic historical account.

8 Pöhlmann 1893–1901, at 1901, 346–348 (and later editions).

9 Lehmann 1989; cf. Lehmann 1970, esp. 401–405, Lehmann 1980.

10 Whitehead 2003, 25–33.

11 The editors would like to thank the anonymous referee for the suggestion to include this topic.

12 Thuc. 3.70–83, 3.85; 4.2; 4.46–48, with Hornblower 1991, 466–491.

13 Thuc. 3.82–83, with McLeod 1979.

14 Gehrke 1985, 88–93, Lintott 1982, 106–109, Fuks 1971.

Small, often mundane grievances, a wounded sense of pride, and quiet (but sometimes justified) desperation set off the disease even before its more dangerous symptoms develop; in a way, Aineias illustrates the beginnings of *stasis* before Thucydides' dramatic events begin, even though he offers some warning examples of devastating consequences, too. Where both authors agree is that disloyalty to the community and war are connected: they fuel each other, through desperation of those suffering aggression and defeat, through opportunism inside the city by the discontented, and through attempts of enemies on the outside to gain advantage by fomenting disputes among the besieged. While it is likely that Aineias knew Thucydides' account,[15] there is no evidence that the *Poliorketika* specifically aims to react to Thucydides' Korkyra episode; but it responds to similar concerns which can only have grown in the decades which had passed since the Athenian historian had written his account.

Aristotle's *Politics* was probably written another two or three decades after Aineias' work. Book v of the *Politics*[16] is the longest ancient work dealing with *stasis*, but its main concern is not the exact process of internal conflict within a *polis*, but the more general mechanisms which cause constitutions to be challenged and overthrown.[17] Aristotle hardly pays any attention to the role of external threats in destabilising *polis* societies, because he is interested in the inherent flaws of those systems. Aristotle and Aineias share the habit of illustrating general points with historical examples,[18] and in both cases, the historical episodes used for this purpose also demonstrate aspects of the phenomenon which are not the main focus of the discussion. Their priorities, however, are almost opposites of each other: in the *Politics*, Aristotle tries to develop a general theory of how *stasis* arises on the basis of inherent flaws within a *polis*'s society and how it might be prevented;[19] he investigates how different forms of government hold up to historical trends and flaws inherent in those systems, and how citizens' desire for equality or personal power shapes states' constitutions.[20] While Aristotle acknowledges the complex multiplicity

15 See my discussion in this volume, p. 78.

16 Aristotle *Pol.* 1301a19–1316b30.

17 Aristotle *Pol.* 1301a19–25; Mulgan 1991, esp. 307–309.

18 Book v of Aristotle's *Politics* contains 67 historical examples (Winterling 1993, 198); for Aineias see my other chapter in this volume, pp. 68–95.

19 Kalimtzis 2000, 103–106, 179–190.

20 Aristot. *Pol.* 1301b5–13, 1301b29–39. Note, however, his discussion of the human factor in triggering factional strife—desire for honour and wealth (1302a32–34) but also envy, ambition (hubris), fear, a reaction to various kinds of injustice or perceived unfairness (1302a35–1303a4). Cf. Polansky 1991, esp. 332–338.

of causes for *stasis*, his main interest is constitutional theory; the historical examples do offer glimpses of particular grievances or events which trigger *stasis*,[21] but even in that context, we rarely hear about the specific actions taken to overthrow a regime.[22] Aineias' focus seems almost the exact opposite of Aristotle's: he emphasises the processes of civic disintegration and illustrates them with stories where occasionally ideological struggles, usually between oligarchs and democrats, become visible.[23] It is important to appreciate that in both cases, the main line of discussion in the text represents the author's priorities and ideas, while the examples demonstrate that both were perfectly aware of other factors, but did not deem them central to their main argument.

The contrast between the discussion of *stasis* in the *Politics* and in the *Poliorketika* illustrates all too clearly how little interest Aineias has in constitutional details. He is wary of people who are unhappy with the political status quo, and it seems that the exact nature of the current government is a lot less important to him than doing everything in his power to avoid constitutional change at a time of external threat. Oligarchic and democratic states could have followed his advice equally, even though they might have found different aspects difficult to put into practice.[24] Lehmann thinks that Aineias assumes a democratic structure which ensures the loyalty of the poor:[25] indeed, there is one passage which suggests to assign the *ochlos*, the masses, to guard certain public spaces,[26] but otherwise the focus is on armed men, presumably hoplites, who would have had political rights in most oligarchic states as well.[27] In fact, some of the measures which curtail citizens' freedoms may have been harder to reconcile with the ideals of democratic states.[28]

Aineias' generic city may be thought to have one of those moderate regimes somewhere close to the blurred boundary between oligarchy and democracy which were probably most common in the classical Greek world, but it does

21 Aristot. *Pol.* 1303b18–1304a18 offers a collection of episodes where *stasis* arises from 'small matters'; see also 1311a37–1312a6 (personal causes that brought down tyrants). External factors play a role in 1302b29–31, 1303a3–11, cf. a general statement on the role of Athens and Sparta in changing constitutions in 1307b20–24.

22 E.g. Aristot. *Pol.* 1307a27–33.

23 Ain. Tac. 11.7–10, 11.10a–12, 11.13–15.

24 Cf. Shipley in this volume, pp. 52–55.

25 Lehmann 1989, 107; 109–110.

26 Ain. Tac. 1.9.

27 Ain. Tac. 1.5–6, 9.1, 38.1–2 (Lehman's examples, alongside a historical episode 11.10–11); cf. Ain.Tac. 9.1. which talks about an assembly 'of soldiers (*stratiōtai*) or citizens'.

28 E.g. Ain. Tac. 10.4–7.

not matter much. The most generic government set-up which almost all *poleis* would have shared is apparently assumed: there are clearly multiple magistrates, *archontes* (ἄρχοντες, also a very generic term), some of whom seem to take on specific, even partly military, functions, such as closing gates and commanding troops within the city.[29] They also supervise or authorise various activities which need special attention during a crisis, such as family gatherings, imports and the employment of mercenaries.[30] Aineias recommends that these leaders should have a bodyguard and avoid dangerous situations; indeed he offers examples where the assassination or capture of magistrates marked the beginning of a coup.[31] Nevertheless, magistrates also need to be watched carefully, since there are other historical episodes where individuals holding high office use their position to overthrow the government or hand over the city to an enemy.[32] Aineias also refers to generals, *stratēgoi* (στρατηγοί); there are several, as one would expect, and they are clearly taking on the main military tasks, such as commanding and organising troops or supervising guard duties. It is possible that these generals are sometimes included when Aineias uses the more generic term *archontes*, since his instructions on securing the gates, where he particularly stresses the importance of clearly defined competences, refer to both *archontes* and *stratēgoi* fulfilling this duty, as if the terms could be used interchangeably.[33] We can say for certain, however, that Aineias is describing a command structure and government setup which is heavily adapted to the crisis, so that the difference between the duties of civil and military officials seems to be blurred.[34]

Apart from the military headquarters, the *stratēgion*, Aineias' generic city is also expected to have a council building (*bouleuterion*) and a *prytaneion*.[35] Most cities with democratic or oligarchic governments would have had a council (*boulē*), and an appropriate space for their meetings: a *bouleuterion* usually has seating for a few dozen up to a group in the low hundreds and space for somebody to address them. *Prytaneia*, usually mainly recorded as venues for public dining, were common even in cities which did not have magistrates of that name:[36] the existence of such a building then tells us even less about a spe-

29 Ain. Tac. 3.6, 18.1–2, 18.21; see Whitehead 1990, 100, on 1.4.

30 Ain. Tac. 10.2, 10.4, 10.5, 10.7, 10.9–10, 13.3.

31 Ain. Tac. 1.4, 17.6 (bodyguards), 11.15, 17.3–4 (examples).

32 Ain. Tac. 11.3–6, 23.7–11.

33 Ain. Tac. 18.3, 18.15, 18.16, 20.1, 20.2 (*strategoi*), 18.1–2, 18.21 (*archontes*).

34 Ain. Tac. 10.4–7, 10.9–11.

35 Ain. Tac. 22.3 (*strategion*), 10.4. (*bouleuterion, prytaneion*).

36 Miller 1978, 4–24; note 9–10 on prytaneia in cities without magistrates called prytaneis.

cific government set-up, but for Aineias, it is clearly a valuable venue to hold and monitor public gatherings. Aineias makes no reference to the activities of specific government bodies in these buildings: the council itself, as one of the central political institutions, is not mentioned at all; its role in decision making is either taken for granted or perhaps assumed to be suspended. Assemblies do not fare any better in the *Poliorketika*. Two historical examples feature a popular assembly, in one case as a means to mobilise people against an oligarchic plot,[37] but in the other the oligarchs call an assembly to gather the people in one place where they and their leaders can easily be brought under control:[38] as a verdict on the usefulness of assemblies in preserving a democratic status quo these examples produce a score draw. In Aineias' generic city, we do not see the *ekklēsia* as a decision making body, either: in just one instance he advises to call an assembly, tellingly 'an assembly of soldiers or citizens'[39] which may suggest openness to oligarchic or democratic traditions in the community. This assembly is merely there to receive orders and information with the explicit expectation that the information will immediately be leaked to the enemy. We have to conclude, therefore, that Aineias would not have trusted an assembly with really important confidential information. Any assemblies of larger groups not sanctioned or ordered by the authorities are definitely considered a threat.[40] In fact, some of the emergency measures put in place to keep the population in check resemble Aristotle's methods of securing a tyranny.[41] Aineias may be worried about people trying to upset the status quo, but in fact, the emergency itself makes standard government procedures almost entirely obsolete and gives all power to the commanding general and a group of officials.[42] No wonder, perhaps, that the exact form of government in this besieged city matters little.

Unlike Thucydides and Aristotle, Aineias never focuses on a theoretical assessment of *stasis*, but his worries about dissenters and what they might do in specific situations are a significant theme in the *Poliorketika*. We can discern from those many references what kind of processes the author had in mind, and what he considered to be their main causes. The term *stasis* (στάσις) never appears in the work, but a simulated internal conflict is described

37 Ain. Tac. 11.8.

38 Ain. Tac. 11.15.

39 Ain. Tac. 9.1–2 ('ἐκκλησιάσαντα τοὺς αὑτοῦ στρατιώτας ἢ πολίτας').

40 Ain. Tac. 10.4–5.

41 Aristotle *Politics* 1313a41–1313b9, cf. Ain. Tac. 10.5.

42 Esp. Ain. Tac. 10.1–11.1. Cf. Shipley in this volume, pp. 53–56.

using the term *stasiasmos* (στασιασμός);[43] a successful revolt of the oligarchs in Korkyra against the democracy is called *epanastasis* (ἐπανάστασις), and later in the chapter the leaders of the people are accused of the same act.[44] More frequently, Aineias talks about plots, *epiboulai* (ἐπιβουλαὶ);[45] one chapter (11) is in fact titled *Epiboulai*, although the chapter titles may not be Aineias' own. Aineias usually thinks of such activities not in the abstract, but in terms of the people involved, for example those 'standing up against' the regime, *antistasiōtai* (ἀντιστασιῶται), who appear twice, in examples from Argos and Herakleia,[46] and, more frequently, plotters, *epibouleuontes* (ἐπιβουλεύοντες).[47] We also encounter traitors, *prodotai* (προδόται), planning treason, *prodosia* (προδοσία), which specifically refers to handing over their city to the enemy.[48] Sometimes these people are described more closely: Aineias is specifically worried about those who are unhappy with the status quo, and people who want change, 'those who want the opposite of what is currently in place'.[49] Another term commonly used in the *Poliorketika* is *neōterizein* (νεωτερίζειν),[50] literally 'to make things new' and 'those who want to make things new' (οἱ νεωτερίζειν βουλόμενοι).[51] The term *neōterizein* and its cognates are generally used in classical Greek to describe the activities of people plotting to overthrow their state or constitution, but they also imply a very specific motivation which, in this case, coincides with other passages where Aineias explains more clearly what motives he has in mind.

What drives Aineias' plotters to want a change of the status quo? We are used to thinking of *stasis* in political terms, efforts to change the political system, as discussed by Aristotle, or perhaps in Thucydides' terms, where changes between oligarchy and democracy may also signify a shift in the alignment with one of the two great powers, Sparta or Athens. Aineias' historical examples expose such motivations, but the *Poliorketika* is no theoretical treatise on *stasis*, and there is no aim to give a clear definition of the wider underlying reasons which drive citizens to do damage to their own community. Aineias leaves the

43 Ain. Tac. 23.3.

44 Ain. Tac. 11.13, 11.15.

45 Ain. Tac. 1.6, 11.2, 11.12, 22.20, 23.7, 29.7, 31.24, 31.33; cf. Losada 1972, 11.

46 Ain. Tac. 11.7, 12.5.

47 Ain. Tac. 2.7, 10.3, 10.15, 11.9, 11.10a, 11.14, 17.2, 17.3, 17.4, 22.20, 23.6; cf. 31.9b.

48 Ain. Tac. 11.3–5, 11.9, 22.7, 31.8, 31.9, 31.25–27; cf. Losada 1972, 6–10.

49 Ain. Tac. 14.1., cf. 10.20. 'τῶν τὰ ἐναντία φρονούντων τοῖς παροῦσι πράγμασι', 'τοῖς μὲν οὖν ἐν τῇ πόλει ὑπεναντία θέλουσιν τοῖς καθεστηκόσι'.

50 Ain. Tac. 10.25, 22.5, 22.6.

51 Ain. Tac. 2.1., 17.5., 22.17, 30.1.

background vague and focuses on the characters who might cause trouble, and with this more practical perspective, he may be providing a much better insight into general perceptions of *stasis* at the time. Much is merely implied or taken for granted: the author assumes that the contemporary reader will recognise some of the characters, mindsets and motivations he is describing.

As we have seen, Aineias' plotters are people who are not satisfied with the status quo, even if it is less clear what exactly this means: he may be talking about the government, but if the siege and the state of martial law he seems to be prescribing went on for some time, we have to assume that dissatisfaction would also increase because of those emergency measures. He is, in fact, expecting that some people will try to leave the city: men intending to become mercenaries and those wanting to sail out with a ship have to ask for permission from a magistrate, but there is no advice to deny all such requests on principle.[52] Few cities would have been able to afford the loss of many men of fighting age, and Aineias is not just keen to prevent desertion during military engagements outside the walls, but he is also concerned with the general morale of his troops, most of whom, we should not forget, were essentially desperate amateurs.[53] Anybody leaving the city was also a potential informer for the enemy, or might tell exiles that an opportunity for a return by force might arise.[54]

Who are the people who might desert and or betray their native city in a crisis? Aineias suspects primarily those who have nothing to lose—men who do not have families or considerable property: for young men particularly, the outlook during a siege on their city would have been very bleak, and the incentives to leave would become greater as the threat of defeat grew. After all, it was not uncommon that, after a city was taken, the victorious forces would kill all the men and sell all the women and children into slavery. Consequently, Aineias has most confidence in those who have something at stake in the community. He recommends that those parts of the city most at risk from attack should be guarded by the wealthiest and those with most prestige in the city because, as he says explicitly, remembering what they had to lose would make them most effective.[55] This is spelled out even more clearly where Aineias describes his ideal gate keeper: in this position, where one man can do fatal

52 Ain. Tac. 10.7–8.

53 Ain. Tac. 23.1, 28.2 (deserters); e.g. 22.24, 22.26, 26.7–10 (morale).

54 Ain. Tac. 10.6. suggests a proclamation which deals with people who leave the city to make contact with exiles.

55 Ain. Tac. 12.15.

damage to the whole community,[56] he wants to see intelligent people capable of healthy suspicion, but first and foremost, he wants men who have families and property to protect.[57] The emphasis on property and prestige comes across as a certain disdain for the poor, but these were indeed advantages which were most dependent on citizenship and therefore on the wellbeing of the community. A traitor might perhaps be lured with financial rewards or at least freedom and safety for himself and his family, but if his city was taken over by the enemy, or if he had to leave the city, it was highly unlikely that he would ever get the chance to be a property owner or an eminent citizen again, even if he was still wealthy, since citizenship and the right to own property were closely guarded privileges and, in most *poleis*, almost impossible for outsiders to obtain.

The worst candidates for gate keeping are, according to Aineias, men 'who because of poverty or the pressure of obligations, or some other difficulties, may be persuaded to join an uprising (*neōterismos*) or instigate a plot themselves.'[58] Debtors in particular represent a danger to community cohesion: in another passage, Aineias refers to the menace of men in dire straits sitting by and waiting for an opportunity.[59] If the debt got out of hand, a debtor might indeed assume that upheaval in the city or even an invasion by an enemy could be a chance for a fresh start, provided they themselves managed to stay alive.[60] To avoid some of these risks, Aineias recommends a reduction or abolition of debt interest, and, in extreme cases, he suggests to cancel debts altogether. Such a measure would of course cause damage to the creditors, and therefore might create even more discontent within the community. Aineias apparently had a solution to this problem, but, unfortunately, he had already explained it in his lost work *Poristikē*.[61] Perhaps his suggestion was similar to that for financing mercenaries, where the wealthy are required to provide money and upkeep for the moment, with a promise to get reimbursed with a tax reduction in the future.[62] Food shortages are also a concern, which means that the community has to try its best to maintain supplies: Aineias suggests incentives for anybody

---

56 See Ain. Tac. 18–20 for numerous examples.

57 Ain. Tac. 5.1.

58 Ain. Tac. 5.1. '… οἵτινες δὶ ἔνδειαν ἢ συναλλαγμάτων ἀνάγκην ἢ δὶ ἄλλην τινὰ ἀπορίαν πεισθεῖεν ὑπό τινων ἢ αὐτοὶ παρακελεύσαιεν ἄν τινας ἐπὶ νεωτερισμῷ.'

59 Ain. Tac. 14.1. 'ὡς πολύ γε φοβερώτατοι ἔφεδροί εἰσιν οἱ τοιοίδε ἄνθρωποι, τούς τε ἐν ἀπορίᾳ ὄντας τῶν ἀναγκαίων εἰς εὐπορίαν καθιστάναι.'

60 Cf. Thuc. 3.81.4.

61 Ain. Tac. 14.1–2.

62 Ain. Tac. 13.4.

managing to import food, and also recommends to provide the basic necessities for those who cannot afford them.[63]

The first impression is almost inevitably that Aineias rates the rich as dependable and honourable and the poor as untrustworthy, an attitude that would not have been particularly unusual among Greek elites of the period.[64] But, in fact, the situation is more complex, because he also identifies the wealthy and influential as most likely to cause trouble within the city. We should not forget that people from a wealthier background might also find themselves in some of the desperate situations Aineias is so worried about, particularly debt. At the same time, if the most prominent people felt that their status or material interests were at stake, they were also the most likely to have the relevant means to trigger a coup, particularly clout, connections and executive experience within the city, as well as acquaintances abroad, which may include friends in the enemy camp.[65] When Aineias suggests to get rid of some of those who 'desire something different from the present situation' he therefore singles out those with a history of leadership and political initiative.[66] Removing fairly prominent people could cause further trouble, so the advice is to find a pretext, for example an embassy or similar public mission abroad, which would get potential plotters out of town with their honour intact. Aineias includes three historical examples where the wealthy (or 'the wealthy and oligarchic', in one case) are staging a coup against a democratic regime;[67] and we have already seen examples which showed magistrates involved in overthrowing a city's government.[68] Stories of democratic coups are more difficult to find, except, perhaps, when Aineias talks about smuggling in weapons to arm those citizens who do not have their own, and there is a story of another group of conspirators who resorted to making shields and armour from wickerwork.[69] But even there, we have to assume that the leaders and instigators of those plots are better situated, prominent men, and, in one case, the leader of a coup is even smuggled

63 Ain. Tac. 10.12 (food imports), 14.1. (supplies for the most needy).

64 Winterling 1993, 190–195, see also Todd 1990, 159–160, 164–167; but note Aristotle *Pol.* 1301b40–1302a2, arguing that there are few well born (εὐγενεῖς) and virtuous (ἀγαθοί) people in a city, but many with wealth.

65 Winterling 1993, 198–199.

66 Ain. Tac. 10.20 'τῶν τὰ ἐναντία φρονούντων τοῖς παροῦσι πράγμασι … καὶ μάλιστ᾽ ἂν ἡγεμόνας τε καὶ αἰτίους γενομένους ἐν τῇ πόλει'.

67 Ain. Tac. 11.7–10, 11.10a–12, 11.13–15—πλούσιοι (καὶ ὀλιγαρχικοί, in 11.13).

68 Ain. Tac. 11.3–6, 23.7–11, see above p. 151.

69 Ain. Tac. 29.4, 29.11–12.

into the city, together with the necessary weapons and armour.[70] Thus, watching the activities of prominent citizens is a crucial aspect of avoiding *stasis* in Aineias' city.

While some plots can be carried out alone, for example by opening the gates or signalling to the enemy from the walls, an actual uprising in the city will need more support, which is why Aineias is so concerned with morale among his men and the desperation of the poor. Everybody in this besieged city is putting up with serious restrictions on their daily lives, and Aineias displays valuable insights into the psychology of his little community and some individuals within it. In particular, he describes personal motivations which seem petty in the grander scheme of things, but which represent desperate decisions of individuals trying to ensure their own survival or that of their family, even if it is against the interest of the whole community. What we get here are insights into activities which are never important enough for final historical narratives of great events, but which give us a better idea of the complexity of *polis* societies: the many viewpoints and interests hidden behind a polis label in a historical narrative, such as '*stasis* in Korkyra' or 'the Mitylenaians revolted'.

In Aineias' *polis* wounded pride is clearly a factor: we are a long way away from the great heroes of the *Iliad*, but here, too, anger about losing face is a crucial factor in sapping loyalty. Where historiography occasionally shows us prominent statesmen motivated by concerns about their personal honour, Aineias takes into account that such values and emotions might motivate every and any man in a *polis*, and potentially with devastating consequences. Thus, when morale is low, Aineias recommends to approach loudly when checking the watches, in order to spare the watchmen the embarrassment of being caught sleeping: there are situations when exacerbating disaffection and despondency is more dangerous than a few men napping on guard duty.[71] Men who cannot be trusted are best not publicly confronted, but quietly removed from positions where they could do damage, and if possible in ways that could be considered a special honour: men who are under suspicion are therefore kept away from festivals by giving them the 'privilege' to celebrate the holiday at home.[72] Fairness is also an issue: for example, the recommendation to make sure that watches are distributed fairly includes advice to reset the water clock at regular intervals to match the seasonal length of days and nights, and if we

70 Ain. Tac. 29.7–8, cf. Winterling 1993, 198.

71 Ain. Tac. 26.7–10.

72 Ain. Tac. 16–17; cf. 10.20.

assume that Aineias is talking about a clock in a public place, he might also be thinking of a recurring public display of ensuring fairness.[73]

Another dangerous example of putting personal interest first was people's desperation to rescue personal belongings and particularly food supplies from outside the walls, even when the situation was already getting dangerous.[74] Aineias makes allowances for evacuating slaves and animals in advance,[75] and for people engaging in last-moment harvesting while the enemy is approaching.[76] The whole city's food supply may depend on it, but this was also a matter of families worried about their own economic survival. The inhabitants of a Greek *polis* under attack could apparently be expected to respond by rushing out to the fields, on their own or in small groups, to rescue or defend their property. Aineias is worried that people might get killed or perhaps captured, but he is also ready to respond to the people's desire to confront the enemy by suggesting properly organised sorties.[77] Nevertheless, all that is left outside, crops, stored supplies and installations which could be of use to the enemy, might have to be destroyed; as he suggests this, Aineias leaves it to the reader's imagination how such measures might be received by the owners of those assets: perhaps he had already discussed these matters elsewhere.[78] There is also a special concern for people whose family members were held as hostages in the enemy camp,[79] accompanied by a hint that the enemy might well kill the hostages in view of the city.[80] Rather than contemplating how to prevent such an atrocity, probably because at times it was unavoidable without surrender, Aineias goes about insulating relatives from the whole conflict, preferably by removing them from the city altogether while there is still a chance to do so. He seems resigned to the idea that hostages' relatives might try something desperate to rescue their family members: the advice here shows a certain understanding for the terrible predicament such people would be in. It also explains more clearly why it was so important to prevent rash actions outside the walls which might allow the enemy to capture more prisoners. Yet again, other ancient sources

73 Ain. Tac. 22.24–25; for problems of interpretation see Whitehead 1990, 158–160; on public water clocks see Armstrong and Camp 1977.

74 Ain. Tac. 10.1–3, 15.2, 16.2–3.

75 Ain. Tac. 10.1–2.

76 Ain. Tac. 7.1–2; cf. Xen. *Hell.* 7.5.14–17.

77 Ain. Tac. 15.3–4, 16.2–3.

78 Ain. Tac. 8.3–5, with a cross-reference to another lost work.

79 Ain. Tac. 10.23–24.

80 This is usually not made explicit in historical accounts, but Diod. 20.54.2–7. offers a particularly gruesome example.

talk about hostage-taking, but in Aineias, we see the effects on individuals and, through them, on the whole *polis*.

This is where Aineias' image of dissent within a *polis* differs most from both Thucydides and Aristotle's ideas of *stasis*: in this generic city under siege, personal motivations matter, and these are often not (or certainly not predominantly) guided by considerations linked to greater political aims. Aineias is willing to accept some of these personal concerns as normal human behaviour rather than malicious intent, even if actions which can arise from such motives are presented as treacherous activities which need to be stopped or prevented. When the odds of survival become less favourable, individuals will re-assess the balance between loyalty to their *polis* and their own personal interests. Aineias talks about *homonoia*, unity of purpose, in the community as the most important antidote to *stasis*. In our terms, what he describes is predominantly about confidence, namely the confidence of every member of the community that collaborating with the *polis's* defence activities is a better bet to preserve the lives and livelihood of themselves and their families than any action they could take on their own, or with a group of dissenters.

Aineias never provides us with a grand theory or narrative of how *stasis* happens: instead his work gathers many different scenarios and insights into the many forms destructive disloyalty might take in a *polis* under pressure, from small acts of sabotage to spectacular attacks which end in bloodshed and destruction. Not all of these could easily be defined as *stasis*, but what we see is a sliding scale of community cohesion failing under pressure, either when individuals lose the will to support their city's defence or, more dangerously, when dissent or discontent fosters groups with common causes or grievances.

Through his advice and some of the examples we get to see a few common scenarios Aineias has in mind at the more dangerous end of this scale, his idea of how a larger plot might turn into outright *stasis*. The most dangerous moments in a city's life, apart from a direct attack by an enemy, are probably mass gatherings, particularly festivals. Such occasions may offer an excuse to bear arms, for example in a procession, or allow plotters to take control of arms which have to be left behind, insufficiently guarded, by some of the participants.[81] At the same time, *stasis* by a large group of conspirators might best be prevented by mass gatherings, too, if those are made up of vigilant citizens and organised properly to disperse the plotters and to deny them an opportunity to overpower the rest of the population.[82] In any case, controlling

---

81 Ain. Tac. 17.1–6.

82 Ain. Tac. 11.7–12.

any kind of gathering is a crucial task of the city officials,[83] and open spaces within the city have to be secured to prevent large groups of people to come together on their own initiative.[84]

Like most Greek *poleis*, Aineias' generic city is probably not very big: the city's force may number in the hundreds rather than in the thousands, so that Aineias has concerns about inviting mercenary forces which outnumber the local army.[85] If mercenaries need to be hired, he suggests a highly decentralised system which will put no single citizen in control of more than two or three mercenaries at a time.[86] But in a fairly small community, the balance of power between different interest groups might be shifted even just by arming locals who do not have armour themselves.[87] All foreign visitors are automatically under suspicion, to the point where they are disarmed on entry and locked inside their inns during night time; even embassies are kept at arm's length from all but carefully selected citizens.[88] Aineias is probably mainly worried about communication with the enemy, which is to be restricted in various ways,[89] but visitors might just add extra support to an internal plot, and at times, perhaps just an appropriate leader was needed, as in the example where a commander is smuggled into a city together with additional weapons.[90]

The most common form of disloyalty in the *Poliorketika* is collaborating with an enemy outside. Conflict within the community, a proper *stasis*, might lead one side to draw on the support of an enemy outside the walls, but the major concern for Aineias is that a single individual with grievances and an opportunity can do immense damage to the whole *polis*. People who plot in groups can be detected and obstructed in various ways, but one desperate man hoping to improve his fortunes when the city is taken is a lot more difficult to detect. Here is yet another issue where where Aineias' attention to individuals and their grievances and selfish calculations plays a crucial role. Every man in the city, including the slaves,[91] could be instrumental in a plot with the enemy. It is worth adding that Aineias perhaps misses a trick when he never even contem-

83 Ain. Tac. 10.4–5.
84 Ain. Tac. 1.9–2.1; 2.7–8.
85 Ain. Tac. 12.1.
86 Ain. Tac. 13.1–4.
87 Ain. Tac. 29.1–12 (two examples); precautions against: e.g. 10.7, 30.1–2.
88 Ain. Tac. 10.9–11, 10.13.
89 Ain. Tac. 10.6.
90 Ain. Tac. 29.7. more support is then let in once the plotters on the inside are ready, 29.10.
91 Ain. Tac. 24.3–9 (a slave who collaborates with the enemy).

plates the idea that women, too, might be potential traitors who collaborate with the enemy. Women do turn up as potential messengers,[92] as part of the cast of various stratagems[93] and, of course, as potential victims who defend their houses by throwing roof tiles;[94] nevertheless, in the *Poliorketika*, they are part of men's plots, but not suspected of being plotters themselves. The sheer size of the main chapters dealing with individual traitors, especially 18–20 and 31, the large number of examples dealing especially with treacherous gate keepers and attempts to communicate with the enemy, shows the importance of the dangers of individual men's disloyalty; cautious tales about such behaviour are presented across the whole work, and include all sorts of offenders and opportunities. The main message is that it takes one man who is desperate, angry or frightened enough to take an opportunity as it arises to take down the whole city. With an enemy waiting outside the walls, an individual's disloyalty might be just as dangerous as a full *stasis* trying to overthrow the status quo. Aineias' attention to the concerns of individuals is therefore crucial to his project.

Homonoia—a common purpose—and the ability to keep calm and organised under pressure are Aineias' best defence against defeat. This kind of community cohesion might still save the city even when the enemy is already within the walls. The best example for this is an episode from Plataia, where the Thebans are already in control of the city, but the locals work together, breaking through walls of houses and using local resources and superior knowledge of the urban topography to overcome the intruders.[95] In such a situation, panic and confusion are especially dangerous, particularly if the defenders cannot tell friend from foe. Plotters might actually try to bring about a panic, pretending that an attack has already happened, in order to help the enemy take the city.[96] Good preparation, sufficient motivation and, most of all, a well-ordered community are key to overcome the most perilous stage of an invasion, when the fight comes down to defending the city itself, street by street.[97] These are also the situations where traitors can do the most damage, if they have already made common cause with the enemy.

92 Ain. Tac. 31.7.

93 Ain. Tac. 4.10–11, 24.7, 40.4–5.

94 Ain. Tac. 2.6, 4.8; cf. 3.6. where Aineias recommends how to organise local watches so that the men can stay close to their houses and keep their children and wives under control.

95 Ain. Tac. 2.3–6.

96 Ain. Tac. 23.6.

97 Ain. Tac. 4.1–5, 39.5, see also 24.6, where a password prevents some confusion.

In summary, a whole range of human behaviour is on display in the *Poliorketika*. Aineias covers deliberate acts of sedition as well as stupidity or selfishness with unintended disastrous consequences, and he takes the potential dangers of both seriously. Planned treachery requires some kind of motivation, from discontent and anger to desperation or self-interest. Combined with the means to carry out a plan, disloyal sentiments can become dangerous as supporters are identified, communications opened, and manpower or equipment secured. Finally, the pivotal point is reached when an opportunity arises: a chance might be easily available as part of the daily routine of the city, but, with a vigilant commander in charge, those dangers might be reduced to a minimum. Aineias offers examples where specific motivations are mentioned, but the actual reason for sedition matters less to him than efforts to restrict access to means and opportunities which could facilitate a plot. At the same time, he emphasises the importance of keeping any kind of discontent among the population to a minimum, as far as that is even possible in a community under severe stress.

Aineias' main weapon against *stasis* and treason is *homonoia*,[98] community cohesion, particularly if it is founded on a confidence in the ability of the *polis* to guarantee the safety its residents. Many problems do not even arise if the people are willing to follow the leadership, and agree on what needs to be done and what sacrifices might be necessary to ensure the community's wellbeing. But this ultimate community asset cannot be taken for granted. At times, securing *homonoia* might entail a balancing act, catering to the needs of interest groups whose concerns might be contradictory, e.g. debtors and lenders, or more generally the wealthy and the poor.[99]

For Aineias' purposes, the most important resource are those inhabitants who agree that security lies in the preservation of the status quo, those who are willing to work together to defend the city, and he assumes that the mass of people can be relied upon to stop a coup.[100] He counts on decent, experienced men with families to defend, people respected by their neighbours and ready to comply with requests from magistrates and commanders.[101] In addition, the community needs order, and in a crisis this can include special measures

98 Ain. Tac. 10.20: "first it should be ascertained whether the citizens are of one mind (in a state of *homonoia*), since this would be of particular benefit during a siege."—"*καὶ πρῶτον ἐπισκεπτέον εἰ ὁμονοοῦσιν οἱ πολῖται, ὡς ἂν ὄντος μεγίστου τούτου ἀγαθοῦ ἐν πολιορκίᾳ*", cf. 14.1.

99 Ain. Tac. 14.1–2.

100 Ain. Tac. 1.9, 11.10a–12.

101 Ain. Tac. 3.4–6; cf. 5.1.

citizens might not tolerate in peace time.[102] Finally, morale is essential: if too many citizens lose hope or assume that the current regime will not be able to save their lives or livelihoods, *homonoia* is unlikely to survive, and neither is the city.

What makes Aineias' work so valuable is that he analyses important aspects of Greek politics and conflict in the context of community life, where individuals and their concerns matter: surviving under siege depends on getting the best out of people. A whole range of damaging behaviours, from individual acts of disloyalty to violent factional strife, are linked to a complex set of human calculations and emotions which matter especially in smaller *poleis* where the actions and personal networks of almost any individual can make a difference. In this context, *stasis* begins to look like a fact of *polis* life—something that can grow, under pressure, out of the perfectly normal attempts of human beings to look after their personal advantage. Under normal circumstances, it might be expected that most people's interests coincide with the wellbeing of the community, but Aineias demonstrates how this situation can shift for some, and he grapples with ways in which loyalty can be maintained, restored or, if necessary, enforced with increased vigilance. In Aineias' small city under siege, it is not philosophical differences about constitutions or the strategic interests of the great power blocks which are the main triggers of *stasis*. In this respect, Aineias' view of *stasis*, even if we have to piece it together from different passages in his work, is a very valuable addition to those other discussions of internal strife which survive from the Classical period, particularly Thucydides and Aristotle. In the *Poliorketika*, we get a chance to look at the failing of *homonoia* from the ground up, encountering a level of *polis* life where the political is intensely personal.

Bibliography

Armstrong, J.E. and Camp, J.M. (1977) 'Notes on a Water Clock in the Athenian Agora', *Hesperia* 46: 147–161.

Eckstein, H. (1964) *Internal War. Problems and Approaches*. Free Press: New York.

Eckstein, H. (1965) 'On the Etiology of Internal Wars', *History and Theory* 4: 133–163.

Fuks, A. (1971) 'Thucydides and the Stasis in Corcyra', *American Journal of Philology* 92: 48–55.

102 Chapter 10 in particular lists a whole range of security measures which look like measures to introduce a kind of martial law to the city, disrupting normal life in a number of ways.

Fuks, A. (1974) 'Pattern and Types of Social-Economic Revolution in Greece from the Fourth to the Second Century BC', *Ancient Society* 5: 51–81.

Fuks, A. (1979–1980) 'Τοῖς ἀπορουμένοις κοινωνεῖν. The Sharing of Property by the Rich with the Poor in Greek Theory and Practice', *Scripta Classica Israelica* 5: 46–55.

Gehrke, H.-J. (1985) *Stasis. Untersuchungen zu den inneren Kriegen in den griechischen Staaten des 5. und 4. Jahrhunderts v. Chr.* Beck, Munich.

Hornblower, S. (1991) *A Commentary on Thucydides, vol. I: Books I–III.* Clarendon Press, Oxford.

Kalimtzis, K. (2000) *Aristotle and Political Enmity and Disease. An Inquiry Into Stasis.* State University of NY Press, Albany. 88–137.

Lehmann, G.A. (1970) 'Dion and Herakleides', *Historia* 19: 401–406.

Lehmann, G.A. (1980) 'Krise and innere Bedrohung der hellenischen Polis bei Aeneas Tacticus', in Eck, W., Galsterer, H., Wolff, H. (eds), *Studien zur antiken Sozialgeschichte. Festschrift für F. Vittinghoff*. Böhlau, Cologne and Vienna. 71–86.

Lehmann, G.A. (1989) 'Aeneas Tacticus und die politisch-soziale Krise der zeitgenössischen Polis-Welt: Einige kritische Anmerkungen', *Index* 17: 105–115.

Lintott, A. (1982) *Violence, Civil Strife and Revolution in the Classical City.* Croom Helm, London and Canberra.

Lonis, R. (1996) 'Poliorcétique et stasis dans la première moitié du IVe siècle av. J.-C.', in Carlier, P. (ed.) *Le IVe siècle av. J.-C. Approches Historiographiques.* De Boccard, Paris. 241–257.

Losada, L.A. (1972) *The Fifth Column in the Peloponnesian War.* Brill, Leiden.

McLeod, C.W. (1979) 'Thucydides on Faction (3.82–83)', *Proceedings of the Cambridge Philological Society* 25: 52–68.

Miller, S.G. (1978) *The Prytaneion: its Function and Architectural Form.* University of California Press, Berkeley.

Mulgan, R.G. (1991) 'Aristotle's Analysis of Oligarchy and Democracy', in Keyt, D. and Miller, F.D. (eds), *A Companion to Aristotle's Politics.* Blackwells, Oxford. 307–322.

Pöhlmann, R.v. (1893–1901) *Geschichte des antiken Sozialismus und Kommunismus.* Beck, Munich.

Polansky, R. (1991) 'Aristotle on Political Change', in Keyt, D. and Miller, F.D. (eds), *A Companion to Aristotle's Politics.* Blackwells, Oxford. 323–345.

Ste Croix, G.E.M. de (1981) *The Class Struggle in the Ancient Greek World.* Duckworth, London.

Todd, S. (1990) 'Lady Chatterley's Lover and the Attic Orators: The Social Composition of the Athenian Jury', *Journal of Hellenic Studies* 110: 146–173.

van Wees, Hans (2008) 'Stasis, Destroyer of Men: Mass, Elite, Political Violence and Security in Archaic Greece', in Brélaz, C. and Ducrey, P. (eds), *Sécurité collective et ordre public dans les sociétés anciennes.* Fondation Hardt, Geneva. 1–48.

Weed, R.L. (2007) *Aristotle on Stasis: A Moral Psychology of Political Conflict*. Logos Verlag, Berlin.

Whitehead, D. (1990) *Aineias the Tactician, How to Survive under Siege*. Translated with Introduction and Commentary. Clarendon Ancient History Series, Oxford.

Winterling, A. (1993) 'Arme und Reiche. Die Struktur der griechischen Polisgesellschaften in Aristoteles' Politik', *Saeculum* 44: 179–205.

CHAPTER 9

Greek *Poleis* and Warfare in the Fourth Century BC: Aineias' *Poliorketika*

*Marco Bettalli**

1 The Otherness of Aineias Tacticus

In the introduction to his excellent commentary on the *Poliorketika*, David Whitehead emphasizes the original perspective that Aineias Tacticus offers us on Greek history, focusing not on Sparta, Athens and the few other great ancient Greek cities, but on the *poleis*, which usually suffered the consequences of such hegemonic cities' political and military aims.[1] I would like to revisit this observation from the point of view of the world of warfare: precisely because the *Poliorketika* adopts a different perspective from that of the main literary sources, it also provides an exceptional account of what warfare was like in fourth-century Greece. For example, almost all the ancient sources—followed, as children follow the Pied Piper of Hamelin, even by modern historians—concentrate at length, if not exclusively, on the battle itself.[2] In the renowned passage of Demosthenes' *Third Philippic* dedicated to changes in the art of warfare,[3] to which we shall return later, one of the most surprising characteristics of the new way of doing battle is apparently the fact that "nothing is resolved in an open field battle". There are no open field battles in Aineias' text, save for a brief mention of how to arrange troops in the event of combat against the enemy. Let us recall the passage (16.7) as it is the only one to describe combat in the open field:

> Wait, in other words, until your relieving force is ready at the appointed place and the enemy already dispersed for plunder: then is the time to bring pressure to bear on them, cutting off their lines of retreat with your

---

* Translation: Emma Thorley, Translation Service, University of Siena.

1 Whitehead (2002) 1–4.

2 Even J.K. Anderson's seminal treatise on warfare in the IV century (1970) focuses on open field battles and warfare in the countryside, citing Aineias only once, and then fleetingly (p. 159).

3 Demosth. 9.47–50.

© KONINKLIJKE BRILL NV, LEIDEN, 2018 | DOI: 10.1163/9789004352858_010

> cavalry, laying ambushes with the picked men, confronting the enemy with the rest of the light-armed troops, and bringing up the heavy infantry in formation not far behind the advance contingents. As to the attack itself, launch it in circumstances where you are under no compulsion to fight but will not be at a disadvantage if you do.[4]
>
> transl. D. WHITEHEAD

Note that this battle is not considered inevitable, but rather one of several possible alternatives; also significant are the mentions of the ambushes to be laid by the special forces and the role of light-armed troops (*kouphoi*), in a context that could be described as dynamic.

Whoever Aineias was—and it is not up to us to resolve the *vexata quaestio* of his identity[5]—his view of war is clearly based on experience and is the result of the reflections of a highly intelligent and capable man. Above all, with the exception of the brief and conventional preamble, his vision is conveyed in a direct manner and is entirely unconditioned by ideological conventions or caution. We could perhaps speak of a '*degré-zéro*' of war writing, which offers us a unique opportunity to enter a mid fourth-century Peloponnesian *polis* and study the measures taken to deal with their recurrent crises, which often led to active fighting. What follows is an attempt to analyze some of the military aspects contained in Aineias' text, which can aid our understanding of the way conflicts were viewed and dealt with in fourth-century Greece. I am well aware that Aineias does not need me to defend him, but I am particularly pleased to write these pages. In fact, I have the strong impression that, although the *Poliorketika* is well known for its description of social turmoil in the *polis* and other aspects that I would define 'folkloristic' (such as the inevitable citation of the women disguised as men in 40.4), it is still underrated from a military viewpoint.[6]

---

4 ἐάνπερ γε σὺ αὐτοῖς εὐκαίρως ἐπιθῇ. ἡτοιμασμένης γάρ σοι τῆς βοηθείας εἰς τὸν παρηγγελμένον τόπον καὶ ἐσπαρμένων ἤδη τῶν πολεμίων πρὸς ἁρπαγήν, οὕτω χρὴ αὐτοῖς προσκεῖσθαι τοῖς μὲν ἱππεῦσιν προκαταλαμβάνοντα τὰς ἀποχωρήσεις, τοῖς δ ἐπιλέκτοις ἐνέδρας ποιούμενον, τοῖς δ ἄλλοις κούφοις ἐπιφαινόμενον αὐτοῖς, τοὺς δ ὁπλίτας ἀθρόους ἐν τάξει ἄγοντα, μὴ πόρρω δὲ τῶν προπεμφθέντων μερῶν. Ἐπιτίθεσο δὲ τοῖς πολεμίοις ἐν οἷς ἄκων μὲν μὴ μαχήσῃ, μαχόμενος δὲ μὴ ἔλασσον ἕξεις τῶν πολεμίων.

5 It will become clear that I have not changed the opinion expressed in Bettalli (1990) 5–6: I believe that Aineias was a mercenary commander of Peloponnesian origin, who may be identified as Aineias of Stymphalos mentioned in Xen. *Hell.* 7.3.1.

6 Cf., for example, van Wees (2004), a text for which I have the greatest admiration, pays very little attention to Aineias' importance as a writer on *militaria*; the recent important summary

2 "There is Always War": A Very Violent Game

Medical books portray all the illnesses of the human body. In other words, in the debate on the omnipresence of war in the Greek world, it may seem somewhat inappropriate to produce as evidence a manual written by a military commander. Moreover, the debate itself seemed to have been silenced, with the majority accepting the affirmation as an inconvenient fact, until the case was reopened by Simon Hornblower and other eminent scholars.[7] In my opinion their arguments are not entirely convincing, although they are sufficiently so to prompt further reflection on the subject. It seems to me that numerous pieces of evidence scattered throughout the *Poliorketika*, not to mention the many examples in other sources, allow us to see the *polis*, or at least the fourth-century *polis*, as a fragile structure that was almost impossible to protect from recurrent bloody outbreaks caused by internal tensions and the continual threat of attack by external forces. In such a situation, war was indeed part of everyday life, which explains the ambiguous attitude shown by Aineias. While he is not oblivious to the desirability of disturbing the normality of everyday life as little as possible and considers the family-related needs of the citizens called to arms,[8] he also does not hesitate in a conscious attempt to militarise everyday life in order to maximise control over a precarious and dangerous situation, as shown in chapter 10 (one of the most frequently cited chapters of the whole text, and not without reason). Aineias does not tell us when (at precisely which point in the escalation of war) this attempt should be put into practice, although it should not be the very moment the enemy arrives; this confirms my interpretation, which tells me there is a very thin line between periods of peace and war. To use the words with which the unforgettable character of Mordo Nahum (who, by coincidence, is Greek!) in Primo Levi's *The Truce* addresses to the young protagonist who has escaped from Auschwitz: "there is always war".[9]

The military context within which Aineias acts is, to a certain extent, surprising in its simplicity. Take the arms, for example: a group of artisans and some wicker is sufficient to transform a collection of citizens into a formidable army (39.11), a circumstance that should give pause for thought to those who see the category

in Cambridge Warfare (2007) contains more citations, but they are often oriented towards the social dimension.

7 Hornblower (2007); cf. Bettalli (2011) in press.

8 3.6, 22.26—arguably some of the most interesting passages in the text.

9 Primo Levi, *La tregua*, Torino (1958); (1989[3]) 189 [*The Truce*, UK, 1969].

of hoplites[10] as operative and essential, also on a practical level. Trickery and scheming abound as in a comedy: impressionable soldiers who suffer from panic attacks (27.1–14), enemies who eat and drink too much (16.5), women who disguise themselves as men (40.4), friends who become enemies (4.1–4), people who pretend they are wounded (11.14) or emerge ready to fight from under a pile of twigs (29.8)—not to mention all the complex but essentially banal stratagems used to force the locks of the city gates (18.3–22), or the wonderful yet impracticable systems for transmitting secret messages over long distances.[11] This may make warfare look like a game but it does not lessen the violence, which is intrinsic to any war and characterizes a society that knows it well and is certainly not surprised by its appearance.[12] Thus, apart from the references to torture that so troubled earlier commentators[13] (6.7), the text is full of cruelty[14] and systematic massacre (4.3–4; 15.8–10); Aineias does not generally seem to show any interest in the preservation of human life.

3 The Citizen/Soldier Revealed and the Selection of Troops

In the three and a half centuries between Archilochos throwing away his shield and the Athenian coward described by Theophrastos,[15] our sources are characterized by a succession of self-proclaimed heroes, for whom *dulce et decorum est pro patria mori*. The *Poliorketika* provides a completely different view. In a context in which the concept of patria itself is dormant, the citizen/soldier—an amateur who is not paid for his services in any way—is stripped of any ideological exaltation, and thus fearful (16.2: "in the grip of understandable terror",[16] comments Aineias), psychologically unstable (26.7–11; 17; 38.4), inefficient (6.1–2), and preoccupied solely with safeguarding his own family and possessions

10 One of the most debated subjects, even recently, on which our text does not provide sufficient evidence for the matter to be discussed here; instead I refer the reader to van Wees (2004).

11 Cf. chapter 31 (*infra*, § 6).

12 The comment (as always, apparently casual) of Jacob Burckhardt comes to mind: "We glimpse an accumulation of bitter suffering in these Greek *poleis* that has rarely been seen in human history." Burckhardt (2010) 83 n. 35.

13 The troubled commentators: Hunter and Handford (1927) 120.

14 Such as the hostages forced to die fighting in the front line during assaults, at 10.23.

15 Archiloc. Fr. 5 West; Theophr., *Char.* 25.3–6.

16 "πεφοβημένων εἰς τοὺς κινδύνους προϊέναι".

(3.6; 7.1; 15.2; 16.2). In short, a disastrous soldier—unlike the mercenary, who is considered far superior (12.2). It goes without saying that Aineias' perspective is by far the most realistic. Precisely because of the inevitably mediocre performance of the citizen/soldier, the question of the recruitment or selection of combatants becomes pressing; Aineias dedicates pages of great interest to this subject, which merit analysis.

The army that the *poleis* place at the disposal of the military commander is—first and foremost—a temporary army. There is no specific call-up procedure, nor is the length of service known: everything is entrusted to the intelligence and the flexibility of the commander, who seems to be guided by the principle of causing "the least possible inconvenience to the smallest number of people"[17] (22.26). Aineias pays no attention to traditional divisions: the cavalry is only mentioned on a few occasions, and in two of these (6.6; 26.4) his choice of words suggests the lack of such a division in many *poleis*; hoplites are only mentioned twice (15.5; 16.7),[18] on the same two occasions as the light-armed troops; and peltasts, which were highly 'fashionable' at the time, are not mentioned at all.[19]

An entire chapter (1) is devoted to the selection of troops with the aim of allocating military tasks to the citizenry, thus providing a truly photographic image of the *polis* through the eyes of a professional soldier.[20] In carrying out this function, the commander must demonstrate an aptitude for understanding his soldiers. Aineias does not waste time underlining this quality, but it brings his leader close to the figure of the ideal commander dwelt on by Xenophon, for example, in book III of the *Memorabilia*.[21] We should mention that Aineias' description of the selection procedures is not clear; nonetheless, we shall seek to render it more linear. First of all, we should point out that, in contrast to the Athenian forces analyzed by Thucydides (2.13) at the outbreak of the Peloponnesian war, Aineias does not seem to make any use of metics or foreigners (the *ochlos* of 1.9 should not be interpreted in this sense), but relies exclusively on the free inhabitants of the city. Thus, for the vast majority of the population only two requirements are specified: physical strength, which seems

17 "ἐλάχιστά τε καὶ ἐλαχίστους τῶν ἀνθρώπων ὀχλεῖν".
18 See above, section 1, pp. 166–167.
19 However, note the mention of *peltai* at 29.6.
20 It is possible but not necessary to hypothesize that he had absolute powers. His relationship with the 'civil' magistracy of the city remains unclear, but is not a subject analyzed in my contribution; nonetheless, see below, note 21.
21 On the selection see, in particular 3.4.5.

to be a prerogative of the younger men,[22] and political trustworthiness,[23] an essential and unavoidable aspect, mentioned frequently by Aineias throughout his work. All those who do not possess these requirements, but who are not clearly identifiable as opponents, are allocated general tasks of limited importance.

Only those who take on positions of responsibility involving leadership functions are also required to be intelligent,[24] these leaders should also be those "with the most to fear from a change of regime"[25] (1.7). Here Aineias alludes to the necessity, made explicit in 5.1 concerning gate keepers, for those who are entrusted with positions of responsibility to offer some form of guarantee, such as property and a family. Epigraphic and literary sources confirm that this was a well-known procedure. Lastly, people who are not only "most judicious", but possess the further qualification of experience in warfare, are destined to 'be attached to the authorities' (1.4).[26] The emphasis here is clearly on intellectual and emotive rather than physical capacities. We can assume that these leaders would be people of advanced years whose political trustworthiness could be taken for granted.

The military commander of the *polis* (maybe his own, but we are not sure), therefore, will first proceed to surround himself by a select group of people—not young, but expert and entirely trustworthy—who will help him organize the troops available. Secondly, he will carry out an inspection in order to separate the healthy and physically strong men from all the others. In the third phase, I would say almost certainly with the help of his newly appointed council of expert citizens, he will separate the trustworthy citizens from those whose loyalty is not entirely certain. Lastly, he will appoint the commanders, favouring those who have most to lose. It is interesting to note that Aineias does not mention any official distinctions or make reference to any katalogos

---

22 Cf. 1.8.

23 1.6. "Loyal and satisfied with the status quo.", "εἶναι δὲ αὐτοὺς εὔνους τε καὶ τοῖς καθεστηκόσι πράγμασιν ἀρεσκομένους".

24 Epitomized by the adjective *phronimos* (cf. also 3.4, 5.1, 15.3), which has a long history related to practical and flexible intelligence, as mentioned in Bettalli (1990) 214, and bibliography.

25 1.7. "καὶ ᾧ ἂν πλεῖστοι κίνδυνοι εἶεν μεταβολῆς γενομένης."

26 1.4. "πρῶτον μὲν οὖν αὐτῶν ἀπονεῖμαι δεῖ τοὺς φρονιμωτάτους τε καὶ ἐμπείρους μάλιστα πολέμου, οἳ περὶ τοὺς ἄρχοντας ἔσονται".

I will not enter into the discussion of to whom Aineias is referring in this expression (*hoi perì tous archontas esontai*): a good *status quaestionis* can be found in Whitehead 2002, 100.

to establish the citizens' economic conditions or military position (horseman, hoplite, etc.): this all remains beyond his scope.

4 A Favourite Subject: Information and Communication

What did it mean to carry out the business of war professionally—and at a high level—in the fourth century, as Aineias did? First and foremost it meant identifying and isolating the business of war as a subject for study and reflection. This approach leads to at least two results: being able to deal with the various situations that will arise in a dispassionate manner; and making improvements in the field of technology or, more simply, in procedures and relationships with subordinates. In Aineias' world, these topics were evidently the subject of frequent discussions and debate, to which some references may be found in the text.[27] In fact, one of the most remarkable characteristics of warfare in the fourth century is precisely the establishment and spread, even the formation from nothing, of this world of soldiers, who were distinct from politicians and aware of being so.[28]

While Aineias is generally well-informed, there is one particular area in which he shows particular interest and admirable assiduity: the field of information and communication. A very 'modern' subject, we may say, and one that was certainly not highly developed in Aineias' day. Nonetheless, using pertinent comparisons we shall seek to demonstrate how this field is fundamental in identifying some epoch-making changes in the art of warfare.

This importance is affirmed by Clausewitz,[29] no less, who defines as 'intelligence' "... every sort of information about the enemy and his country—the basis, in short, of our own plans and operations". As we know, for a long time the Greeks were unaware of the importance of intelligence.[30]

---

27 Cf. for example 22.13; 25.1; 26.6.

28 The consolidation of the politics/war rift, resulting in the creation of 'professional' generals, such as our Aineias, is a well-known subject which we shall not go into here. Lengauer (1979) was a pioneer of these studies; Tritle (1992) (to be taken into consideration but not followed) was against this view, which contrasted the v to the iv century.

29 C. von Clausewitz, *Della guerra*, trad. it. Milano 1970, 84. [Carl von Clausewitz, *On War*. Princeton University Press (1989)].

30 Most recent references on the subject: Richmond (1998); Russell (1999); Sheldon (2003) (mainly referring to the Roman world).

Concerning the factors considered essential to ensure a successful outcome in battle, there is a clear shift of focus in the *Poliorketika* from the notion of strength/courage, typical of this tradition, to the notion of knowledge/information. Throughout his manual, Aineias places great emphasis on the need for knowledge and, of even greater importance, on the *ways* of transmitting knowledge.

Above all, knowledge of one's own side is essential; not by chance does the text commence, following the brief introduction, with an affirmation of the peremptory need for *skepsasthai*—careful observation, examination—of all aspects of the *polis* that the military commander has to defend: the size of the male population, the topographic features of the *polis* and the consequent necessity of organizing guards and patrols (1.1).[31] The importance of gathering knowledge, or 'intelligence', also concerns the area surrounding the city: at 16.20. Aineias notes that, thanks to this knowledge, "The difference, in fact, between your respective circumstances will be just as great as if they ⟨had to fight⟩ at night while you were in daylight."[32] For the same reasons, the movements of one's own troops in the surrounding area must be followed very attentively: all information must be reported to the command station inside the city "from as far out as possible" (4.6). This need for control also refers to citizens who are not involved in military actions but are working in the fields of the surrounding countryside (7.2). Inside the city, as we see from chapter 10, the commander must know about and control not only all arms (10.7, 10.9) and mercenaries present in the city (10.7), but also everyday aspects such as festivals, the activities of seers (10.4), letters sent or received (10.6), and even the occupants of hotels (10.9).

Aineias obviously deems knowledge of the enemy, in as far as this is possible, as equally decisive, although the presence of external enemies in the *Poliorketika* is well known for being limited and not particularly invasive. Chapter 6, dedicated to *hemeroskopia*, concerns the information that well-trained scouts can obtain about the composition, plans and movements of the enemy troops.

However, the field in which Aineias is most innovative is the related one of communication, which constitutes the most remarkable characteristic of the *Poliorketika*. Aineias appears to be fully aware that the real problem is not so much obtaining information, as transmitting it. He could not go beyond the technological capabilities of his era, which to us seem extremely limited, but

31 Socrates gives similar advice to the young Glaucon, who wishes to gain a position of command in Athens, in Xen. *Mem.* 3.6.8–10.

32 16.20. "διενέγκοι γὰρ ἂν ἑκατέροις ὅσον εἰ τοῖς μὲν νυκτὸς μάχεσθαι εἰλήχει τοῖς δὲ ἡμέρας".

it is fundamentally important that he even posed the question of communication, coming up with solutions that, although they may seem primitive and inefficient to us, are nonetheless signs of an interest (we could even say a passion) that was far ahead of his time. We shall look below at some significant passages on this subject.[33]

On the theme of transmitting information, and therefore communication, Aineias explains the use of lanterns and other long-distance signalling devices in great detail in some passages.[34] However, the most complex system is not described, probably having been left to the *Paraskeuastikè biblos*, which no longer exists, but we know about it from Polybius (10.44). This system sought to increase the *quantity* of information transmitted over long distances during daytime, apparently reaching a theoretical maximum of 24 messages.[35] Lastly, we cannot but cite the famous and extremely long chapter 31, dedicated to secret messages, which contains descriptions of two real cryptographic systems (31.17–19; 31.30–31) which—despite their simplicity and, at least in the first case, complete uselessness from a practical viewpoint—are two of the few examples of the kind to have existed in the ancient world.[36]

We can therefore maintain that, in the pages of his manual (which, we should not forget, is only one part of a more systematic and comprehensive work), Aineias' leitmotiv is the need to create an information network to control both declared and potential friends and enemies, which is capable of transmitting and exchanging the information obtained with the greatest efficiency, i.e. as much news as possible with the greatest secrecy and speed.[37]

---

33 Aineias is aware of the fundamental distinction between true and false news, just as he recognizes the importance of controlling the spread of news in terms of the soldiers' morale. Cf., for example, the interesting ch. 27 on the spread of panic attacks, often caused by uncontrolled news, and ch. IX, which points out that carefully filtered news of the citizens' excellent organization could dissuade enemies from attacking.

34 Cf. in particular 6.4–7; 7.2–4; 22.21–22; 26.12–14.

35 Cf. Hershbell (1978), which contains the best description of Aineias' 'telegraph', in a context that pays close attention to the oral/written dynamic, which is a good starting point on which to base reflections on the topic.

36 Cf. Bettalli (1990) 302–305. The subject does not seem to be particularly popular with scholars, although I believe it is worthy of careful reconsideration. Among the few articles published in recent years, we suggest Richmond (1998) and Debidour (2006).

37 The question of the speed of execution recurs frequently in the text: cf. for example 3.6; 6.5–6; 9.1; 15.2.

5 Aineias, Demosthenes, Philip and the Art of Warfare

It is not easy to demonstrate that this interest in the world of information/communication was *peculiar* to the fourth century, or in any case that it developed at an accelerated rate in the first half of the century, after the Peloponnesian war. On the one hand, to use a very famous example, Aischylos had already described an impressive system of relayed fire signals used to transmit the news of the fall of Troy to Mycenae (*Ag.* 281–316); moreover, it is plausible that many battles of the Peloponnesian war were won thanks to the use of informers, spies and other such figures (the first that comes to mind being the Battle of Sphakteria). On the other hand, there are plenty of examples from the same century of complete disinterest towards such subjects, not to mention an amateurism that Aineias would undoubtedly have disapproved of.[38] The number of sources is definitely limited, and the few that exist are generally inclined to favour approaches in keeping with the hoplite ideology. As we well know, this ideology was connected with a very simple kind of warfare, based on field battles, and paid little attention to knowledge of the enemy's troops and movements.

Nonetheless, it seems clear to me that Aineias' manual testifies to his new awareness of issues such as rapidity, secrecy, knowledge of the land and intelligence: all of which are aspects of a more varied, dynamic and altogether more complex kind of warfare. Contemporary sources can provide further insight into this question: we refer not to Xenophon, the undisputed authority on fourth-century warfare, whose agreement with Aineias has been highlighted on several occasions,[39] but to the Demosthenic corpus, and in particular some passages of the *demegoriai* which focus primarily on the art of warfare.

Demosthenes is active in, and reflects upon, a context that is chronologically contiguous to Aineias (only a few years later) but entirely different: as an orator he analyzes the chaotic workings of assembly-based democracy, which causes delays, indecision and reconsiderations in a framework that is completely unfavourable to the effective conduct of war. In this context, it is no coincidence that Aineias never mentions any form of collective decision-making mechanism.[40] Nonetheless, some aspects recall the needs put forward

38 Cf. for example Xen. *Hell.* 5.1.20–22; 7.5.10; although there are many others.

39 One of his last works, the *Hipparchikòs*, competes with the *Poliorketikà* for the title of the oldest existing military manual: cf. Petrocelli (2001); Vela Tejada (2004).

40 At most he mentions the typically democratic institution of drawing lots adopted, for security reasons, to allocate a section of the walls to each tribe (3.1) or to identify the strategists responsible for the bolts of the city gates (20.2).

by Aineias himself: limiting ourselves, for now, to the splendid pages of the *First Philippic*,[41] we see Demosthenes suggest that citizens should be used as *epoptas*, as supervisors of operations, and, at the same time, as informers of the assembly, which otherwise would never know anything about what was going on (4.25) or find out far too late. In contrast Philip, king of Macedonia, manages to prevail because he is always extremely well-informed about everything and can anticipate the movements of the Athenians and the enemy in general.[42] It is also helpful to mention another passage of the oration (4.35–36), in which the concept of war as a pre-eminently anomic place is contested: a place, that is, in which it is impossible to be efficiently organized, in contrast to the organization the Athenians manage to achieve for the city's religious festivals, which are by nature predictable as they are included in a precise calendar of events. Aineias seems, indeed, to be trying to apply organization and a solid framework to war, which is by its very nature unpredictable.

Speaking of the changes in the way battles took place in the fourth century, it is impossible to ignore the figure of Philip II. He is often evoked in relation to the *Poliorketika*, which appeared in the mid-fifties, when Philip was about to come onto the scene: a circumstance that could have rendered Aineias' work obsolete as soon as it came out. In actual fact, it seems to me that Aineias anticipates at least a few of the innovations that are attributed to Philip: a mode of warfare based on the acquisition of information, rapidity of action, flexibility, and the use of troops containing as few inefficient and unreliable citizens as possible (even of hoplite status) and as many good soldiers as could be recruited from among the mercenaries.[43]

Philip is therefore not the subverter of the traditional rules of warfare, and is not responsible for ruining the game that the Greeks had loved to play for centuries. Demosthenes adopts this ideological view in the famous passage of the *Third Philippic* (9.47–50), in which he analyzes the changes that have taken place in the art of warfare during his time, including some aspects that are

41 Demosth. 4 probably written in 351 BC.

42 Cf., e.g., 4.18, a passage that recalls the psychological situation suggested by the *Poliorketikà*, ch. 9: if Philip finds out how well organized and prepared the Athenians are, he will stay where he is.

43 It is not my task to discuss the relationship between citizens and mercenaries in Aineias. However, in this context, I would like to recall one of Philips's great predecessors (in many aspects)—Jason of Pherae—who (as reported by Polydamas of Pharsalus to the Spartan assembly a few years before the *Poliorketika* was written) placed great emphasis on the performance of mercenaries, which was far superior to that of the citizens: cf. Xen. *Hell.* 6.1.5. Aineias seems to agree with this position (cf. 12.2).

already present in Aineias, as mentioned above.[44] The Macedonian king is in fact deemed responsible for putting into practice and developing a whole series of innovations which were already familiar to those conducting war and which could not be effectively implemented due to the social and military framework of the *polis*: in this sense, Aineias does not appear behind the times at all, but rather a soldier full of ideas who probably realizes how difficult it is to conduct war in the outdated world in which he lives.[45]

6 Sieges

The final chapters of Aineias' work (32–40) oblige us to mention the subject of siege as dealt with by Aineias, at least briefly.[46] Again (see previous section), a commonly accepted viewpoint maintains that our author has in some way been unlucky, as the art of *poliorketikà* underwent such significant developments with the appearance of Philip and especially Alexander,[47] that Aineias' text quickly became an outdated compilation of no practical interest. I have never fully accepted this hypothesis[48] and believe that I can now counter it even more firmly. Aineias has excellent knowledge of the means that were effectively used at that time, demonstrating himself to be an expert on city fortifications: glancing through the final chapters, we find that he is perfectly familiar with battering rams (32.3–7), knows catapults, which were introduced into Greece no more than a dozen years before the text was written[49] and, although he does not propose their use for defence, he lists a plau-

44 Note that Demosthenes, albeit in a very aggressive passage against his arch-enemy, introduces the changes that have taken place in the conduct of warfare with the verbs *kekinesthai* and *epidedokenai* (κεκινῆσθαι, κἀπιδεδωκέναι, in 9.47) which—somewhat contradictorily—do not evoke a negative opinion of the changes.

45 For the more traditional position, which sees Aineias as an author outdated by events and therefore the last representative of an era, rather than as an innovator, cf., e.g. Bengtson (1962) 468. I myself was much closer to this position in my commentary: cf. Bettalli 1990, 34–36.

46 For some reason that is not entirely clear, sieges in the Greek world have been the subject of very few studies in the last few decades. For a general introduction, see the recent summary by Kern 1999.

47 On this subject see Keyser (1994).

48 Cf. Bettalli (1990) 312–315, and references.

49 If it is true that they were imported into the Greek world by the expedition that Dionysus sent to save Sparta in 368, on that occasion Archidamos would have pronounced the saying reported by Plut. *Mor.* 219a 8 ("Heracles, the valour of men is lost!"), suggesting

sible series of counter-measures against attacks using the new *mechanema*. He talks extensively about ladders (36) and tunnels (37), and theorizes about the use of reserve troops to ensure the continuous replacement of soldiers on the perimeter walls when the city is attacked (38.1–2). Lastly, he is well aware of the phenomenon of *Geländemauern* (40), and was probably familiar with the recent examples of Megalopolis and Messene.

None of this became obsolete—not only in the decades that followed, but at any time in the ancient world! At the very most we could speak of an increase in the size of the *mechanemata* and the spectacularity of the sieges, but without completely changing the scene. This is one of the reasons for the persistence of Aineias' fortunes until the Byzantine era, although we should also mention the considerable traditionalism, not to say static nature, of warfare in the ancient world. Our mistake (and the reason why Aineias is considered outdated) lies in seeing the art of siege as an inexorable *continuum* of progress, and placing too much emphasis on any invention, despite the fact that their practical impact is often difficult to gauge. Not to mention the fact that the siege of Plataia described by Thucydides (2.75–77, in the year 429) is much more highly 'evolved' than similar episodes that took place fifty years later.

I would like to call attention to a simple fact in the *Poliorketika*, but one which has not been sufficiently underlined: assaults on city walls were very common at the time, despite the costs they entailed in terms of human lives. It is difficult to say whether this was because defence strategies were often unrealizable due to the large sums of money they required, because the walls of the *poleis* (especially the smaller ones) were anything but insurmountable or, lastly, because the assaults themselves led to (frequently critical) treachery within the city walls: perhaps it was a combination of all these circumstances.

7 Conclusions

Let us read again a well-known passage from Isokrates' Archidamos, written in around 366, not long before the Poliorketika. Archidamos, the son of Agesilaos, speaks in defence of the positive role the Spartans played in the Peloponnese, a role that was placed in doubt, if not refuted, by the Battle of Leuktra and the expeditions of Epaminondas in subsequent years.

---

great sadness for the non-technological world that is disappearing, albeit very slowly. On the subject of the introduction of the *mechanemata* to the Greek world, cf. Pimouguet-Pedarros (2000).

> (64–65) Also I think that not only the people of the Peloponnesus in general but even the adherents of democracy, whom we consider to be especially unfriendly to us, are already yearning for our protection. For by revolting from us they have gained nothing of what they anticipated; on the contrary, they have got just the opposite of freedom; for having slain the best of their citizens, they are now in the power of the worst; instead of securing self-government, they have been plunged into misgovernment of many terrible kinds; accustomed as they have been in the past to march with us against others, they now behold the rest taking the field against themselves; and the war of factions, of whose existence in other territories they used to know only by report, they now see waged almost every day in their own states [...]. (67–68) They feel such distrust and such hatred of one another that they fear their fellow-citizens more than the enemy; instead of preserving the spirit of accord and mutual helpfulness which they enjoyed under our rule, they have become so unsocial that those who own property had rather throw their possessions into the sea that lend aid to the needy, while those who are in poorer circumstances would less gladly find a treasure than seize the possessions of the rich; having ceased sacrificing victims at the altars they slaughter one another there instead; and more people are in exile now from a single city than before from the whole of the Peloponnesus [...].[50]
>
> LOEB translation

We will not dwell on the ideological vision suggested by these words (the Spartans as defenders of the landowners, as the Romans will be a century and a half later). What interests us is the description of the Peloponnese in the

---

50 οἶμαι δὲ καὶ τὸν ἄλλον ὄχλον τὸν ἐν Πελοποννήσῳ καὶ τὸν δῆμον, ὃν οἰόμεθα μάλιστα πολεμεῖν ἡμῖν, ποθεῖν ἤδη τὴν ἡμετέραν ἐπιμέλειαν. οὐδὲν γὰρ αὐτοῖς ἀποστᾶσι γέγονεν ὧν προσεδόκησαν, ἀλλ' ἀντὶ μὲν τῆς ἐλευθερίας τοὐναντίον ἀποβέβηκεν (ἀπολέσαντες γὰρ αὑτῶν τοὺς βελτίστους ἐπὶ τοῖς χειρίστοις τῶν πολιτῶν γεγόνασιν), ἀντὶ δὲ τῆς αὐτονομίας εἰς πολλὰς καὶ δεινὰς ἀνομίας ἐμπεπτώκασιν, εἰθισμένοι δὲ τὸν ἄλλον χρόνον μεθ' ἡμῶν ἐφ' ἑτέρους ἰέναι, νῦν τοὺς ἄλλους ὁρῶσιν ἐφ' αὑτοὺς στρατευομένους, καὶ τὰς στάσεις, ἃς ἐπυνθάνοντο πρότερον παρ' ἑτέροις οὔσας, νῦν παρ' αὑτοῖς ὀλίγου δεῖν καθ' ἑκάστην τὴν ἡμέραν γιγνομένας [...]

ὕτω δ' ἀπίστως τὰ πρὸς σφᾶς αὐτοὺς καὶ δυσμενῶς ἔχουσιν, ὥστε μᾶλλον τοὺς πολίτας ἢ τοὺς πολεμίους δεδίασιν: ἀντὶ δὲ τῆς ἐφ' ἡμῶν ὁμονοίας καὶ τῆς παρ' ἀλλήλων εὐπορίας εἰς τοσαύτην ἀμιξίαν ἐληλύθασιν, ὥσθ' οἱ μὲν κεκτημένοι τὰς οὐσίας ἥδιον ἂν εἰς τὴν θάλατταν τὰ σφέτερ' αὐτῶν ἐκβάλοιεν ἢ τοῖς δεομένοις ἐπαρκέσειαν, οἱ δὲ καταδεέστερον πράττοντες οὐδ' ἂν εὑρεῖν δέξαιντο μᾶλλον ἢ τὰ τῶν ἐχόντων ἀφελέσθαι: [68] καταλύσαντες δὲ τὰς θυσίας ἐπὶ τῶν βωμῶν σφάττουσιν ἀλλήλους: πλείους δὲ φεύγουσι νῦν ἐκ μιᾶς πόλεως ἢ πρότερον ἐξ ἁπάσης τῆς Πελοποννήσου.

years after Leuktra, which is powerful despite the excessive rhetoric, as this is the field of action of Aineias: this is his terrain.[51] Here, the Peloponnese is losing the position of importance it had previously attained—in other words, it is becoming outdated and unimportant, while the new Greek world is taking shape elsewhere, in the centre and north of the country. In this context, it is impossible to expect Aineias to represent the vanguard of a new way of conducting war, or of change at all. However, he can be seen as a rare example of an intelligent, flexible, up-to-date commander; one who knows warfare in all its aspects and how common it is, but at the same time is aware of the need to safeguard oases of calm and everyday life whenever possible.

Reading him again twenty years on, I am increasingly convinced that this author is essential to our understanding of war and society in the *poleis* of the fourth century BC.

Bibliography

Anderson, J.K. (1970) *Military Theory and Practice in the Age of Xenophon*. University of California Press, Berkeley.

Bengston, H. (1962) 'Die Griechische Polis bei Aeneas Tacticus', *Historia* 11: 458–468.

Bettalli, M. (1990) *Enea Tattico, La difesa di una città assediata* (*Poliorketika*). Introduzione, traduzione e commento. ETS Editrice, Pisa.

Bettalli, M. (2011) 'Guerre tra polemologi. Dodici anni di studi sulla guerra nel mondo greco antico, 1998–2009', *Quaderni di Storia* 73: 161–231.

Burckhardt, J. (2010) *Storia della civiltà greca, vol. 1: I Greci e il loro mito—La polis*. Bollati Boringhieri, Turin.

Debidour, M. (2006) 'Le secret et les messages secrets dans la Poliorcétique d'Énée le Tacticien', in Olivier, H., Giovannelli-Jouanna, P. and Bérard, F. (eds), *Ruses, secrets et mensonges chez les historiens grecs et latins*. Université Jean Moulin Lyon 3, Lyon. 213–241.

Hershbell, J.P. (1978) 'The Ancient Telegraph: War and Literacy', in Havelock, E.A. and Hershbell J.P. (eds) *Communication Arts in the Ancient World*. Hastings House, New York. 81–92.

Hornblower, S. (2007) 'Warfare in Ancient Literature: The Paradox of War', in Sabin, P., van Wees, H. and Whitby, M. (eds) *The Cambridge History of Greek and Roman Warfare vol. 1*. Cambridge University Press, Cambridge. 22–53.

51 I am aware that this interpretation, which takes Aineias to be Peloponnesian, is not completely incontrovertible (cf. *supra*, note 4).

Hunter, L.W. and Handford, S.A. (1927), *Αἰνείου Πολιορκητικά. Aineias on Siegecraft*. Clarendon Press, Oxford.

Kern, P.B. (1999) *Ancient Siege Warfare*. Indiana University Press, Indiana.

Keyser, P.T. (1994) 'The Use of Artillery by Philip II and Alexander the Great', *Ancient World* 25: 27–59.

Lengauer, W. (1979) *Greek Commanders in the 5th and 4th Centuries BC. Politics and Ideology: a Study of Militarism*. Wydawnictwa Uniwersytetu Warszawskiego, Warsaw.

Petrocelli, C. (2001) *Senofonte, Ipparchico. Manuale per il comandante di cavalleria*. Edipuglia, Bari.

Pimouguet-Pedarros, I. (2000) 'L'apparition des premiers engins balistiques dans le monde grec et hellénisé: un état de la question', *Revue des études anciennes* 102: 5–26.

Richmond, J.A. (1998) 'Spies in Ancient Greece' *Greece and Rome* 45: 1–18.

Russell, F.S. (1999) *Information Gathering in Classical Greece*. University of Michigan Press, Ann Arbor.

Sheldon, R.M. (2003) *Espionage in Ancient World. An Annotated Bibliography of Books and Articles in Western Languages*. McFarland, Jefferson.

Tritle, L. (1992) 'Virtue and Progress in Classical Athens: the Myth of the Professional General' *Ancient World* 23: 71–89.

van Wees. H. (2004) *Greek Warfare: Myths and Realities*. Duckworth, London.

Vela Tejada, J. (2004) 'Warfare, History and Literature in the Archaic and Classical Periods: The Development of Greek Military Treatises' *Historia* 53: 129–146.

Whitehead, D. (2002), *Aineias the Tactician, How to Survive under Siege*. Translated with introduction and commentary Updated Edition. Bristol Classical Press and Duckworth, Bristol and London.

CHAPTER 10

Light Infantry and Leadership in Aineias

Nick Barley

The subject of Aineias Tacticus' work is ostensibly that of siege warfare; *Poliorketika*, or *How to Survive Under Siege*, could suggest a work focused on useful methods and tricks to prevent walls being breached and guards being tricked; we would expect talk of battlements, supplies, and trickery, and we are not disappointed. However Aineias' conception of siege is much broader than this; he does not limit himself to discussing the physical defence of city walls, and devotes a significant amount of his text to taking the battle to the enemy in well-organised surprise attacks using mixed forces co-ordinated from the city. Taken as a whole, Aineias' text presents us with a systematic analysis of the ways in which an invasion could be discouraged and, if necessary, repulsed. In this manner his work goes beyond simply describing how the walls of a besieged city can be defended, becoming a treatise on the effective defence of the *polis* as a whole: its territory, its people, and its sense of geographic, cultural, and historical self-identity. Aineias considers the physical defence of walls and gates as a last resort, preferring that an invasion be met on the very borders of a state's territory.[1] The reasons for this are twofold: first, local topographical knowledge lends a huge tactical advantage to the defenders, which can only be taken if the enemy are met in the field within the borders in some manner; secondly, Aineias' main concern, the potential for revolt from within, is at its most critical and dangerous point when a city is actively besieged.[2]

Aineias Tacticus' surviving work is such an invaluable collection of advice, anecdote, and practical theory, all presented as 'standard' responses to invasion or threat, that his evidence has the power to dramatically change the way we view Classical Greek warfare. True, the *Poliorketika* is a work of the mid-fourth century, but Aineias does not present any of his advice as being innovative or part of some technical or intellectual 'revolution' of warfare or generalship, the existence of which is frequently advanced by modern scholarship,[3] and his

1 Ain. Tac. 8.1; 16.16.

2 Ain. Tac. 2.3–6; 16.19.

3 As summarised by Wheeler: (2007) 213–223.

 | DOI: 10.1163/9789004352858_011

definition of 'siege' is so broad that his advice should be read as a commentary on the art of warfare in general. This advice focuses heavily on the use of light infantry and ad-hoc forces, as well as on the important role of the general and his subordinate commanders, while the systematic and effective use of reconnaissance, manoeuvre, and communication is presented as desirable and within the grasp of what we may risk calling an 'average' *polis*. These elements all combine to produce a detailed response to the presence of enemy forces which is notable for its reluctance to engage in any form of conventional or direct combat. This system thus differs greatly from current mainstream thought, especially with regard to the use and effectiveness of what van Wees termed 'The Other Warriors', those non-hoplite forces which have traditionally been viewed as peripheral to the main event.[4] Effective and organised light infantry was known to the Greeks at the time of the Persian Wars,[5] was deployed to significant effect at various times during the Peloponnesian War,[6] and throughout the Corinthian War.[7]

It is important to note that light infantry was only effective in the field when both organised and, generally, in combination with a phalanx that could serve as a checking unit or a 'phalanx in being', but when organised and intelligently commanded they represented a significant threat to enemy troops of all kinds.[8] Flexible, unconcerned with keeping anything more than the most rudimentary of formations, able to operate in virtually any territory and enjoying a significant advantage in speed and manoeuvrability over a phalanx, light infantry introduced a deadly dimension to battle: range.[9] This range threatened both the physical and ideological cohesiveness of an unaccompanied or insuf-

4 Van Wees (2004) 61–76.

5 See, e.g. Herodotos 9.21–23. How and Wells (1912) 295; Lazenby (1993) 222–223; van Wees (2004) 181.

6 E.g. Thuc 1.106. Kitto (1966) 271; Hornblower (1991) 166. The effective deployment of light infantry was difficult for Classical Greek generals to achieve, yet here we see it used in combination with a pursuing hoplite force to exploit advantageous terrain while in unfamiliar territory, the very model of an effective engagement as recommended by Aineias Tacticus! See also the various engagements of the Athenian general Demosthenes and his immediate counterpart Brasidas. It is surely no coincidence that these generals were able to use, and react to, light infantry, and were also highly effective.

7 E.g. Xen. *Hell* 4.5.13–18.

8 E.g. Hdt 9.20–24; Thuc 4.124–128; Xen. *Hell* 2.4.32–34. Many more examples will be discussed in a future paper.

9 They were thus able to inflict damage without suffering in kind, assuming the enemy was unable to deploy effective counter-measures. See Adcock (1957) 15: "… this is the most artful thing in battle, to harm the enemy and preserve your own life."

ficiently protected hoplite phalanx by forcing a response, generally a retreat or the dispatching of a small unit of hoplites (organised by year-group in the Spartan system) to chase down the attackers.[10] Thus the only way to offer resistance to a unit of light infantry was to weaken the central strength and force multiplier of the hoplite, the phalanx itself. Why, then, was light infantry not deployed earlier, to a greater extent, and to greater effect throughout the Classical Period? Commentators point to an apparently prevailing system of ethics that underpinned warfare in the Classical Period;[11] they note that such infantry required a significant degree of training,[12] or argue that light infantry forces were simply ineffective for states, especially those with large hoplite populations, to develop and maintain.[13] This is not the place to deal with these arguments in detail, suffice to say here that none of them are defensible, nor can they be combined to present a coherent argument. I prefer the view that light infantry was not developed because its effective use required a combination of skills and tactical nuance that was simply beyond the majority of generals to deploy and for citizen troops to maintain on a large scale; there was no doctrinal or moral bar to their use, just practical reasons wholly apart from the theoretical strengths they held against heavy infantry. Despite these objections to the use of light infantry it is of significant interest that Aineias Tacticus not only recommends the use of skirmishing forces, scouting forces, and infantry equipped with ranged weapons, but that he assumes his putative city will have access to them. His concern is not with the training of such soldiers, but in the effective co-ordination of their skill set with other military elements available to the commander of a city. His advice not only discounts any form of 'playing by the rules' but actively encourages the commander to engineer scenarios when the enemy can be attacked from range and while intoxicated.[14] Aineais Tacticus' advice is down-to-earth and practical, indicating those pointers a pro-

---

10 Thuc 4.125, 5.68; Xen. *Hel* 6.4.12, 17. Van Wees (2004) 98, 277. Although this appears to have been the initial Spartan response to the Athenian light infantry during the battle of Sphacteria, Thucydides does not record an order from the Spartan commander to that effect. Lazenby (2004) 98. Wylie (1992) 86 refers to the tactic as "… orthodox Spartan tactics."

11 Ober (1996) 56. See Krentz (2002) 23–37 for the argument that Ober's 'rules' are simply circumstance and are all broken frequently. Hanson (2000a) 36. See also Anderson (1970) 149; Lazenby (1991) 101; Wheeler (2007a) 212.

12 Hanson (2000) 31; Hunt (2007) 134–135. van Wees (2004) 89–97 for discussion of the lack of training required for hoplite forces.

13 Hunt (2007) 135.

14 Ain. Tac. 16.5–7—see below.

fessional, or at least highly experienced, soldier considered most vital to give to those facing the existential threat of invasion; his is not a historical work, but a guide, and as such we must pay close attention to his surviving work if we are to offer some balance to the modern focus on hoplite warfare and the pitched battle.

This image of warfare generated by Aineias Tacticus' focus on ambush, communication, and leadership is not found in other, almost contemporary, primary sources, which consistently portray the warfare of the age as being an affair of heavy infantry; the hoplite dominated the battlefields of Greece and the great pitched battles of this time were primarily hoplite encounters.[15] Our primary sources focus on these encounters, and it is easy to see why: battles such as Nemea, Mantinea, and Koronea were massive hegemony-shifting events which involved the major powers of the time. Modern scholarship has followed this evidence, and several excellent accounts of Classical Greek warfare have appeared over the course of the last half century; discussion of warfare and the pitched battle is as attractive to modern audiences as it must have been to the citizens of the various *poleis* of the Greek world.[16]

An account of 'average'-sized *polis* warfare is thus difficult to construct; evidence for such encounters in the major historians is scarce and a simple scaling down of pitched battles is fraught with difficulty. The large numbers of hoplites and the presence of many allied contingents generally placed limitations on the feasibility and extent of tactical manoeuvring in pitched battles; it was extremely difficult to co-ordinate large armies effectively and as such tactics remained relatively simple. However, there would be a point at which this no longer applied, and when dealing with smaller forces the situation could be quite different; with no tactical or logistical necessity for a challenge to, or acceptance of, pitched battle, engagements between the forces of average-sized *poleis* could have had their emphasis on manoeuvrability and the creating of situational advantage, rather than a simple head-on clash.[17] The lack of direct evidence in the historians regarding this form of warfare is unfortunate; atti-

15 Pritchett (1974) 117: "The citizen-hoplite, not the mercenary peltast, won or lost the day at Leuktra, Mantineia and Chaironeia."

16 See especially Anderson (1970); Cartledge (1977); Salmon (1977); Holladay (1978); Krentz (1985), (1994), (2002); Cawkwell (1989); Hanson (ed.) (1991), (2000); Dawson (1996); Goldsworthy (1997); van Wees (2004); Lendon (2005); Rawlings (2007); Sabin, van Wees and Whitby (2007) (eds); Schwartz (2009).

17 The forming of small elite units would suggest that there was a place, and recognized need, for both skill with weapons and ability to manoeuvre on the battlefields of ancient Greece.

tudes to warfare cast a great deal of light on how the Greeks viewed their history, society, and moral values.[18] It is in this context that we approach the relatively underused Aineias Tacticus, the first of the ancient tacticians, to see if his *Poliorketika* can aid the reconstruction of the 'everyday' warfare that must have occurred between average-sized *poleis*.[19]

Much of his advice is aimed at keeping people busy and involved, to such an extent that they have no chance to begin a revolt; measures should be taken to prevent large-scale meetings and only those satisfied with the status quo should be allowed positions of responsibility.[20] Such internal measures form an important part of Aineias' advice, and are solidly connected to his ideal defence of the *polis*, namely that one should occupy the outer territories in such a way that the dangers of invasion become unacceptable to the enemy.[21] Indeed, a simple demonstration of organisation and willingness to resist could well be enough to discourage potential attackers: in 378 BC the discipline and resolve of Chabrias' mercenaries was enough to dissuade Agesilaos from attacking them. Agesilaos was put off his attack by the ease with which Chabrias' orders were followed; upon hearing the command Chabrias' men went down on one knee, rested their shields against the other, and awaited the charge of the Spartan hoplites.[22]

Aineias presents us with four areas of increasing proximity to the physical and theoretical centre of the *polis*; as a ring is penetrated or surrendered the potential danger to the *polis* is increased, with the corresponding reduction in tactical flexibility constituting a double blow. The areas to be defended

---

18 See, e.g. Connor (1998), Dawson (1996) 47–99. Aeschylus was able to characterise the Persian wars as a clash between Persian bow and Dorian spear: *Persae*, 85–86; 147–149; 817. On moral values, or lack thereof, in Greek warfare see: Ober (1996) 53–71, against whom Krentz (2000) 167–178, (2002) 23–39. Also Pritchett (1974) 186–189; Hanson (1991) 3–5, (2000) 9–18; van Wees (2004) 126–127, 132–138; Hornblower (2007) 40–42; Wheeler (2007) 188–192, 199–202.

19 E.g. Aineias is graced with a single mention in Hanson's *The Western Way of War* where he is referred to as projecting an "... especially idiosyncratic viewpoint" Hanson (2000) 48; although he is upgraded to a "... pragmatic guide ..." in Hanson (2007) 3. He appears only six times in Hanson (ed.) (1991) and eight times in van Wees (2004). It is to be admitted that these excellent works are not monographs on the practice and development of siege warfare, but then, as argued below, neither is Aineias' *Poliorketika*.

20 Ain. Tac. 1.5–8; 5.1; 10.19–21. At 10.17 Aineias recommends that the death penalty should be imposed on mercenaries found to be damaging morale.

21 Ain. Tac. 16.16–18.

22 Diod. Sic. 15.32–33; Xen. *Anab.* 1.5.11–14. The position indicated an immediate readiness for action while saving the men from having to adopt an energy-sapping stand to attention.

can be defined as: territory, city walls, internal defence, and morale. He gives appropriate advice for how to resist and dissuade further aggression at each of these stages, with the focus throughout being one of effective scouting, reliance on local geographical knowledge, the use of swift and effective communication, and co-ordinated attacks utilising different troop types. It is clear that Aineias' primary concern is the quelling of both doubt and ambition amongst certain sections of the *polis* through a combination of demonstrable superiority in command and action, and through the careful monitoring and deployment of suspect individuals.[23]

By making use of natural terrain features it should be possible to deny certain routes to the enemy and discourage others from being taken; the active and permanent garrisoning of passes is not advised, but it is desirable to establish a military presence at the first opportunity and as far away from the city as possible.[24] Given the importance Aineias attaches to rapid communication and flexibility the most logical and effective force to undertake the occupation of suitable routes would be one made up of light troops; such a force could respond rapidly to a threat, could skirmish and harass the enemy, and would be able to operate unsupported by heavy infantry or cavalry if necessary.[25] It would also be able to scout, establish effective and quick communications, and could disrupt an invasion to such an extent that the enemy would be forced to withdraw. The use of scouts would be highly important at this stage, with Aineias advising that only those experienced in war should be selected for such duty, as a panicked or misinformed report could be seriously damaging.[26]

Should the enemy manage to penetrate the outer territory, or an effective defence prove impossible to mount, then a defence of the local territory should be undertaken; again, because of their speed and flexibility, light infantry would be the ideal force to offer resistance at this scale. At 8.1–4 Aineias advocates a 'scorched earth' policy be undertaken if the invading forces are too large to be immediately met in the field: he advises that rivers be enlarged, useful building material be concealed, ambushes be laid, and crops be disposed of; all tasks

23 On which see further Pretzler in this volume, 156.

24 E.g. Ain. Tac. 8.1; 16.9, 16.

25 See below for the importance of hoplite phalanxes accompanying lightly armed forces but not engaging the enemy. Although effective this was not an essential element in the use of light infantry; the Athenian general Demosthenes was defeated in 426 by an Aetolian force which lacked heavy infantry support. Thuc. 3.97–98; see Roisman (1993) for a detailed discussion of Demosthenes' military capabilities, or lack thereof. Also Best (1969) 188–189.

26 Ain. Tac. 6.1–3. The emphasis on experience suggests these scouts are not simply hoplites serving in a scouting capacity, but are perhaps specialists. Pritchett (1974) 188–189.

more suited to light infantry than cavalry or hoplites.[27] The setting of traps, such as the undermining of potential engine placements or drawing the enemy into an ambush inside the city gate would require a flexible and swift force, as would the disposing of crops.[28] We are not informed exactly how the crops should be disposed of; Aineias tells us he has already set out this information in his *Preparations* and will not repeat it.[29] Perhaps all livestock and stored food would be rapidly moved behind the walls, while small groups of men would attempt to spoil or fire any standing crops that could not be harvested due to the timing of the invasion or the proximity of the enemy. The deliberate burning of crops seems unlikely; such an action would be both time consuming and damaging to the morale of the city, two elements Aineias strongly and constantly advises against. Indeed, the deliberate destruction of all crops could simply force the enemy into directly attacking the walls of the city itself while also depriving the general of vital time and room to manoeuvre; the enemy would be need to disperse their forces in order to burn or otherwise damage crops, a situation that could be exploited by a general with the skills Aineias details.

Should it prove impossible to engage the enemy at a distance from the city, then the physical structure and immediate terrain features of the city itself come into play, with Aineias recommending that advantageous positions be occupied and used as bases for attack and retreat.[30] The emphasis remains on attack, although with the city in such close proximity Aineias seems to be recommending a shift from the flowing counter-attack useful in the outer territories to a relatively compact attempt to draw the enemy into position and attack them from the city. Although it is clear that light infantry would be able to undertake some of these roles well, the limited space and concentration of enemy forces could make this rather more difficult, and heavy infantry would come to the fore of the defence.[31] However Aineias concentrates heavily on the

27 See Krentz (1997) 68–69 for the forceful arguments that if the Peloponnesians sought a decisive engagement then they should have invaded Attica with an army that did not overwhelmingly outnumber the Athenian force. That Pericles was still subject to abuse for refusing to engage the Peloponnesian force in defence of the Acharnian's land demonstrates the dangerous depth of feeling the ravaging of land could engender. For the targeted nature of Spartan agricultural devastation see Garlan (1975) 110–112; Foxhall (1993) 142–143; van Wees (2004) 123–124.

28 Ain. Tac. 32.8; 39.2; 8.4. van Wees (2004) 121.

29 Ain. Tac. 8.4.

30 Ain. Tac. 16.18.

31 Peltasts required a considerable amount of room to manoeuvre if they were to be used

advantages to be gained from local knowledge and familiarity with terrain, so it would seem he still regards this position as a flexible one; perhaps he is thinking of a situation such as the Plataian break out in 428, except on a larger scale and with the focus on causing the enemy to withdraw, not on escape to a friendly territory.[32] The break out by Plataian forces after nearly two years of siege was innovative, effective, and was based on many of the principles that Aineias advocates, namely: stealth, speed, surprise, ruse, local knowledge, and use of a mixed force. Although the break out was successful, it caused a minimum of damage to the Peloponnesian force, so perhaps something rather similar to Brasidas' ambush of Kleon at Amphipolis is a more appropriate example.[33] By launching a surprise attack from the city when the Athenian army was unwisely withdrawing, Brasidas was able to secure a comprehensive victory. Constant observation of the enemy, combined with Brasidas' desire to engage what he saw as an army that was light on morale and effective leadership, resulted in 600 Athenian casualties.[34] Before the engagement Thucydides has Brasidas deliver a speech, of which Aineias would undoubtedly approve, in which he praises the use of unorthodox methods and distances himself from any set 'form' of engagement, stating that "These are the tricks of war and win great acclaim when the enemy is completely fooled to the maximum benefit of one's own side."[35]

effectively, although they would still prove highly useful at this point there would come a time where using them for operations other than the sort mentioned at 39.2 would prove highly dangerous.

32 Thuc. 3.20–24.

33 Thuc. 5.8–11.

34 Thucydides' account suggests something of a 'perfect storm' here: Kleon's hoplites were chafing at his lack of drive and initiative and he felt forced into moving his army close to Amphipolis in order to quell the disquiet. Although his force included 300 cavalry he does not appear to have had light infantry support, as evidenced by his request to Polles of Odomantia for Thracian mercenary support; these would most likely have been in the form of peltasts and other skirmishing forces. When Kleon realised that Brasidas' forces were massing to sally from the city he opted to fall back, but Brasidas launched a surprise attack on the withdrawing Athenians which allowed his large numbers of peltasts to be used to their best advantage. See Anderson (1965) 1–4 and Best (1969) 29–35. Aineias would probably prefer the general to survive such actions however.

35 Thucy. 5.9. However see 4.126 where Brasidas suggests the skirmishers his army faces are cowardly and undisciplined; his speech appears more motivational than rhetorical as he closes with the warning that the enemy still pose a significant threat if they are underestimated.

Aineias seems to present three more levels of proximity and danger which should be considered: defence of the city walls, internal defence, and factors regarding the mental health and morale of the city. Regarding the first two he has little to say about light infantry, other than that missiles launched from the walls would prove effective against both troops and machines, and that the indiscriminate throwing of stones should be discouraged.[36] However, the last, and most important, of the danger areas which Aineias presents, i.e. morale, is discussed and commented on throughout the work.

Morale and the danger of betrayal from within form the overriding topics of concern for Aineias: overt threats such as an enemy army could be recognised and responded to well before they posed a physical threat to the structure and integrity of the *polis*, however attempting to respond to a fifth column or revolt was highly dangerous and distracting. Aineias therefore concentrates his attention on the prevention of revolt, advising, amongst other things, that people be kept busy, prevented from meeting in large groups, and that the general should see that he performs certain duties in person, both to ensure that they are performed correctly and to improve morale by his presence.[37] The importance of morale to Greek armies has been well documented, with the need for the general to inspire his men by leading by example being constantly emphasised; the general's role was one of battle planner followed by inspirational figurehead rather than one of tactical innovator in the face of battlefield developments. Despite a trend towards more cerebral command in the fourth century the general was still actively involved in the battle; Kleombrotos fell at Leuktra while Epaminondas died at Mantinea, and, according to Diodoros, the Athenians were not impressed by the general Lysikles' survival at the battle of Chaironeia.[38]

Xenophon recognised the importance of the general to an army's morale and believed that the general should exceed his troops in bravery while setting an example for enduring hardship for the good of his army.[39] He should also be hardworking, respectful, cheerful, dignified and live a simple life, he should know when to reward and when to punish his men, being neither harsh nor

36 Ain. Tac. 32.5; 38.6.

37 Ain. Tac. 10.4; 17.1–6; 38.4. The importance of keeping people visibly busy is something which comes out very strongly throughout Aineias' work: see below, n. 55.

38 Xen. *Hell.* 6.4.13–14; 7.5.24–25; Diodoros Siculus, 16.88.2. See Lengauer (1979), Wheeler (1991) 136–139 and (2007) 213–223 for discussion of the battlefield role of the general and the development from leader to manager; Hanson (2000) 107–116 for an analysis that favours a much more direct role for the general, even into the fourth century.

39 Xen. *Hell.* 6.2.16–23; *Ages.* 5.1–3.

easy but steering a balanced course, hitting "… a paternalist mean between the excessive severity of a Klearchos (*Anab.* 2.6.9–14) and the excessive leniency of a Proxenos (*Anab.* 2.6.20) …"[40] He should be a persuasive speaker, but one who is able to back up his words with powerful actions and cunning strategies.[41] Wheeler's comment that "… the general and his army's morale became inextricably intertwined …" is indeed true, and is well demonstrated by the examples of the Spartan commander of mercenaries Mnasippos on the one hand, and Agesilaos on the other.[42] Not only was Mnasippos keeping his mercenaries two months in arrears, but he dealt with a complaint of a lack of food by striking those captains who brought up the issue. As Xenophon pointed out, this was hardly helpful behaviour, and was certainly not the inspiring example expected of a Spartan commander; Plutarch was of the opinion that Spartan forces were so successful because they had "… learned to respect rather than fear their leaders."[43] Unsurprisingly, Mnasippos did not survive.[44] He contrasts starkly with Xenophon's perfect example: Agesilaos. He was a man who apparently shared in his men's danger and discomfort at every level, becoming their leader not only by his position but also by his excellence as a soldier.[45] The Athenian mercenary commander Iphikrates led in a similar fashion; by sharing in the hardship suffered by his men he provided an inspiring example. Polyainos tells us that when his men were hungry and ill-equipped during a winter campaign, Iphikrates not only took off his shoes, but put on summer clothes.[46] Iphikrates was also apparently a stern disciplinarian; after killing a sentry he found sleeping he is reported to have said "I left him as I found him."[47]

Aineias may not have approved of Iphikrates' excessively harsh methods, but the sentiments offered by Xenophon gel well with the image of generalship that Aineias' text generates: the general should be sensible and energetic, as well as being personally concerned with the maintenance of the status quo.[48] He must be a capable man, able to calm down and organise citizens when they panic at reports of threats to their property, and must maintain a personal presence

40 Cartledge (1987) 207.
41 Cartledge (1987) 206–207.
42 Wheeler (1991) 145.
43 Plut. *Mor.* 217A; 227D; *Apoph. Lac.* 63.4.
44 Xen. *Hell.* 6.2.16–23.
45 Xen. *Ages.* 6.47.
46 Poly. 3.9.34; Xen. *Anab.* 3.4.47–48.
47 Front. *Strat.* 3.12.2.
48 Ain. Tac. 1.6–8.

amongst the citizenry as far as safety allows.[49] What comes out most clearly in Aineias is that the general must be a cerebral man, a man who understands the importance of clear lines of communication and good local knowledge, combined with the imagination to engage in an energetic defence of territory using a mixed force. Indeed, in comparing an army to a human body, Iphikrates remarked that the phalanx was the trunk, the light infantry the hands, the cavalry the feet, and the general the head, and he went on to say that "when the other parts are missing, the army is lame and disabled, but when the general is killed then the entire army is worthless."[50] Although Iphikrates has created a convincing figure here, he has left out one vital component: blood, which could be categorized as the overall morale of the army. If the general was unable to mount a visible and energetic defence, and the enemy were allowed to begin an active siege, then the possibility of the figure bleeding to death before taking any significant wounds comes into play. Iphikrates was also able to state that "I am not a cavalryman, a hoplite, an archer, or a peltast, but one who knows how to command all of these."[51] Aineias would most likely approve of this sentiment; his general maintains morale and order by being able to offer the right words, actions, praise, or punishment as the situation demands—he does not simply take his place in the ranks to fight.

So Aineias' general is a leader, more than a fighter, and as such he operates as the focus of both citizen and military morale. According to Aineias this general should plan, organise, analyse information, and prepare the countryside and city for enemy incursion; he should not rush to engage and he should certainly not risk his life in combat unless part of the city is directly threatened and his presence would bolster the defences.[52] Aineias is concerned for the safety of the general, and at no point does he advise that the general actually leave the safety of the city, rather he should be stationed at the agora, or, if this is a weak position, at the most defensible position of the city and have easy access to both runners and a trumpeter.[53] However the general should still be a highly visible individual, and should see to a number of tasks personally, especially the closing of the gates and the undertaking of patrols if morale is especially low.[54] Finally, in the event of the city being actively besieged, the general should

49 Ain. Tac. 14.1–15.4, 22.2, 26.10.

50 Poly. 3.9.22.

51 Plut. *Mor.* 187B.

52 Ain. Tac. 38.2; nowhere else in the text does Aineias advocate the general actually joining the combat.

53 Ain. Tac. 22.2–3.

54 Ain. Tac. 18.22; 20.1. The general should do this before dining. Aineias appears to regard the

form a relief force with a group of 'picked men', and be alert to any section of the wall in danger of falling.[55]

This is a logical and systematic system: the first concern should be to organise and co-ordinate the defence of the outer territories by sending and receiving messages and building up an overall image of the invading forces and their strategy. If the outer territories cannot be blocked or the attempts at delay or dissuasion fail then the general should prepare for a defence of the local territory, and ensure that discipline and morale remain high by becoming actively involved with the everyday tasks of the guards, patrols and gatekeepers. Finally, if all attempts at external resistance fail then the general's role partly reverts to that of his fifth-century predecessors; he becomes a rallying figure, an inspirational fighter leading a selected unit of men to bolster morale and raise hopes. However, although the general is in overall command of the defence of the city, Aineias does not expect him to act alone, and offers a significant piece of advice by which subordinate officers could help the city survive while under siege, namely the creation of street level commanders.

The discussion of these commanders comes in the second half of an analysis of how best to post guards; Aineias first advises that in the event of a panic or an unexpected attack the tribes of the city should be assigned by lot a section of wall to guard.[56] The need for an immediate response and for the population to be occupied in the event of a siege is something emphasised several times in the work; the authorities must take control, energise, and disperse the people if disorder was to be avoided.[57] Although this would seem an adequate initial response to an unexpected attack, Aineias points out that a well prepared general would have gone one step further and, in peacetime, appointed the most competent and judicious men of each street to act as local commanders.[58] Thus, in the event of attack, the men of each street would be able to rally around a central and well-known figure before making their

time immediately after consuming food as a dangerous one, advising at 16.5 to attack after the enemy had taken food and drink and at 26.2 to ensure that the first watch begins its patrol before consuming food. The advice is well given, with several accounts of ruses and tricks involving food, or the times at which food was conventionally consumed. Pritchett (1974) 158, N.7; van Wees (2004) 132–133.

55 Ain Tac. 38.2.

56 Ain. Tac. 3.1.

57 E.g. Ain. Tac. 3.1; 8.1; 9.1; 10.13; 15.2; 16.2–9 (where Aineias qualifies his advice by reminding the reader that although speed is of the essence, organisation and preparedness remain more important); 26.1.

58 Ain. Tac. 3.4–6.

way to a previously determined location; those from the streets close to the agora would congregate there, those close to the theatre would congregate there, and so forth. The potential for panic would be reduced, because citizens would be able to simply 'follow the script', would rapidly find themselves under authoritative command, and would muster close enough to their homes and family to provide reassurance to their household.[59] Aineias advises that this system be adopted as it provides an immediate and effective response should anything occur during the night, however it is clear that it would be equally effective at other times, and also carries with it several other distinct advantages which fit well with Aineias' view of how to prepare the state for siege.

By taking command down to the street level the general could maximise the advantage to be gained from local geographical knowledge and could gain a simpler and more accurate count of his military forces. Street-commanders also highlight the level of preparedness, caution, and attention to detail that Aineias suggests be adopted; the general should do his utmost to keep the enemy away from the walls of the city, but should also be realistic and have procedures in place in case the walls are breached. To prevent panic and confusion a rendezvous point should be established to which friendly units can retreat, thus preventing them from being mistaken for invading forces and also preventing the spread of panic; once again, a contingency is in place for unexpected events which would disrupt the normal chain of command.[60] A final suggestion from Aineias finishes off this section; open spaces within the city which serve no useful purpose should have ditches dug around them and generally be made as inaccessible as possible.[61] Although Aineias recommends this be done in order to reduce the possibility of hostile gatherings, such an action would also aid the defence of the city if it were breached by funnelling the enemy into natural defensive features and reducing the amount of terrain needing to be held. This action, when combined with the street-commanders and organised rendezvous points, suggests that Aineias does not view the loss of the city walls as end of the siege, simply the beginning of the final phase; however his core philosophy remains the same in this phase as with the defence of territory—organisation, advantage of terrain, and preparedness are vital if the state is to survive. This points to a vital element of Aineias' approach

59 The number of street commanders would also help to avoid a "Mindarus dead, men starving, we don't know what to do" moment. Xen. *Hell.* 1.1.23.

60 Ain. Tac. 39.5.

61 Ain. Tac. 2.1–2.

to warfare, and one which is especially important to both light infantry and effective leadership: communications.

Effective and constant communications are emphasised throughout Aineias' text, both as a means of keeping updated on the movements of the enemy, and as a means of ensuring co-ordination between the various elements of the defence.[62] Thus he emphasises the importance of maintaining communications between different battlefield elements while in the field, and makes frequent reference throughout his work to the use of scouts and the various ways in which swift and reliable messages could be sent.[63] This focus on scouting and information gathering is an essential element of Aineias' conception of warfare, and a task which would seem best suited to lightly armed forces; indeed the use of experienced and reliable scouts is heavily emphasised in chapter 6 where Aineias advises that outposts be placed at strategic points around the city. These outposts must be placed on high ground to ensure their signals can be accurately read in the city, and in the absence of such ground a system of runners should be set up to allow messages to reach the city in good time.[64] The accuracy and reliability of the information thus transmitted is of great concern to Aineias; incorrect or hastily judged messages could easily begin a panic in the city, potentially providing the catalyst to open revolt which Aineias is at such great pains to avoid. In order to avoid such panics these stations should be manned by at least three experienced men, who should be able to recognise and report on enemy dispositions and movements, be able to recognise feints and diversionary tactics, and be able to accurately judge the strength of the enemy force and their line of march.[65] This concern for strategic awareness is echoed throughout the work; indeed Aineias assumes both that the presence of enemy forces will be reported before they reach the borders of the territory, and that the enemy will have an equal, if not greater, concern and capacity for information.[66] A further, and equally important, assumption made by Aineias is that the state will have a ready supply of men with the tactical and strategic know-how to recognise the important movements and actions undertaken by an invading force, and that while the presence of these men is taken for granted, the presence of cavalry is not.[67] The point is sim-

62 Ain. Tac. 4.1–12; 6.4–7; 7.1–4; 15.5–7; 16.16–17; 22.21–24; 26.12–14; 27.1–4.

63 Ain. Tac. 15.4.

64 Ain. Tac. 6.1–6.

65 Ain. Tac. 6.1–3.

66 Ain. Tac. 9.1. The enemy's *capacity* for gathering such information would be increased due to the presence of informers and malcontents within the city itself.

67 Ain. Tac. 6.6.

ple but important: not all states were geographically or economically suited to maintaining cavalry or breeding horses, however all states had an interest in defence to some degree and generally had a violent or turbulent history. The natural consequence of this, combined with Aineias' frequent allusions to the advantage of knowing the local terrain, suggests that the city would have a pool of experienced individuals who could be trusted to perform the vital roles of scouting and information gathering as recommended.[68] An example of particular interest can be found in Thucydides, involving the destruction of a Chaonian force advancing on the city of Stratos in 429. The Chaonians were advancing on Stratos with two Peloponnesian contingents and, keen to claim the credit for taking the city themselves, were marching some distance ahead of their Peloponnesian allies. The men of Stratos observed this, and set a number of ambushes to coincide with an assault launched from the city itself; the attack was successful, with the Chaonian force being defeated without the Peloponnesian force realising it had even been engaged.[69] The Stratians followed up this success with a sensible ranged attack against the now amalgamated Peloponnesian and Chaonian force; they were still outnumbered and so used their skill with the sling to delay and harass the enemy, who retreated under cover of darkness.[70] Effective observation, knowledge of local terrain, and a swift and well organised attack using lightly armed forces resulted in enough damage being inflicted to force a withdrawal; an almost demonstration of how to survive under siege. Perhaps this is representative of warfare between smaller states, a style of warfare that was the very opposite of the large pitched battles that dominate the pages of Herodotos, Thucydides, and Xenophon; multiple small engagements, ambushes, and delaying tactics were all to be aggressively undertaken, and a full attack should only be launched if a distinct advantage in information, terrain, and organisation was held.

The importance Aineias attaches to secure, swift, and frequent communications indicates a potential understanding of the way light infantry functioned best; as part of a flowing and sustained attack that required constant communication and a detailed knowledge of local terrain to be successful. Although the damage thus inflicted may have been relatively minor when compared to the inexorable forward grind of a well-disciplined phalanx, the flexibility and manoeuvrability of light infantry allowed them to attack again and again until

68 The frequent references to the advantages of intimate local geographical knowledge would seem to discount the possibility of these scouts being exclusively mercenaries.

69 Thucydides 2.81.

70 Thucydides 2.82; such was the effectiveness of the Stratian slinging that the Peloponnesians were forced to move around their camps in armour.

their target broke. The impact such tactics could have upon hoplite forces was well demonstrated by Iphikrates, who achieved notable success using mercenary peltasts with supporting hoplite forces in 390. While engaged in skirmishing and harassing operations around Arkadia, Iphikrates' forces gained a fearsome reputation, with the men of Arkadia refusing to engage them, and being taunted by the Spartans for fearing "... the peltasts, just as children fear hobgoblins."[71] The discretion of the men of Arkadia proved rather more appropriate than the bluster and confidence of the Spartans, when the peltasts of Iphikrates engaged and destroyed a regiment of Spartan hoplites. Iphikrates' tactics were simple: his peltasts were to advance to missile range, hurl their javelins and retreat before any pursuers were able to catch them. The plan worked perfectly; following the death of several men the Spartan commander ordered the first ten year classes to pursue and drive off the peltasts.[72] The young Spartans, unable to catch the lightly armed peltasts, returned to the ranks of the phalanx, taking more casualties as the peltasts took this opportunity to strike again. The Spartans, completely on the defensive, retreated to a small hill where the sight of a confident, organised, and fresh phalanx marching on their position broke whatever will to fight still remained.[73]

Aineias recognises the advantages light infantry possessed, and his overall conception of how to survive under siege depends heavily on their deployment, both as scouting forces and as forces capable of direct military action in their own right.[74] Light infantry were by definition lacking in armour, a factor which has played some part in their modern dismissal as being unsuited to engaging a phalanx, especially on the flat plains where most battles were apparently fought. However, weight and quantity of armour is no reasonable measure by which to evaluate any military unit's effectiveness, many other factors must also be considered if a reasonable and fair conclusion is to be reached. Although light infantry were weak in terms of physical protection, they were strong in three other attributes to which Aineias attaches a great deal of importance: speed, mobility and flexibility. Indeed, many of Aineias' points regarding the need for alert and vigilant watchmen may have been a response to the effectiveness of peltasts rushing and occupying vulnerable gatehouses or other positions, a tactic put to good use by Brasidas during the 420s. These attributes also made light infantry perfect forces to carry out a number of tasks

---

71 Xen. *Hell.* 4.4.16.

72 Xen. *Hell.* 4.5.14–15.

73 Xen. *Hell.* 4.5.16–18.

74 E.g. Ain.Tac. 6.5, 9.1.

which Aineias advises, such as occupying high ground and gathering intelligence, and also constituted an important defensive system which took the place of the panoply of the hoplite in providing protection against attack; if the archer could not be reached then the archer could not be harmed, by a hoplite at least.[75]

In chapter 15 Aineias advises that cavalry and lightly armed forces must march in the vanguard of any relieving or counter-attacking force sent from the city; these elements should effectively and thoroughly scout the ground ahead, occupy suitable elevated ground, and function as a screening force to prevent enemy units from scouting or attacking the hoplites.[76] A high degree of organisation and caution is advised; the various elements must be constituted rapidly but not at the expense of good order, and should remain close enough to each other for quick communication and physical assistance to be rendered when necessary.[77] Caution on the march is emphasised further in the next chapter, with Aineias explaining that intelligent enemies will not simply march into a territory expecting a pitched battle, but will have been vigilant in their reconnaissance and will most likely have sent peltasts in advance to stage ambushes.[78] In order to engage a force such as this effectively and safely, Aineias advises stealth and patience—two attributes rarely associated with Greek warfare. The commander should wait until the enemy became complacent (and preferably drunk!) before setting his own ambushes with picked units of light infantry, and engaging with the remainder of his light forces.[79] The ambushing troops would most likely have been peltasts—they held an advantage over other forms of lightly armed due to the nature of their style of engagement; they were more flexible and probably much more damaging than other forms of light troops because they operated so close to the enemy and, thanks to their shield and skirmishing nature, were potentially able to enter close com-

---

75 The lesson given in Xen. *Cyr.* 7.4.15 that not all the slingers in the world could stand up to a phalanx in close combat is easily reversed; not all the hoplites in the world could face a unit of slingers that could avoid close combat.

76 15.5. Also 27.15 and 28.5 where Aineias advises that reconnaissance should be undertaken before the night-time guards are stood down and that nobody should be let out of the town before reconnaissance of the immediate neighbourhood had been undertaken. See Xen. *Hell.* 7.1.15–17 for an example of an attack timed to coincide with the end of the night watch and the rousing of the army.

77 Ain. Tac. 15.4.

78 Ain. Tac. 16.4 The 'strongest forces.'

79 Ain. Tac. 16.5–7.

bat.[80] Peltasts were used by Iphikrates in this manner in an ambush of Peloponnesian forces in 389; by travelling at night, occupying mountainous terrain, and using his transport ships as decoys Iphikrates' troops were able to attack the Peloponnesians as they crossed hilly terrain, killing 250 Peloponnesian hoplites and around twelve Spartan governors.[81] This incident, overshadowed by Iphikrates' more famous victory at Lechaion, bears close resemblance to much of what Aineias has to say about the use of light troops, ruse, stealth, intelligence, and patience. Iphikrates had used troop carriers to gain a tactical advantage and ensure his men were fresh; had used ruse and misdirection to confuse the enemy; occupied high ground; and had successfully ambushed the enemy on his own terms and to his own advantage. The Peloponnesian force also serves as an example of how *not* to move through territory; Anaxibios, the commander, was full of confidence because he was marching with a friendly force through friendly territory with the result that, when attacked, it truly appeared that the Peloponnesians were fighting in the dark, while the peltasts of Iphikrates were fighting in the light.[82]

To complete his outline of this attack Aineias also assigns a role to the heavy infantry and cavalry. The hoplites should move up behind the light elements, although a direct combat role is not specifically assigned to them; perhaps Aineias has a 'checking' role in mind for them.[83] An Athenian hoplite force was present on the island of Sphakteria in 425 and was never actually engaged in close combat; instead it served as a lure to the Spartan hoplites and as a checking force, preventing the Spartans from being able to pursue the light infantry which too vigorously.[84] Equally a hoplite force commanded by Kallias accompanied Iphikrates' peltasts when they attacked a Spartan *mora* near Lechaion in 390. Kallias did not engage, but remained close enough to

---

80 Best (1969) 19–20. The equipment and style of engagement of peltasts did not preclude them from fighting hand-to-hand. Indeed, if individual hoplites could be drawn far enough away from the protection of the phalanx then the peltasts' advantage in speed and numbers would best be expressed in a rapid switch to close-quarters battle.

81 Xenophon *Hellenica* 4.8.36–39.

82 Ain. Tac. 16.19; Xen. *Hipp.* 8.3. Also Xen. *Hell.* 2.4.17 where Thrasybulus rouses his troops by stating that a barrage of javelins will force the enemy will to take shelter, and will be as easy as blind men to strike down. Thucydides emphasises the desperate situation of the Spartans on Sphacteria in 425 by describing the visual and aural disturbances caused by large numbers of skirmishers operating in a relatively small space: great clouds of dust obscured their vision and the constant shouting of the skirmishers drowned out the commands of the Spartan officers. Thuc. 4.34.

83 Ain. Tac. 16.7.

84 Thuc. 4.32–39.

the Spartan phalanx to limit their response to the skirmishing attacks of the peltasts; the *mora*, having retreated to a hill, was finally broken at the sight of Kallias' phalanx advancing.[85] Finally, cavalry should be used to cut off the enemy lines of retreat. Aineias has presented us with a highly organised and sophisticated method of engagement which makes use of all the major unit types of the day: cavalry, heavy infantry, and light infantry. This force must not rush to engage but should be cautious, gather information, and attack on its own terms, preferably when the enemy is dispersed and lacking in discipline. A further level of sophistication is added when Aineias discusses the use of naval transport to aid attacks on an enemy force.

The *Poliorketika* ends with the enigmatic statement that "A fleet may be equipped in two ways …"[86] However, Aineias does discuss the use of naval forces in three passages in the *Poliorketika*, two of which emphasise his advice, given throughout the text, that the general is better off relying on speed and surprise than brute force and that forces should not be committed unless they have an advantage. First we are advised that if boats are available then they should be used to transport troops behind retreating enemies, thus both reducing the possibility of the pursuing troops being ambushed and ensuring that they arrive fitter and fresher than those retreating.[87] Secondly a fleet of boats should be used to deposit troops behind the enemy in order to ensure complete surprise and maximum impact when the dispersed enemy forces are finally engaged.[88] So in addition to the other troop types outlined above we see marines being put to use; the hypothetical enemy is now being shot at from range by light troops, ambushed by peltasts, chased down by cavalry, threatened by hoplites, and attacked from behind by marines—a flowing melange of styles and equipment that adds a much needed pinch of spicy realism to the relatively standard affairs of pitched battles.

Light infantry could be, but often were not, used in Greek warfare because they required high levels of communication, cerebral generalship, and local geographical knowledge to be used effectively. As such it is highly intriguing to find that a Greek author recommends the combined use of communication, cerebral generalship, and local geographical knowledge as useful tools for a state threatened with invasion or actually under attack. The nature of Aineias' military experience may account for this; he presents such a rounded, practical,

85 Xen. *Hell.* 4.5.13–18.

86 Ain. Tac. 40.8. "Ναυτικοῦ δὲ στρατεύματος δύο εἰσὶ στόλοι …"

87 Ain.Tac. 16.13.

88 Ain. Tac. 16.21–22.

and unbiased view of warfare that he must have had direct experience with the use and command of light infantry, both as stand-alone forces and in combination with hoplites and cavalry. Perhaps Aineias is pointing out that quantity is not all that matters; that although large states could afford to focus on acquiring vast navies or huge hoplite armies, even they could be defeated or dissuaded from attacking smaller states in the face of an organised resistance. This resistance, made up of an effective, rapidly moving force that comprised many elements would make use of ambush, traps, and ruses to try to ensure that the potential danger to the morale of the populace was minimised, and the enemy kept as far away from the city walls as possible.

Aineias' work is conventionally referred to as a manual on how to survive under siege. This title is accurate and to the point: he relates many ways in which the general can defend his city from external and internal attack, and offers examples and analysis as to how this has been attempted, successfully or otherwise, in the past. However the manual is rather more than this: a significant amount of the advice Aineias offers is to do with taking the battle to the enemy; indeed, Aineias would probably agree with the statement that the best way to survive under siege is to ensure that the enemy cannot commence one. The rings of danger and proximity to the city I have identified demonstrate that Aineias recognises at least three phases of an invasion during which a city should mount a highly active external defence. This point is forcefully made by Aineias in his preface, where he states that:

> When men leave their country and engage in warfare and encounter perils beyond their own frontiers, and disaster occurs by land or sea, the survivors still have their own country and city and fatherland between them and utter destruction.[89]

By engaging the enemy at the greatest possible distance from the city itself, it was still possible to have native soil, however little, between the soldiers and utter extinction, and Aineias recommends that the general make the most of this window of opportunity by undertaking an aggressive defence. This defence should not consist of a challenge to or acceptance of any form of 'pitched battle', but should be multifaceted, utilising naturally strong defensive points, constant skirmishing operations, and retreat to the next ring if the enemy

89 "Ὅσοις τῶν ἀνθρώπων ἐκ τῆς αὑτῶν ὁρμωμένοις χώρας ὑπερόριοί τε ἀγῶνες καὶ κίνδυνοι συμβαίνουσιν, ἄν τι σφάλμα γένηται κατὰ γῆν ἢ κατὰ θάλασσαν, ὑπολείπεται τοῖς περιγιγνομένοις αὐτῶν οἰκεία τε χώρα καὶ πόλις καὶ πατρίς, ὥστε οὐκ ἂν ἄρδην πάντες ἀναιρεθείησαν."

is determined enough to try and force the matter; indeed, the last thing the general should do is seek a direct and final confrontation as the loss of such an event could lead to an immediate physical siege of the city itself.[90] Perhaps Aineias recognised that a siege is a matter of mentality rather than physical structure; an army in the field could be besieged as readily as a city, and the way to do this was to utilise combined forces which made use of the local terrain. An invading force would thus find itself being delayed, distracted, and under attack from the moment it entered the very outer territory of the city; an organised display of resistance at this point could prove valuable, demonstrating to the enemy that the city was prepared and willing to resist. However the objective here was not to utterly defeat the enemy or to risk friendly forces in non-essential engagements; rather the objective was to inflict sufficient damage, or give the impression of an ability to do so, to make the enemy withdraw. Aineias does not appear concerned with received standards of glory or heroism; he does not want the city to go 'all in' and invite the enemy to engage in combat, rather the city's forces should harass, annoy, and fall back to prepared positions if necessary. In advocating this Aineias demonstrates his military credentials and gains entry into a supposedly select club of individuals from antiquity: those who were concerned with outcomes rather than methods and who were able to see beyond the limitations of the hoplite way of war to the usefulness of combined forces. From Aineias' treatise we can reconstruct a relatively detailed picture of 'everyday' warfare between average sized *poleis*. This picture demonstrates that a narrow focus on the hoplite and the pitched battle is one that is simply not representative of warfare in Greece, and that by engaging further with this most unfairly ignored of sources our understanding of the techniques, purpose, and remarkable sophistication of Greek warfare can be greatly improved.

Bibliography

Anderson, J.K. (1970) *Military Theory and Practice in the Age of Xenophon*. University of California Press, Berkeley.

Best, J. (1969) *Thracian Peltasts and their Influence on Greek Warfare*. Wolters-Noordhoff, Groningen.

Bianco, E. (1997) 'Ificrate, ῥήτωρ καὶ στρατηγός', *Miscellanea Greca e Romana* 21: 179–201.

90 Whitehead (1991) 24: "He is certainly no prisoner of the pitched-battle mentality …"; van Wees (2004) 123–124.

Cartledge, P. (1977) 'Hoplites and Heroes: Sparta's Contribution to the Technique of Ancient Warfare', *Journal of Hellenic Studies* 97: 11–27.

Cawkwell, G. (1989) 'Orthodoxy and Hoplites', *Classical Quarterly* 39: 375–389.

Dawson, D. (1996) *The Origins of Western Warfare: Militarism and Morality in the Ancient World.* Westview Press, Boulder.

Ducrey, P. (1986) *Warfare in Ancient Greece.* Schoken Books, New York.

Fagan, G. and Trundle, M. (eds) (2010) *New Perspectives on Ancient Warfare.* Brill, Leiden.

Gabriel, R. and Metz, K. (1991) *From Sumer to Rome: The Military Capabilities of Ancient Armies.* Greenwood Press, Westport.

Garlan, Y. (1975) *War in the Ancient World. A Social History* (translated by J. Lloyd). Chatto and Windus, London.

Goldsworthy, A. (1997) 'The *Othismos*, Myths and Heresies: the Nature of Hoplite Battle', *War in History* 4: 1–26.

Hamel, D. (1998) *Athenian Generals: Military Authority in the Classical Period.* Brill, Leiden.

Hanson, V.D. (ed.) (1991) *Hoplites: the Classical Greek Battle Experience.* Routledge, London.

Hanson, V.D. (2000) *The Western Way of War.* University of California Press, Berkeley.

Holladay, A. (1982) 'Hoplites and Heresies', *Journal of Hellenic Studies* 102: 94–103

Hornblower, S. (2007) 'Warfare in Ancient Literature: The Paradox of War', In Sabin, P., van Wees, H. and Whitby, M. (eds) (2007) *The Cambridge History of Greek and Roman Warfare vol. 1.* Cambridge University Press, Cambridge. 147–185.

Jarva, E. (1996) *Archaiologica on Archaic Greek Body Armour.* Studia Archaeologica Septentrionalia 3.

Kern, P. (1988) 'Military Technology and Ethical Values in Ancient Greek Warfare', *War and Society* 6.2: 1–20.

Kern, P. (1999) *Ancient Siege Warfare.* Indiana University Press, Bloomington.

Konecny, A. (2001) 'Κατέκοψεν τὴν μόραν Ἰφικράτης . Das Gefecht bei Lechaion im Frühsommer 390 v. Chr.', *Chiron* 31: 79–127.

Krentz, P. (1985) 'The Nature of Hoplite Battle', *Classical Antiquity* 4: 50–61.

Krentz, P. (1991) 'The Salpinx in Greek Warfare' in Hanson V. (ed.) *Hoplites. The Classical Greek Battle Experience.* Routledge, London.

Krentz, P. (1994) 'Continuing the *Othismos* on *Othismos*', *Ancient History Bulletin* 8: 45–49.

Krentz, P. (2000) 'Deception in Archaic and Classical Greek Wafare', In van Wees (ed.) *War and Violence in Ancient Greece.* Duckworth, London. 167–2000.

Krentz, P. (2002) 'Fighting by the Rules: the Invention of the Hoplite Agon', *Hesperia* 71: 23–39.

Krentz, P. (2007) 'War', in Sabin, P., van Wees, H. and Whitby, M. (eds) *The Cambridge*

History of Greek and Roman Warfare vol. 1. Cambridge University Press, Cambridge. 147–185.

Lendon, J. (2005) *Soldiers and Ghosts: A History of Battle in Classical Antiquity*. Yale University Press, Yale.

Lengauer, W. (1979) *Greek Commanders in the 5th and 4th Centuries BC. Politics and Ideology: a Study of Militarism*. Wydawnictwa Uniwersytetu Warszawskiego, Warsaw.

Lissarrague, F. (1994) *L'autre Guerrier: Archers, Peltastes, Cavaliers dans L'imagerie Attique*. La Decouverte, Paris.

Ober, J. (1996) *The Athenian Revolution: Essays on Ancient Greek Democracy and Political Theory*. Princeton University Press, Princeton.

Parke, H. (1933) *Greek Mercenary Soldiers*. Clarendon Press, Oxford.

Pritchett, W.K. (1971) *Ancient Greek Military Practices*, vol. 1. University of California Press, Berkeley.

Pritchett, W.K. (1974) *The Greek State at War*, vol. 2. University of California Press, Berkeley.

Pritchett, W.K. (2007) *The Ancient Greeks at War*. Manchester University Press, Manchester.

Roisman, J. (1993) *The General Demosthenes and His Use of Military Surprise* (*Historia* Einzelschriften 78). Franz Steiner Verlag, Stuttgart.

Roisman, J. (1998) 'General Demosthenes Goes to Lebanon', *Classical Bulletin* 74: 55–63.

Salmon, J. (1997) 'Political Hoplites?' *Journal of Hellenic Studies* 97: 84–101.

Schwartz, A. (2009) *Reinstating the Hoplite* (*Historia* Einzelschriften 207). Franz Steiner Verlag, Stuttgart.

Storch, R. (2001) 'The Silence is Deafening: Persian Arrows Did Not Inspire the Greek Charge at Marathon', *Acta Antiqua Acedemiae Scientiraum* 38: 381–394.

Thorne, J. (2001) 'Warfare and Agriculture: The Economic Impact of Devastation in Classical Greece', *Greek, Roman and Byzantine Studies* 42: 225–253.

van Wees, H. (ed.) (2000) *War and Violence in Ancient Greece*. Duckworth, London.

van Wees, H. (2004) *Greek Warfare: Myths and Realities*. Duckworth, London.

Wheeler, E. (1987) 'Ephorus and the Prohibition of Missiles', *Transactions of the American Philological Association* 117: 157–182.

Wheeler, E. (1988a) *Stratagem and the Vocabulary of Military Trickery* (*Mnemosyne* Supplement 108). Brill, Leiden.

Wheeler, E. (1988b) 'The Modern Legality of Frontinus' Stratagems', *Militärgeschichtliche Mitteilungen* 44: 7–29.

Wheeler, E. (1991) 'The General as Hoplite', in Hanson (1991) 121–170.

Wheeler, E. (2001) 'Firepower: Missile Weapons and the "Face of Battle"', In Dabrowa (2001) 169–184.

Whitehead, D. (1990) *Aineias the Tactician, How to Survive under Siege. Translated with introduction and commentary*. Clarendon Ancient History Series, Oxford.

Whitehead, D. (2003) 'Polyaenus on Iphicrates', *Classical Quarterly* 53: 613–616.

Whitehead, D. (2008) 'Fact and Fantasy in Greek Military Writers', *Acta Antiqua Academiae Scientiarum* 48: 139–155.

Williams, M. (2004) Philopoimen's Special Forces: Peltasts and a New Kind of Greek Light-Armed Warfare (Livy 35.27). *Historia* 53: 257–277.

Woodcock, C. (1928) 'Demosthenes, Son of Alcisthenes', *Harvard Studies in Classical Philology* 39: 93–108.

Wylie, G. (1993) 'Demosthenes the General—Protagonist in a Greek Tragedy?' *Greece and Rome* 40: 20–30.

CHAPTER 11

Mercenaries in Aineias Tacticus

James Roy

Whether or not Aineias Tacticus is to be identified with the Arcadian federal *strategos* Aineias of Stymphalos, it is generally agreed that he was an experienced soldier[1] writing around the middle of the fourth century.[2] In that period mercenaries were widely employed in the Greek world, and Aineias clearly assumes that they are a normal feature of Greek warfare. Yet detailed evidence on how mercenaries were recruited and employed is scarce, apart from the quite exceptional insights offered by Xenophon's *Anabasis*, and it is worth analysing what Aineias has to offer on the terms of mercenary service. In addition, the considerable number of men willing to serve as mercenaries in fourth-century Greece has been seen as an aspect of social and economic tensions in Greek society, and mercenary armies have been seen as a threat to the stability of Greek communities.[3] Aineias was notoriously concerned with the dangers arising from internal divisions and possible treachery in cities threatened by war, and it is therefore worth examining what threat he thought that mercenaries might pose.

The text of Aineias as it survives is incomplete, and in what survives Aineias distinguishes between mercenaries and other troops only when the distinction is necessary for his purposes.[4] He takes it for granted that a city will have its own citizen troops (e.g. 1.4–7), and very probably allied forces as well (e.g. 3.3, 12.1). Some of what he says about soldiers in general would no doubt apply also to mercenaries serving alongside these other troops. Nonetheless there are enough specific references to mercenaries to reveal something of Aineias' thinking about them.

These references to mercenaries fall into two categories: on the one hand historical incidents are cited in which mercenaries played a part, and on the other there are recommendations on how mercenaries should be managed by a city employing them.

1 Bettalli 4, 26.

2 Bettalli (1990) 5; Whitehead (2001) 8–10.

3 E.g. Marinovic (1988) 197–269; cf. Bettalli (2006a) and (2006b).

4 Marinovic (1988) 219.

 | DOI: 10.1163/9789004352858_012

In the first category there are eight instances (10.21–22, 11.7–10, 12.5, 18.8–11, 18.13–19, 23.7–11, 24.3–9, 28.5; mercenaries are also implied in 5.2, 11.3–6, and 11.13). Apart from an incident in an unnamed city for which no historical context can be deduced (23.7–11), these examples probably all fall in the first half of the fourth century.[5] One or two dates are admittedly somewhat uncertain: for instance, the attempt by wealthy men at Argos to take control of the city using mercenaries (11.7–10) has been dated to 386–371 by David, approved by Whitehead, but Bettalli has suggested 415 as a possibility. At any rate none of the cited examples is obviously taken from an earlier written[6] source: see Pretzler in this volume[7] on ways in which information was transmitted. In all cases the reference is to a mercenary force acting as a unit, or to a detachment from such a force. The reference to a few mercenaries in the employment of Dionysios I of Syracuse (10.21–22) is clearly in the context of the large mercenary force under his command, and the thirty mercenaries used by Charidemos of Oreos to gain entry to Ilion (24.3–9) were presumably drawn from a larger force. In these examples Aineias offers no complaint about the behaviour of the mercenaries themselves: they are simply professional soldiers doing a job, admittedly sometimes for politically dubious employers. The only doubtful case is at 24.8. There Aineias says that thirty mercenaries who were successfully smuggled into Ilion then killed the gate-keeper, set about other *xenas praxeis* and took control of the gates: Charidemos' main force was immediately admitted through the gates and captured the town. The phrase *xenas praxeis* is odd, and has been taken to be pejorative—'barbarous acts' in the Loeb translation—but it has also been taken to refer to normal mercenary practice.[8] The latter meaning is required by the context: Aineias is clearly recounting, fairly briefly, a businesslike and successful proceeding to secure control of the gates; there was no time to waste on gratuitous savagery. The only occasion on which the mercenaries were a danger to the community was at Herakleia, where a mercenary force made it possible for their own mercenary commander to take control of the city (12.5). The commander, whom Aineias does not name, was Klearchos and the incident was notorious, but the conclusion that Aineias draws is not that the mercenaries were depraved or vicious but that it was a fatal error to employ a mercenary force more powerful than the citizen army of the polis. Nor did such dangers arise only from mercenaries: a force of allies had behaved similarly at Chalkedon (12.3). Aineias therefore insists heavily on the need to have a citizen

5 Hunter and Handford (1927) xxxvi–xxxvii.

6 David (1986); Whitehead (2001) 130; Bettalli (1990) 248–249.

7 Pretzler in this volume, 76–77.

8 E.g. Hunter and Handford (1927).

army larger and stronger than any mercenary force that is employed (12.2, 4). He does not however at any point suggest that mercenaries are unsatisfactory when kept under proper control.[9]

Aineias' proposals on how a city should manage any mercenaries that it employs are clearly intended to establish secure control. No-one may recruit mercenaries nor may mercenaries accept service save in the presence of the city authorities (10.7). The safest way to recruit mercenaries is for wealthy men each individually to recruit one, two, or three mercenaries, according to their personal resources (13.1). Each mercenary should then receive pay and maintenance partly from the citizen who has recruited him and partly from the city (13.2). This passage is far from clear, but it seems possible that pay was to come from the city and maintenance from the citizen recruiting.[10] Gabba 1979 has drawn attention to a passage in Plautus' *Aulularia* (526–531), a play set in Athens: in these lines there is talk of a soldier turning up at a rich man's house to receive money, along with various other people, mainly tradesmen, who also want payment; it is made clear that the soldier has not eaten that day (and also that he has to leave again without getting his money). Gabba suggests that the reference is to a mercenary receiving maintenance from a wealthy citizen, and that the practice would be common in the Greek world. According to Aineias the wealthy citizen was to receive eventually reimbursement of any money spent on the mercenary, through a reduction in taxes (13.4). The mercenary was also to be billeted in the house of the man who recruited him (13.3). This citizen apparently had wider responsibility for his mercenary. If a mercenary failed to report for guard duty, his superior officer was to replace him by someone else, at whatever cost could be negotiated: the cost was to be borne by the *proxenos* (22.29). It seems likely that the *proxenos* was the citizen who had recruited the mercenary, as Gauthier suggested,[11] and that suggests that the citizen recruiting had responsibility for the mercenary that went beyond providing maintenance and a billet.

The mercenaries were to be grouped in *lochoi*, and for each *lochos* a reliable citizen was to be appointed as *lochagos* (13.1). The mercenaries were to assemble in their *lochos* to carry out their normal duties, and also any other instructions given by the city's magistrates (13.3). Another officer, the *taxiarchos*, appears at 22.29: little is said about his functions, save that he is to punish any mercenary who fails to appear for guard duty, but he was evidently supe-

9 See Bettalli (2006a) 57.

10 Bettalli (1990) 254–255; Whitehead (2001) 135–136.

11 Gauthier (1972) 52–53. Bettalli (1990) 284–285; Whitehead (2001) 161–162.

rior in rank to the *lochagoi*, and therefore presumably a citizen commanding a number of *lochoi* of mercenaries.

These recommendations leave many questions unanswered, though in some cases reasonable assumptions can be made. Aineias says nothing for instance about how the mercenaries are to be selected, or how their technical competence is to be checked, but any necessary scrutiny could have been made by a responsible citizen with suitable experience when mercenaries were being recruited in the presence of the city authorities. Aineias equally says nothing about the appropriate payment for a mercenary, or about any bonuses that might be paid, or about any distribution of booty or money from the sale of booty. Aineias' main concern is defence against a siege, but he envisages sorties against the enemy (e.g. 16.1–20), and in them booty in the form of weapons or supplies might have been captured. Since the mercenaries' loyalty would presumably be influenced by the rate of remuneration, some comment from Aineias might have been expected, but he may have supposed that there were normal rates of pay and normal arrangements for booty, and that they would apply. It is not clear, however, how Aineias expected a city to find mercenaries to recruit. It is clear in Xenophon's *Anabasis* that some *lochagoi* recruited men for their *lochos* and then the complete *lochos*, officer and men, took service as a unit, and the same practice appears at *Hellenica* 4.2.5.[12] However, Aineias' proposals would make such recruitment by *lochos* impossible: he is very clear that the city was to divide the mercenaries into *lochoi* once they were recruited, and that the *lochagoi* were to be trustworthy citizens of the polis. Aineias' suggestions would work only if mercenaries seeking work turned up piecemeal at the city wanting to recruit them; they leave no scope for taking on already existing units with their own mercenary officers. Such piecemeal recruitment in turn raises the question of how the mercenaries would know where work was available: conceivably news might get about by word of mouth that war was likely at some city, or that a city had started recruiting, but Aineias, for whom concern for security is paramount, can hardly have wanted an uncontrolled number of experienced but unemployed mercenaries to drift towards a city that might be under danger of attack. It is true that Isokrates wrote of large numbers of men adrift in the Greek world, many seeking employment as mercenaries,[13] and when he wrote his *Philip* in 346 he claimed that it was possible to find as many soldiers as one wished among this wandering population, and that it was no longer necessary to send out recruiting agents as had been done in the

12 Lee (2004) 291–292.

13 E.g. Isoc. 4.168.

past.[14] Despite some rhetorical exaggeration[15] Isokrates depicts social conditions that would have provided a city with recruits as Aineias wished, if the number of potential mercenaries turning up at a city could have been controlled in some way that Aineias does not mention.[16] There is no need to suppose that Aineias' proposal that wealthy citizens should each engage one, two, or three mercenaries would mean that each such citizen would have to find his own mercenary or mercenaries: at the moment of recruitment each mercenary could have been assigned by the city authorities to a suitable citizen.

Aineias sees possible problems in choosing passwords if a city recruits mercenaries from different areas speaking different forms of Greek (24.3): the same problem could arise with allies from different regions (24.1–2, cf. 15). These concerns do not tell us much about the scope for recruitment, since they do not necessarily imply a great geographical range of origin among the recruits; differences in local dialects could have given rise to the problem.

There are however other difficulties. At 24.7 Aineias relates how Charidemos of Oreos succeeded in smuggling thirty mercenaries into Ilion by disguising them as prisoners; to make the disguise more convincing he had them accompanied by women and children. These women and children are generally taken to be 'camp-followers of the mercenaries',[17] and typical of a mercenary force. Aineias was therefore aware that mercenary forces were often accompanied by women and children, but he says nothing of this elsewhere in his text. Was the wealthy citizen who recruited and billeted a mercenary also expected to billet the mercenary's woman and children? Did the welfare of the women and children have any influence on the mercenary's loyalty, or did it affect the security of the city? Indeed were these women and children to be allowed to remain in the city? Aineias recommends (10.1) that slaves and livestock be removed to a neighbouring state before an invasion (though 10.5 implies that domestic slaves would remain in the city, as Whitehead[18] notes). The recommendation was presumably intended to reduce the number of mouths to feed, and clearly sending away the mercenaries' followers would also have saved supplies. However, it might not have pleased the mercenaries, and it is not clear what the women and children would then have lived on.

---

14 Isoc. 5.96.

15 Marinovic (1988) 237–269.

16 See the full analysis by Bouchet (2010) of the varying ways—not all negative—in which Isokrates depicts mercenaries.

17 Whitehead (2001) 168. See also Marinovic (1988) 225; Bettalli (1990) 288.

18 Whitehead (2001) 116. See also Bettalli (1990) 18 n. 57.

Aineias was aware that some mercenaries might commit offences requiring disciplinary action, though he says nothing to suggest that maintaining discipline would be a problem in a well-managed mercenary force. As noted above, if a mercenary failed to report for guard-duty, his *proxenos*, evidently the citizen recruiting him, had to pay the cost of a replacement (22.29): Aineias then specifies further that on the following day the *taxiarchos* was to impose on the defaulting mercenary the usual punishment. Clearly punishment was appropriate, but Aineias does not suggest that such an incident needed more than some purely routine measures. However, there were to be lighter penalties applicable to less serious offences and extreme penalties for two very major offences, and the way in which these penalties were to be made known needs examination.

Aineias devotes a fairly long passage (10.1–19) to a series of measures that need to be formally proclaimed by herald. Those concerning mercenaries are the last in the list (10.18–19). In a *stratopedon* of mercenaries a proclamation is to be made by a herald of the penalties that apply to various offences by a mercenary. First, any mercenary who wishes to leave is to be given the opportunity to do so immediately, but subsequently anyone caught trying to leave will be sold into slavery. For less serious offences the penalty will be imprisonment or a fine, according to prevailing custom (or possibly law—'*nomos*'). However "any manifest harm to military operations ..., by undermining morale in camp, is a capital offence."[19] Firstly, it is striking that the mercenary's terms of service allowed punishment even by enslavement or death, admittedly for very serious offences. Secondly, however, the proclamation is to be made in a *stratopedon* (10.18). Bettalli rightly points out that 'stratopedon' need not mean 'camp' since the word can mean simply 'army',[20] but Whitehead translates it as 'camp'.[21] However, whatever meaning is supposed here, the mercenaries are clearly all gathered in a single body to hear the proclamation; nowhere else does Aineias suggest that the mercenaries should ever come together as a single body, and indeed elsewhere he recommends that the mercenaries be divided into *lochoi* and that they operate as separate *lochoi*. No doubt in the face of an all-out assault on the city the city's defenders might have been gathered into a single force to resist: but the proclamation was clearly to be made before any such desperate situation could arise, at a time when a mercenary who wished to depart could still be allowed to do so.

19 Translation: Whitehead (2001) 55.

20 Bettalli (1990) 245.

21 Whitehead (2001) 55.

When one examines Aineias' text in search of evidence for the conditions of mercenary service in his day, some conclusions emerge, but also some inconsistencies and some problems. He does not see mercenaries as bad soldiers, or as politically dangerous, provided that they are securely controlled. Individual mercenaries may need to be punished for offences, but mercenaries in themselves do not appear to be any less reliable, in Aineias' view, than citizens or allies, both also potential sources of trouble. Mercenaries are a normal feature of a city's military force, at least in periods of danger, and they are evidently readily available for recruitment. Exactly how they are to be found and recruited is not made clear, save that they should not be taken on in pre-existing units with their own officers, and there is no mention at all of what is to be done with the women and children who, as Aineias knew, often accompanied mercenary forces. Clearly some account must be taken of the fact that the work, in its present form, is incomplete, and also of the fact that Aineias was a pioneer in producing a manual of military practice—in fact several manuals, though the others are lost. Nonetheless it is notable that, when Aineias devotes a section (13.1–4) to the recruitment and maintenance of mercenaries, he says nothing of important aspects of recruitment and providing for camp-followers.

It is also notable that his recommendations for how mercenaries should be recruited and maintained differ from what is seen in the historical examples that he cites. In them the mercenary force is recognisable as a large unit, and indeed Aineias treats a city's mercenaries as a military unit when he says that they should be assembled together to hear a proclamation on disciplinary measures. Save in that passage, however, he proposes that a city should recruit mercenaries individually and then assign them to *lochoi* under the command of reliable citizen officers, and that the mercenaries should then operate separately in the several *lochoi*. In addition each mercenary is to be attached to a wealthy citizen who will pay him money for his keep (probably), provide him with a billet in his house, and have some general responsibility for him. Such methods would certainly give a city more control over its mercenaries, but, as scholars have pointed out,[22] there is no evidence that it was ever put into practice. Aineias himself, for all his military experience, does not seek to supply any.

22 Parke (1993) 95; Bettalli (1990) 253.

Bibliography

Bettalli, M. (1990) *Enea Tattico. La difesa di una città assediata (Poliorketika). Introduzione, traduzione e commento*. ETS editrice, Pisa.

Bettalli, M. (2006a) 'Hoi ton Hellenon aporoi: I mercenari del mondo Greco classico tra violenza, emarginazione e integrazione', in Urso G. (ed.) *Terror et pavor. Violenza, intimidazione, clandestinità nel mondo antico. Atti del convegno internazionale Cividale del Friuli, 22–24 settembre 2005*. Edizioni, Pisa. 55–64.

Bettalli, M. (2006b) 'L'immagine del mercenario nella Grecia del IV secolo a.C.', vol. 1 in Viaggiolo M.A. (ed.) *Guerra e pace in Sicilia e nel Mediterraneo antico (VIII–III sec. a.C.). Arte, prassi e teoria della pace e della guerra. Atti delle quinte giornate internazionali di studi sull'area elima e la Sicilia occidentale nel contesto mediterraneo, Erice, 12–15 ottobre 2003*. Edizioni della Normale, Pisa. 19–28.

Bouchet, C. (2010) 'Isocrate et la question des mercenaires', *Studia Humaniora Tartuensia* 11.A.1: 1–25.

David E. (1986) 'Aeneas Tacticus 11.7–10 and the Argive Revolution of 370 BC', *American Journal of Philology* 107: 343–349.

Gabba, E. (1979) 'Sul miles impransus dell'Aulularia di Plauto', *Rendiconti dell'Istituto Lombardo* 113: 408–414.

Gauthier, P. (1972) *Symbola. Les étrangers et la justice dans les cités grecques*. Annales de l'Est, Université de Nancy, Nancy.

Hunter, L.W. and Handford, S.A. (1927), *Αἰνείου Πολιορκητικά. Aineias on Siegecraft*. Clarendon Press, Oxford.

Lee, J.W.I. (2004) 'The Lochos in Xenophon's Anabasis', in Tuplin C. (ed.) *Xenophon and his world. Papers from a conference held in Liverpool in July 1999*. Franz Steiner, Stuttgart. 289–317.

Marinovic, L.P. (1988) *Le mercenariat grec au IVe siècle avant notre ère et la crise de la polis*, translated by J. Garlan and Y. Garlan. Annales Littéraires de l'Université de Besançon, Paris.

Whitehead, D. (2002), *Aineias the Tactician, How to Survive under Siege. Translated with introduction and commentary* Updated Edition. Bristol Classical Press and Duckworth, Bristol and London.

CHAPTER 12

Raiders from the Sea: The Maritime Context of the *Poliorketika*

*Philip de Souza**

The aim of this chapter is to examine the likely maritime context of the archetypal Greek *polis* that Aineias the Tactician envisages as the setting for his advice in the *Poliorketika*, and to assess the maritime aspects of some of the military activities he describes or recommends.

When discussing ancient warfare we must take care not to impose anachronistic modern concepts of strategy and tactics on the ancient world, which is what historians have traditionally done, especially with regard to naval aspects. Recent studies have emphasized the importance of non-strategic and non-political objectives in ancient Greek warfare, particularly the raids for honour, status and booty that are a feature of the Homeric poems.[1] Vincent Gabrielsen has argued that, from about 750 BC onwards, across the ancient Mediterranean world, warfare was dominated by what he terms the "raid mentality".[2] This refers to the widespread assumption of the right, to some extent even the obligation, of the strongest men in a community, to acquire slaves and booty through force of arms. In the Greek world in the Homeric period (c. 750–650 BC) the aims of raiding were limited to killing and despoiling opponents in order to achieve personal wealth and status. During the next few centuries, however, the Greeks developed more sophisticated political communities, with centralized governments that had the capacity to raise large military forces for inter-state warfare. Although it became possible for these states to seek long-term, strategic objectives, such as the domination of territory and the exaction of tribute, nevertheless the "raid mentality" remained a key element of ancient Greek warfare.[3]

* I owe thanks to all the participants in the 2010 Aeneas Tacticus conference for their comments and suggestions, but special thanks are due to Maria Pretzler for her help, encouragement and perseverance.

1 See, for example *Il.* 11. 669–761; *Od.* 9.39.52, 14.222–234. For modern scholarship see van Wees (1992); de Souza (1999) 17–24; van Wees (2004) 22–33, 66–70, 202–206.

2 Gabrielsen (2001).

3 Gabrielsen (2001) 222–226; (2007) 248–264. See also de Souza (1999) 25–36; Pritchett (1991) 312–348.

 | DOI: 10.1163/9789004352858_013

In relation to naval warfare, it has been assumed that the triremes, quadriremes and larger vessels that were developed in the sixth, fifth and fourth centuries BC were primarily designed and deployed to fight at sea, in order to control maritime trade and supply routes, or deny control of them to the enemy.[4] Recent scholarship has challenged this view, emphasizing the fact that, while ship-to-ship combat did occur, sometimes with considerable frequency, the typical mode of deployment for these ships was as troop carriers, transporting fighting men in large numbers to attack targets on land.[5] The major advantage that the use of such ships offered was that they enabled fleets to operate across long distances far more swiftly than land-bound forces. Athenian triremes of the Classical period, for which the most detailed information survives, normally had a crew of 200. This included a captain (*trierarchos*), commanding a handful of officers and a dozen or so sailors (*hyperesiai*), plus ten heavily armed hoplite infantrymen (*epibatai*) and a few archers.[6] However, the most numerous category of men in the crew were the "oarsmen" (*nautai*). In cases where another fleet was encountered and there was no alternative to combat at sea, and on voyages when the sails provided insufficient motive power, they rowed the ships by pulling on long wooden oars, but this was by no means their sole function. They could also be used as soldiers, in the general category of "light troops" (*psiloi* or *kouphoi*). Typically, they would ravage enemy territory, assault and plunder lightly defended settlements, build fortifications and construct siege works. If it came to a pitched battle on land they would be deployed as skirmishers, providing missile fire with javelins, slings and hand-thrown rocks.[7] The aims of raiding were to damage the military and economic capability of the enemy, to force them to commit soldiers to coastal defence and to damage their morale, whilst at the same time improving the funds and resources of the raiders, and boosting their morale. Our main sources for the wars of the Classical period, Thucydides, Xenophon and Diodorus Siculus, tend to focus on campaigns that involve pitched battles and long sieges, but their narratives do contain many examples of the raid mentality in action.

---

4 The two most egregious examples are Rodgers (1937) 3–10, 192–194, and Starr (1989) 3–6, 35–49.

5 de Souza (2013); (2010) 84–90; Krentz (2007) 172; van Wees (2004) 62–64, 222–224. For a recent restatement of the traditional view see Hale (2009).

6 Morrison, Coates and Rankov (2000) 107–126. More hoplites or other troops might be added when the need for them was anticipated.

7 On the uses of light-armed troops see Barley in this volume, pp. 182–205; Pritchett (1991) 61–67. On the ideological issues involved see van Wees (2004) 62–65; Trundle (2010).

The raid-based form of naval warfare featured prominently in conflicts of the fifth and fourth centuries BC, especially the Peloponnesian War (431–404 BC). Thucydides describes in detail how the Athenians showed that the coastlines of mainland Greece and the islands were particularly vulnerable to such seaborne raids. A raiding strategy was advocated by Perikles (Thuc. 1.142–143), and put into action by him early on in the war in spectacular fashion with a 100-ship raid on the Eastern Peloponnese in 430 BC (Thuc. 2.56). Athenian ships used Naupaktos, where they settled the rebellious Messenian helots in 456 BC (Thuc. 1.103), as a base from which to raid Spartan allies in Akarnania in 429 BC (Thuc. 2.102–103), and again in 425 BC (Thuc. 4.49). An Athenian fleet of 60 ships raided the island of Melos in 426 BC (Thuc. 3.91), while another fleet of 30 ships attacked Leukas (Thuc. 3.94). In 425 a fleet of 80 ships made a raid on the Corinthia (Thuc. 4.42–45). In 424 BC an Athenian amphibious strike force of 60 warships, with 2,000 hoplites and a small cavalry troop, transported on specially adapted triremes, took control of the strategically important island of Kythera. They attacked Asine and several other places on the coast of Lakonia itself. Their tactic was to establish a temporary camp from which to ravage the countryside for several days before moving on to another position (Thuc. 4.52–54). Spartan attempts to deal with the problem by stationing garrisons in likely target settlements were thwarted by the Athenians' mobility, and by the fact that the Athenians could deploy more than 6,000 light-armed oarsmen, backed up by their hoplites and cavalry, making their amphibious strike force too strong for most of the garrisons to challenge (Thuc. 4.55–56).

The frequency of maritime raids did not decline in the later stages of the Peloponnesian War and continued to be a feature of Greek city-state warfare in the first half of the fourth century BC, as the following selection of examples shows. In 392 BC the Athenian *strategos* Thrasyboulos ravaged Lesbos with a fleet of 40 triremes (Diod. 14.94). In 389 BC a series of raids and counter raids by small fleets characterised the operations in the Hellespontine area of Anaxibios, the Spartan *harmost* of Abydos, and the Athenian *strategos* Iphikrates, whose fleet of eight ships carried 2,000 peltasts (Xen. *Hell.* 4.8.35). In the same year the Aiginetans repeatedly raided the coast of Attica at the behest of the Spartan *harmost* Eteonikos (Xen. *Hell.* 5.1.1); this was still a serious drain on Athenian resources and morale in 374 BC (Xen. *Hell.* 6.2.1). Xenophon says that it was in order to relieve pressure on the Thebans that the Athenians sent a fleet of 60 ships under Timotheos to ravage the Peloponnese in 375 BC (Xen. Hell. 5.4.62–63). Diodoros briefly outlines raids in 369 BC by Jason of Pherai and the Athenians under Chares (Diod. 15.95). The key point here is that Aineias lived and wrote at a time in which the raid mentality was still central to Greek military thinking and the Greek city-states routinely practiced this form of warfare.

The Maritime Context of the City under Attack

From the very beginning of his work it is clear that the maritime aspect of warfare is familiar and commonplace for Aineias. Travel by ship is implied by the opening sentence, in which Aineias emphasizes the overwhelming importance of his theme, the successful defence of one's homeland, by contrasting it with the lesser importance of attacks against enemy territory:

> When men leave their own territory to meet combat and danger beyond its borders, the survivors of any disaster which strikes them, on land or at sea, still have their native soil and state and fatherland between them and utter extinction.[8]
>
> Preface 1

A few sections later he indicates that the typical city whose defence he is writing about is a maritime one, either on the sea, or with a harbour nearby:

> In time of war, then, and when the enemy is nearby, the first requirement is that those sent out from the city on any business, on land or at sea, be furnished with both daytime and night-time signals for communicating with those still in the city. This is so that the latter can tell friend from foe, if enemies put in a sudden appearance.[9]
>
> 4.5

The possible topicality of this recommendation is illustrated by a stratagem in the first century AD collection of Sextus Julius Frontinus. The commander in question is Iphikrates, who was active in the first half of the fourth century, and thus would seem to be an exact contemporary of Aineias.[10] According to

---

8 Ain. Tac. Preface 1 "Ὅσοις τῶν ἀνθρώπων ἐκ τῆς αὑτῶν ὁρμωμένοις χώρας ὑπερόριοί τε ἀγῶνες καὶ κίνδυνοι συμβαίνουσιν, ἄν τι σφάλμα γένηται κατὰ γῆν ἢ κατὰ θάλασσαν, ὑπολείπεται τοῖς περιγιγνομένοις αὐτῶν οἰκεία τε χώρα καὶ πόλις καὶ πατρίς, ὥστε οὐκ ἂν ἄρδην πάντες ἀναιρεθείησαν."

Unless otherwise indicated, all Aineias quotations are from David Whitehead's excellent English translation.

9 Ain. Tac. 4.5. "πολεμοῦντα οὖν χρὴ καὶ ἐγγὺς ὄντων τῶν πολεμίων, πρῶτον μὲν τὰ ἀποστελλόμενα ἐκ τῆς πόλεως κατὰ γῆν ἢ κατὰ θάλατταν ἐπί τινα πρᾶξιν πρὸς τοὺς ὑπομένοντας μετὰ συσσήμων ἀποστέλλεσθαι καὶ ἡμερινῶν καὶ νυκτερινῶν, ἵνα μὴ ἀγνοῶσι πολεμίων αὐτοῖς ἐπιφαινομένων, εἰ φίλιοι ἢ πολέμιοί εἰσιν".

10 See Bettalli (1990) 4–10; Whitehead (1990) 4–13; Lane Fox in this volume, p. 45.

Frontinus, Iphikrates fitted out the ships of his own fleet in a similar fashion to those of the enemy, and sailed into the harbour of a particular city where he did not trust inhabitants' loyalties. On being very warmly welcomed, he thereby uncovered their treachery and proceeded to ransack the city, presumably in punishment for their treachery, and as a warning to others (Frontin. *Str.* 4.7.23).

The historical example that Aineias chooses to illustrate his point is, however, one of the earliest of all those that he cites.[11] It involves the kind of raiding and counter raiding that has been described above, in this case clearly intended to produce captives, either for ransom, or sale as slaves:

> When Peisistratos was general at Athens, for instance, he received word of the approach of some men from Megara in ships to attempt a night attack on the Athenian women during their celebration of the Thesmophoria at Eleusis. On hearing this Peisistratos laid an ambush for them. When the men from Megara had disembarked—unnoticed, they imagined—and were some way from the sea, Peisistratos burst from his ambush and overpowered them, killing most of them and also capturing the ships in which they had come. These he then immediately filled with his own soldiers; and then, taking from among the women those best suited to sailing with them, he landed at Megara late in the day and at some distance from the city itself. On sighting the ships sailing in, many of the Megarians, including their officials duly gathered to watch the arrival, as they naturally supposed, of a large body of female captives. ⟨Then the Athenian soldiers were ordered⟩ to disembark with daggers and stab some of the Megarians, but carry off to the ships as many as possible of the most distinguished of them; which they did.[12]
>
> Ain. Tac. 4.8–11

11 On Aineias' preferences and usage with regard to historical exempla see Pretzler (pp. 68–95) and Shipley (p. 50) in this volume.

12 Ain. Tac. 4.8–11 "Πεισιστράτῳ γὰρ Ἀθηναίων στρατηγοῦντι ἐξηγγέλθη ὅτι οἱ ἐκ Μεγάρων οἱ ἐπιχειροῖεν ἀφικόμενοι πλοίοις ἐπιθέσθαι νυκτὸς ταῖς τῶν Ἀθηναίων γυναιξὶν θεσμοφόρια ἀγούσαις ἐν Ἐλευσῖνι: ὁ δὲ Πεισίστρατος ἀκούσας προενήδρευσεν. ἐπεὶ δὲ οἱ ἐκ τῶν Μεγάρων ὡς λεληθότες ἀπέβησαν καὶ ἀπὸ τῆς θαλάττης ἐγένοντο, ἐξαναστὰς ὁ Πεισίστρατος τῶν ἐνεδρευθέντων τε ἀνδρῶν ἐκράτησεν καὶ διέφθειρεν τοὺς πλείστους, καὶ τῶν πλοίων οἷς ἀφίκοντο ἐγκρατὴς ἐγένετο. ἔπειτα παραχρῆμα τοῖς ἑαυτοῦ στρατιώταις πληρώσας τὰ πλοῖα ἔλαβε τῶν γυναικῶν τὰς ἐπιτηδειοτάτας συμπλεῦσαι, καὶ κατήγετο εἰς τὰ Μέγαρα ὀψὲ ἀπωτέρω τῆς πόλεως. κατιδόντες οὖν τὰ πλοῖα προσπλέοντα ἀπήντων πολλοὶ τῶν Μεγαρέων, αἵ τε συναρχίαι καὶ οἱ ἄλλοι, θεώμενοι ὡς εἰκὸς αἰχμαλώτους ἀγομένας ὡς πλείστας ... καὶ μετ̓ ἐγχειριδίων ἀποβάντες τοὺς μὲν καταβαλεῖν, ὅσους δὲ ἂν δύναιντο τῶν ἐπιφανεστάτων συναρπάζειν εἰς τὰ πλοῖα."

No other surviving source describes this event, although Herodotos, following a different anecdote about Peisistratos, this time involving a feigned injury, does mention that he won his reputation as a military leader while commanding a force against the Megarians (Hdt. 1.59.4). Aineias' account is also similar to an anecdote related by Plutarch in his *Life of Solon*. Plutarch, describing the rise of Solon to a position of prominence in Athenian politics, makes Peisistratos a subordinate of Solon in broadly similar deception of the Megarians. In this version it is a false messenger from Solon who fools the Megarians, based on Salamis, into making the attack. Solon then substitutes some young men armed with daggers for the Athenian women, with the result that the Megarians are all killed when they impetuously leap out of their ships to attack what they think is a group of defenceless women (Plut. *Solon* 8). While this version clearly differs in many details from the story related by Aineias, both are told from an Athenian perspective, and both imply that the Athenians were, at the time the event took place, vulnerable to seaborne raids by the Megarians. Thus, while we simply cannot tell how far Aineias was adapting or enlarging upon the account he found in Herodotos, or making use of other accounts in sources no longer available to us, we can see that his version depends for its relevance and impact on the basic premise that seaborne raids on *polis* territory were a feature of Greek warfare.

Aineias shows that the vulnerability that results from a *polis* being close to the sea is a major concern of his when he refers to countermeasures that he has already described in an earlier work, now lost:

> As regards the number and nature of ruses to be employed against enemies landing on sandy or rocky shores; what kinds of barriers should be in readiness against them at the harbours of the city and its territory, to prevent them from sailing in or else, if they have done so already, to bar their exit … … there is a full treatment in my book *Preparations*.[13]
>
> Ain. Tac. 8.2; 8.5

It is important to note here that Aineias does not suggest meeting the invader's ships at sea. The most likely reason for this is that he assumes that there will be insufficient warning available to the *polis* under attack for it to deploy warships to intercept the enemy. He does, however, anticipate that the *polis* may have

13 Ain. Tac. 8.2; 8.5. "Πρός τε τὰς ἀποβάσεις τῶν πολεμίων εἰς τὰ ψαμμώδη καὶ στερεὰ ὅσα καὶ οἷα χρὴ κατασκευάζεσθαι δολώματα τοῖς ἀποβαίνουσι, τοῖς τε ἐν τῇ χώρᾳ καὶ τῇ πόλει λιμέσιν οἷα εἰς τούτους δεῖ φράγματα παρασκευάζεσθαι πρὸς τὸ μὴ εἰσπλεῖν ἢ τὰ εἰσπλεύσαντα μὴ δύνασθαι ἐκπλεῦσαι, … γέγραπται δὲ τελέως περὶ αὐτῶν ἐν τῇ Παρασκευαστικῇ βίβλῳ."

its own warships. The ways in which he recommends they should be used are discussed below.

Awareness of both the importance and the possible exploitation of the normal maritime activities of the *polis* is evident in two specific measures that Aineias recommends to be taken when the enemy's arrival is imminent:

> Allow no citizen, or immigrant to leave by sea without an identity token; and see that vessels have been instructed in advance to drop anchor only by those gates which the order then proceeds to specify.
>
> Ain. Tac. 10.8

> … When the city is running short of grain or oil or something else, incentives proportionate to the size of his cargo—and a crown in his honour—should be on offer to anyone who imports them; and the ship's owner (*naukleros*)[14] should have beaching and launching facilities.[15]
>
> Ain. Tac. 10.12

Here Aineias assumes that seaborne mercantile traffic is an everyday occurrence, and that the supplies for the beleaguered *polis* can easily be imported by sea. The reference to beaching and launching facilities may simply indicate that they should be available, but could imply that, for this vital service, they should be exempted from the normal charges. A good degree of familiarity with the routines and customs of a maritime polis is also implicit in the following recommendation relating to the need to minimize opportunities for internal conflict:

> In a state lacking unanimity and full of mutual suspicion, foresight and caution is needed when crowds go out to see torch races, horse races … not to mention occasions when the community beaches its ships or buries its dead.[16]
>
> Ain. Tac. 17.1

---

14 Whitehead translates *naukleros* as "ship's captain", but the word more usually refers to the owner of the vessel, who may or may not be its captain; Reed (2003) 12–13. Bon (*armateur*) and Vela Tejada (*armador*) make the same distinction; see also Bettalli (1990) ad loc.

15 Ain. Tac. 10.8; 10.12: "ἐκπλεῖν μηδένα ἀστῶν μηδὲ μέτοικον ἄνευ συμβόλου, τά τε πλοῖα προπαρηγγέλθαι ὁρμίζεσθαι καθ ἃς πύλας ἐν τοῖς ἐχομένοις ῥηθήσεται. … καὶ ὧν ἂν σπανίζῃ ἡ πόλις, σίτου ἢ ἐλαίου ἢ ἄλλου τινός, τῷ εἰσάγοντι κατὰ πλῆθος τῶν εἰσαγομένων τόκους προκεῖσθαι καὶ στέφανον δίδοσθαι εἰς τιμήν, τῷ δὲ ναυκλήρῳ ἀνολκὴν καὶ καθολκήν."

16 Ain. Tac. 17.1. "Ἐν δὲ μὴ ὁμονοούσῃ πόλει καὶ ὑπόπτως πρὸς ἀλλήλους ἐχόντων χρὴ προνοοῦντα

In later sections Aineias cautions his readers to beware of the danger of clandestine infiltrations of their city from the sea. He gives an example involving a group of *polis* officials who used the pretext of leading a force out of the city to deal with enemy raiders to rendezvous with a mercenary force:

> But in fact they went to a spot where mercenaries whom they had secretly brought in by sea were ready and waiting ... They then used the mercenaries to seize control of the city ...[17]
>
> 23.11

The section of the *Poliorketika* dealing with the need to keep gates closed draws attention to the particular vulnerability of gates that give access onto a harbour: "Furthermore, ships are not to be moored in front of the gates but at a distance." (28.1) The point is developed further when Aineias discusses the many ways in which arms might be smuggled into a city so that a faction can seize control:

> One further point: do not ignore any ships, whether large or small, which sail in and drop anchor nearby, at night or in daytime. The harbour-guardians and dispatch-officers should go on board and see the cargo for themselves—bearing in mind that when the men of Sikyon neglected such precautions they suffered a great calamity.[18]
>
> 29.12

The reference to the men of Sikyon seems to be a recent, notorious episode. The capture of Sikyon is briefly alluded to by Xenophon in his narrative of the events following the rout of Spartan forces by the Thebans at Pellene 369 BC (Xen. *Hell.* 7.1.18).[19] Two slightly different versions of the stratagem used by the Theban commander Pammenes are given by Frontinus and the mid second century

---

εὐλαβεῖσθαι τὰς μετ᾽ ὄχλου ἐξόδους ἐπὶ θεωρίαν λαμπάδος καὶ ἱπποδρομίας ... ἔτι καὶ περὶ τὰς πανδήμους νεωλκίας καὶ τὰς συνεκφορὰς τῶν τελευτησάντων."

17 Ain. Tac. 23.11. "πορευθέντες δὲ εἰς τόπον ὅπου ἦν αὐτοῖς ξενικὸν προητοιμασμένον κρυφαίως κομισθέντες κατὰ θάλατταν, ... τὴν δὲ πόλιν καταλαβόντες τοῖς ξένοις".

18 Ain. Tac. 29.12. "ἀλλὰ μὴν οὐδὲ τῶν κατὰ θάλατταν προσορμιζομένων πλοίων νυκτὸς καὶ ἡμέρας οὔτε μεγάλων οὔτε μικρῶν ἀδιασκέπτως ἔχειν, ἀλλὰ ἐμβαίνοντας τοὺς λιμενοφύλακάς τε καὶ ἀποστολέας ἰδεῖν αὐτοὺ; τὰ ἀγώγιμα, ἐνθυμουμένους ὅτι καὶ Σικυώνιοι ἀμελήσαντες τῶν τοιούτων μεγάλα ἐσφάλησαν".

19 See Bettalli (1990) ad loc; Whitehead (1990) ad loc.

AD writer Polyainos (*Frontin.* Str. 3.2.10; Polyainos *Strat.* 5.16.3). In both cases success is achieved by infiltrating an apparently harmless merchant ship into Sikyon's harbour as a troop-carrier.

Making War by Sea

The first way in which Aineias suggests a *polis* might make use of its own warships comes in the context of pursuing enemy raiders who have plundered the territory of the polis and are heading back to their own base:

> If ships (*ploia*) are available, the best way of keeping your soldiers fresh is to make the pursuit by sea. This will ensure, amongst other things necessary for success, that provided your voyage goes unobserved by the enemy you will arrive before they do.[20]
>
> 16.13

The caveat that such a measure will be most effective if the ships are unobserved reinforces the impression gained from the lack of any reference to intercepting invaders at sea that Aineias assumes that it is relatively easy to avoid detection when approaching a *polis* by sea until it is too late for a warning to be effective. There are various ways in which this might be achieved. Skilful use of islands, mountains and headlands to conceal an approach, or deliberately sailing out of sight of land would both be tricky, but not impossible for navigators well used to the waters in which they were operating. Night voyages are also possible, again provided the course is well known. Alternatively ships could take an indirect, but sufficiently speedy route to the anticipated destination of the retreating enemy. Each of these would be far more likely if the "enemy" in question were from, or operating on behalf of, a nearby *polis*. This scenario does seem to be what Aineias has in mind, and it is in keeping with the well known tendency of the Classical Greek *poleis* to be frequently at war with their neighbours.

There are two obvious ways in which we might interpret the phrase translated by Whitehead as "keeping your soldiers fresh". Aineias could be suggesting that transporting the men by sea will allow them to travel without any physical

20 Ain. Tac. 16.13. "ἄριστον δ, ἵνα νεοκμῆσιν τοῖς στρατιώταις χρήσῃ, ὑπαρχόντων γε πλοίων, κατὰ θάλατταν τὴν δίωξιν ποιεῖσθαι: τό τε γὰρ φθάσαι καὶ τὰ ἄλλα εἰς τὸ δέον σοι συμβήσεται, ἐὰν μὴ κατοφθῇς πλέων ὑπ αὐτῶν".

effort, and thus arrive rested. This implies that they will not be rowing the ships themselves, but that they will be passengers. This interpretation is strengthened by Aineias' use of the word *ploia*, which is the standard Classical Greek term for any sort of ships, but is not the specific term for purpose-built warships. The implied distinction is made clearer by his use of the standard word for warship, *naus*, a few sentences later (16.22), when he advocates the deployment of a seaborne force when making counter-attacks against invaders:[21] Alternatively, if he is assuming that a multi-level, oared warship such as a trireme is being used, then, besides the obvious expediency of using sails when the winds allow, he could be suggesting that different levels of oarsmen take turns at rowing.[22]

Aineias clearly does have the use of purpose-built warships in mind for the next naval tactic he recommends:

> If you do have a fleet (*nautikon*) the warships (*tas naus*) should be ready manned. Putting some of your men in warships will prove no less aggravating to the enemy, provided the ships sail along their seaboard near the coast roads, with the aim of making a landing in their rear to coincide with your own harassing by land. In that way you can attack the enemy when they are least prepared for it and take them by surprise.[23]
>
> 16.21–22

Here Aineias is still ostensibly discussing counter-attacks against an invader, but the methods he advocates could equally well be those of an aggressive maritime raid that is not a direct response to an enemy incursion.

There is no doubt that Aineias was fully aware of the considerable commitment in terms of resources and infrastructure that operating even a small fleet of warships required. As part of his recommendations to be wary of and ready

---

21 On the ancient Greek vocabulary for ships see Casson (1971) 157 and (1986) 169; Morrison and Williams (1968) 244.

22 For discussion of the performance of the replica trireme *Olympias* under oar and sail, and the various possible permutations see Morrison, Coates and Rankov (2000) 236–267. For observations on the comparison between the trials of this vessel to ancient *testimonia* see Whitehead (2012).

23 Ain. Tac. 16.21–22 "ὑπάρχοντος δέ σοι ναυτικοῦ, πεπληρῶσθαι τὰς ναῦς: οὐ γὰρ ἐλάσσω τοὺς πολεμίους ὀχλήσουσιν οἱ ἐν ταῖς ναυσίν, ἐάν γε παραπλέον ὑπάρχῃ πρὸς τὰ ἐπιθαλάσσια χωρία καὶ τὰς παρὰ θάλασσαν ὁδούς, ἵν̓ ὑπὸ ὑμῶν τε ὀχλῶνται καὶ ὑπὸ τῶν ἐκ τῶν νηῶν κατὰ νώτου αὐτοῖς τὴν ἀπόβασιν ποιουμένων. ταῦτα δὲ οὕτως πράττοντες ἀπαρασκευαστοτάτοις ἂν τοῖς πολεμίοις ἐπιθεῖσθε καὶ παρὰ δόξαν τὰ ἀφ ὑμῶν αὐτοῖς ἀποβαίη".

for the acts of internal dissidents, he relates an instance of betrayal by a *polis* official:

> Just before the betrayal of Chios, for instance, the authorities were tricked by one of their number who was party to the plot. He persuaded his colleagues that as it was peacetime they ought to have the harbour-boom hauled ashore, dried out, and given a new coating of pitch, the old warships' tackle sold, and repairs made to the leaking roofs of the shipsheds, to the adjacent stoa, and to the adjoining tower where the officials had their quarters. This was an excuse for ladders to be ready for those who were planning to seize the shipsheds and the stoa and the tower.[24]
>
> 11.3

Aineias first refers to a "boom" for closing off a harbour, which, given that it has to be dried out, would probably be made of stout rope. It would have required strong, possibly fortified, fixing points. Next he refers to the "tackle" of warships, which could include sails, lines, ropes, oars, oar-loops and several other items.[25] However, it is likely that the most expensive piece of naval infrastructure referred to in this short passage would be the shipsheds (*neoria*), which were essential for the protection and maintenance of warships. Shipsheds were necessary because the hulls of ancient, wooden warships were only lightly coated with pitch and paint. Unless they were regularly removed from the water, dried, careened, repaired and re-coated they became waterlogged, encrusted with barnacles and riddled with the burrows of marine borers. *Poleis* that had fleets of warships, had, therefore, to invest in well constructed roofed slipways, or shipsheds, where this work could be carried out. The shipsheds of Piraeus, the fortified port of Athens, are the best known, and the most costly, but there were many more throughout the Greek world. Smaller *poleis* might have only a handful of permanent shipsheds, but even these represented a sig-

---

24 On the possible date and context of this apparently notorious event see Whitehead (1990) ad loc.

Ain. Tac. 11.3. "Χίου γὰρ μελλούσης προδίδοσθαι, τῶν ἀρχόντων τις, συμπροδιδοὺς καὶ ἀπατῶν τοὺς συνάρχοντας ἔπεισε λέγων, ὡς, ἐπειδὴ εἰρήνη εἴη, τοῦ τε λιμένος τὸ κλεῖθρον εἰς γῆν ἀνασπάσαντας ξηρᾶναι δεῖ καὶ πισσαλοιφῆσαι καὶ τὰ παλαιὰ τῶν νεῶν ἄρμενα ἀποδόσθαι, τῶν τε νεωρίων ἐπεσκευάσθαι τὰ στάζοντα καὶ τὴν ἐχομένην αὐτῶν στοὰν καὶ τὸν πύργον, ἐν ᾧ διητῶντο οἱ ἄρχοντες, ἐχόμενον τῆς στοᾶς, ἵνα ἐκ προφάσεως κλίμακες προπορισθῶσιν τοῖς μέλλουσι καταλαμβάνειν τὰ νεώρια καὶ τὴν στοὰν καὶ τὸν πύργον."

25 See Casson (1971; 1986) 235–239; Morrison, Coates and Rankov (2000) 168–178.

nificant financial commitment in order to have warships ready for action, as Aineias envisages.[26]

Shipsheds, even those built on natural rock foundations, required a great deal of wood for their construction, so they, and the ships and equipment they contained, were extremely vulnerable to fire. Hence Aineias includes them in his list of locations where internal dissidents might choose to start fires:

> And another thing: think twice before ever venturing out of town at night with a crowd. This is because, at such moments, conspirators both inside the community and beyond are up to their tricks, seeking to lure one out with ruses like fire-signals or setting ablaze a shipshed or gymnasium ⟨or⟩ major public shrine ...[27]
>
> 23.6

There are no recorded examples from the Classical period of shipsheds being set ablaze by spies or traitors, though in 369 BC the Thebans attacked Gytheion, where the Spartans had their shipsheds (Xen. *Hell.* 6.5.32). This is perhaps an indication of how carefully guarded they were.[28]

Having reviewed these passages and considered how the measures and tactics that Aineias describes, or advocates fit into the wider context of ancient Greek naval warfare, one issue remains—the statement that he makes at the very end of the *Poliorketika*:

> As regards grainless rations, shortages during a siege, and the means of making water fit to drink, these are matters discussed in my book *Preparations*; so having dealt with them I pass on to the organization of a fleet. There are two purposes to fleet operations ...[29]
>
> 40.8

---

26 For brief surveys see Blackman (1968); (2008) 654–661. For more details see Blackman and Rankov (2013); Lóven (2011).

27 Ain. Tac. 23.6. "Οὐ μὴν οὐδ ἐν ταῖς νυξὶν ἀσκέπτως μετ ὄχλου ἐκπορευτέον, οἷα καὶ ἐν τοῖς καιροῖς τῶν ἐπιβουλευόντων τινὲς τεχνάζουσιν, οἱ μὲν ἔσω τῆς πόλεως οἱ δὲ καὶ ἔξωθεν, προσάγεσθαι βουλόμενοι τοιοῖσδε ἀπατήμασι, πυρσεύσαντές τι ἢ ἐμπρήσαντες νεώριον ἢ γυμνάσιον ἢ ἱερὸν πάνδημον ...".

28 For a later example, dated to c. 205 BC see Polyainos *Strat.* 5.17.2, where Herakleides of Tarentum, acting on behalf of Philip V of Macedon, sets fire to Rhodian shipsheds, destroying thirteen of them.

29 Ain. Tac. 40.8. "Περὶ δὲ τροφῆς ἀσίτου καὶ ὧν σπάνις ἐν πολιορκίᾳ καὶ ὑδάτων ὡς δεῖ πότιμα ποιεῖν, ἐν τῇ Παρασκευαστικῇ βύβλῳ δεδήλωται. ἐπεὶ δὲ ταῦθ ἡμῖν εἴρηται, περὶ ναυτικῆς τάξεως δίειμι. Ναυτικοῦ δὲ στρατεύματος δύο εἰσὶ στόλοι ...".

For the last, incomplete sentence I have modified Whitehead's translation to give what I think may be the intended meaning.[30] It offers a hint of the contents of the further section of Aineias' work that is apparently introduced here, but that, for unknown reasons, has not been preserved. Based on those sections discussed above where Aineias refers to the military use of ships or warships (4.8; 8.2; 16.13; 16.21–22), and bearing in mind the frequency with which Classical Greek warships were used for raids, I tentatively suggest that the two purposes Aineias went on to elaborate were, on the one hand pursuit or ambush of enemy raiders, and on the other hand direct attacks, or as they might be thought of, retaliatory counter-raids, on enemy territory. The two different purposes would require different tactics and also different ratios of light to heavy infantry in the ships crews, with more of the latter for an operation intended to assault a potentially well defended target, such as a city. At best this is a speculative hypothesis. The incomplete nature of the *Poliorketika* precludes a more definitive one.

Bibliography

Bettalli, M. (1990) *Enea Tattico. La difesa di una città assediata (Poliorketika). Introduzione, traduzione e commento.* ETS editrice, Pisa.

Blackman, D.J. (1968) 'The Shipsheds', in Morrison, J. and Williams, R.T. (eds) *Greek Oared Ships*, Cambridge University Press, Cambridge. 181–192.

Blackman, D.J. (2008) 'Sea Transport, Part 2: Harbors', in Oleson, J.P. (ed.) *The Oxford Handbook of Engineering and Technology in the Classical World.* Oxford University Press, New York and Oxford. 638–670.

Blackman, D.J. and Rankov, B. (2013) *Shipsheds of the Ancient Mediterranean.* Cambridge University Press, Cambridge.

Casson, L. (1971) *Ships and Seamanship in the Ancient World* (second edition 1986). Princeton University Press, Princeton.

Dain, A. and Bon, A.-M. (1967) *Énée le Tacticien: Poliorcétique.* Les Belles Lettres, Paris.

30 On the translation of *stoloi* as "purposes" see *LSJ* sv. It is worth noting some of the alternative translations that scholars have put forward, because they suggest different structures and contents for the missing section. Whitehead himself translates the passage thus: "There are two ways of equipping a naval force …" In his commentary he notes that earlier scholars have preferred to see Aineias as referring to two ways of using a fleet, or two divisions within a fleet. Whitehead (1990) 97. Bon followed the latter course: "L'armée de mer comporte deux flottes …" Dain and Bon (1967) 91; so did Vela Tejada, "Las flotas de la armada son dos …" Vela Tejada (1991) 256. See also Bettalli (1990) ad loc.

de Souza, P. (1999) *Piracy in the Graeco-Roman World*. Cambridge University Press, Cambridge.

de Souza, P. (2010) 'Raids on the Coast of Attica', in Sekunda, N. (ed.) *Ergasteria: Works in Honour of John Ellis Jones on his 80th Birthday*. Gdansk University, Gdansk.

de Souza, P. (2013) 'War at Sea', in Campbell, B. and Tritle, L. (eds) *The Oxford Handbook of War in the Classical World*. Oxford University Press, Oxford and New York. 369–394.

Gabrielsen, V. (2001) 'Economic Activity, Maritime Trade and Piracy in the Hellenistic Aegean', *Revue des Études Anciennes* 103: 219–240.

Gabrielsen, V. (2007) 'Warfare and the State', in Sabin, P. van Wees, H. and Whitby, M. (eds) *The Cambridge History of Greek and Roman Warfare*, vol. I. Cambridge University Press, Cambridge. 248–272.

Hale, J.R. (2009) *Lords of the Sea. The Epic Story of the Athenian Navy and the Birth of Democracy*. Viking, New York.

Krentz, P. (2007) 'Archaic and Classical Greek War' in Sabin, P. van Wees, H. and Whitby, M. (eds) *The Cambridge History of Greek and Roman Warfare*, vol. I. Cambridge University Press, Cambridge. 147–185.

Lovén, B. (2011) *The Ancient Harbours of the Piraeus* (two vols). Monographs of the Danish Institute at Athens, Aarhus.

Morrison, J.S. and Williams, R.T. (1968) *Greek Oared Ships 900–322 BC*. Cambridge University Press, Cambridge.

Morrison, J.S., Coates, J.F. and Rankov, N.B. (2000) *The Athenian Trireme. The History and Reconstruction of an Ancient Greek Warship*. Cambridge University Press, Cambridge.

Pritchett, W.K. (1991) *The Greek State at War. Part V*. University of California Press, Berkeley.

Reed, C.M. (2003) *Maritime Traders in the Ancient Greek World*. University of Cambridge Press, Cambridge.

Rogers, W.L. (1937) *Greek and Roman Naval Warfare*. Naval Institute Press, Annapolis.

Starr, C.G. (1989) *The Influence of Sea Power on Ancient History*. Oxford University Press, New York and Oxford.

Trundle, M. (2010) 'Light Troops in Classical Athens', in Pritchard, D.M. (ed.) *War, Democracy and Culture in Classical Athens*. Cambridge University Press, Cambridge. 139–160.

van Wees, H. (1992) *Status Warriors: War, Violence and Society in Homer and History*. J.C. Gieben, Amsterdam.

van Wees, H. (2004) *Greek Warfare: Myths and Realities*. Duckworth, London.

Vela Tejada, J. (1991) *Eneas el Táctico. Poliorcética. La estrategia militar griega en el siglo IV a.C.* Ministerio de Defesa, Madrid.

Whitehead, D. (1990) *Aineias the Tactician, How to Survive under Siege. Translated with introduction and commentary*. Clarendon Ancient History Series, Oxford.

Whitehead, I. (2012) 'Triereis Under Oar and Sail', in B. Rankov (ed.) *Trireme Olympias. The Final Report. Sea Trials 1992–1994. Conference Papers 1998*. Oxford University Press, Oxford. 155–160.

CHAPTER 13

Notes from Home: A Perspective on Fourth-century Defence from Arkadian Stymphalos

Ben Gourley

A few years ago in writing on the defences at ancient Stymphalos, my colleague Hector Williams and I suggested that the small Arkadian town might serve as a "potential paradigm" for the study of Greek fortifications in the age of fourth-century defensive warfare.[1] Nowadays, anyone who pays a visit to the mountain site is likely to encounter little more than a few foundations scattered around the valley bottom in a picturesque natural situation; they could be forgiven for wondering 'why such a grand claim?' It is true that unlike many of the great preserved stone fortifications of contemporary date, such as those of Messene or Aigosthena where visitors can wander into a tower chamber and immediately imagine themselves in the world of an ancient soldier, Stymphalos does not offer much on first inspection. Looking beyond immediate impressions, however, it seems that due to a number of fortunate circumstances of survival (both historical and archaeological), this small town has more to offer on the subject of defensive warfare than it would first appear. To start with, Stymphalos has the rather peculiar responsibility of being the likely home of Aineias the Tactician,[2] the creator of the unique vision of small town warfare in fourth-century Greece that is the *Poliorketika*. Less famous and more grounded in the material realities of fourth-century defences than its famous son, this city also happens to have been archaeologically investigated in some detail. The fortifications in particular have been studied extensively through systematic archaeological investigations carried out under the auspices of the University of British Columbia and the Canadian Archaeological Institute in Athens, and thanks to this, we have a holistic understanding of the archaeological context of fourth-century civic defences.[3]

1 Williams and Gourley (2005) 215.

2 For Aineias' likely origins see Whitehead (1990) 10–13.

3 Archaeological work at Stymphalos has taken place under the aegis of the Canadian Archaeological Institute at Athens with funding from the Social Sciences and Humanities Research Council of Canada, the University of British Columbia, and private donors. We acknowledge with thanks the contributions of many colleagues, students and volunteers on the project

 | DOI: 10.1163/9789004352858_014

Reconciling these two pieces of information—the text and the material remains—into one coherent vision of the social and material organisation of the fourth-century defence is not an easy task. Of course the association of Aineias with the city introduces quite reasonably high expectations of the archaeology at the site. The obvious question arises: do the archaeological remains embody at least some physical evidence of principles advocated by that writer in his military treatise?

Chronologically, the timeframe is tantalising. The fortifications at Stymphalos are almost certainly a product of the rapid developments in defensive architecture of the early fourth century and the body of archaeological evidence suggests a date in the second quarter of that century.[4] Aineias is understood to have compiled his surviving manuscript around that time, though his association with the city remains a piece of detective work, and at best a 'most likely' scenario rather than a historical fact.[5] In the many references to places around the Greek world that do crop up in his manuscript, Stymphalos does not appear at any time. The absence is curious, and is the obvious missing link that would make it possible to talk with any certainty about the defences as having had some influence on the author's military training.

One of the problems in looking at the text and surviving remains together is the incongruous nature of the evidence before us. The archaeology, though providing clarity on how and when things were built, is very limited in its ability to disentangle the complexity of the social and technological act of defending a town and territory. On the other hand, in Aineias' text, we are instructed on several aspects of the social, psychological and organisational sides of defensive warfare. It is very seldom that one gets a sense of the physicality of a defensive system as a whole. His recommendations are primarily concerned with the management of populations in time of danger, and are administered in general terms. There are some superficial parallels, but the two stand far apart when considering the specific aspects of the evidence.

For this reason I would like to look at Aineias and his relation to Stymphalos not with an eye towards direct comparison, but more as a means of finding what is complementary between the two. Because most of the readers of this volume will be more familiar with Aineias than the site of Stymphalos, I think

---

over the past two decades and the support of the Greek Archaeological Service, the Archaeological Society of Athens (which originally held the rights to the site) and the people of Stymphalia.

4 For excavation reports see Williams (1983–1985) and Williams et al. (1995–2003).

5 Whitehead (1990) 12.

it is best to focus on the city itself and what it has to offer. The fact that we know quite a bit about its defensive context allows us to make some general observations about the physicality of defence in a small *polis* in this period. I leave it to the reader to form their own opinion of whether it synchronises with Aineias or not. Whether or not the author was from Stymphalos is in many respects irrelevant: the city is representative of the kind of place with which Aineias would be familiar, and in that sense the archaeology provides a useful tool for grounding the text in a physical environment. If it is possible for us to imagine that Aineias did have *this particular* city in mind when writing, then all the better.

Ancient Stymphalos

Occupation History

Human activity in the Stymphalian valley can be traced as far back as the Neolithic period, but very little of the early history of the region is known in any detail.[6] The occasional surface finds, many of which are in the collections of the local natural history museum, suggest a sustained human presence in the region for the following millennia, but the first obvious evidence for organised settlement is that of the ancient town dating to the early fourth century. Some isolated Bronze Age material has been found on the acropolis of the fourth-century town and in a regional context Knauss has suggested extensive Mycenaean dams and canal works for water management and reclamation of arable land in the valley.[7] Archaeological evidence from the town itself shows a foundation date of c. 375–350, and although it is clear from a handful of sources, including Pindar and Homer, that a pre-fourth-century Stymphalos, did exist (see discussion below),[8] the circumstances surrounding the construction of this new town are obscure. Both Pausanias and Strabo refer to events at this time: Strabo documents a siege in the early fourth century by Iphikrates, and Pausanias intriguingly records that the city had been moved from its original location at an undisclosed date.[9] Archaeological evidence for the fifth and early fourth century is sparse but includes coinage and ceramic clearly associated

6 Hope Simpson and Dickenson (1979) 84; Howell (1970) 97–98.

7 Knauss (1990) 1–52; Knauss (1991) 364.

8 Stymphalos is listed as a contributor in the Catalogue of Ships: Hom. *Il* 2.605. Pindar celebrates Olympian victors from Stympalos in the 460s: *Olympian* 6.

9 Strab. 8.8.4; Paus. 8.22.1.

with the fifth century found in deeper deposits of excavations in the lower town in association with domestic structures.[10]

For the centuries following refoundation our knowledge is much clearer. Late classical Stymphalos flourished, particularly in the early third century.[11] It has been suggested that prosperity was fuelled at least in part by participation of the city's inhabitants in foreign campaigns as mercenary soldiers.[12] There is some evidence for such a claim—Xenophon in particular refers to Stymphalians amongst foreign troops in both the *Anabasis* and *Hellenica*[13]—although it seems unlikely that Stymphalos was any different from other Arkadian small towns in this respect. Certainly, the numismatic record for the third century suggests a wide international pool of coinage in circulation in the town in the third and second centuries.

The second century seems to have be a transitional one in the valley and town. It has been suggested that the city suffered somewhat at the time of the Roman destruction of Corinth in 146 BC, with key elements of the civic infrastructure such as the fortifications and the Acropolis Sanctuary going out of use at that time.[14] Whether or not Stymphalos was affected by nearby events,

10 Bronze Age pottery was identified on the acropolis and reported by Hope Simpson and Dickenson (1979) 84. Williams reports possibly BA sherds in excavations in the lower town in Williams et al. (2002) 171. A handful of Bronze Age lithics were found on the acropolis during excavation of a structure of unknown function: Williams et al. (2002) 171. Several fifth-century coins including those of the Arkadian League dating to 465–455 have been studied by Robert Weir: see, Weir (2002) in e-journal *Labyrinth* 79; and Williams et al. (2002) 136.

11 Many of the monuments in the agora date from this period and the fortifications were overhauled.

See A. Orlandos (1924–1930) for monuments in the agora, and for the renovation of the fortifications, see below, and also Williams and Gourley (2005) 213–259.

12 Williams and Gourley (2005) 215, and note 5 p. 215. Weir also suggests mercenary activity as contributing to pool of coinage circulating in the town: Weir (2000).

13 Stymphalians in foreign armies: Xen *Hell.* 7.3.1, 2; *Anab.* 1.1 Perhaps one of the more significant events of the late fourth century is reported by Diod. 19.63.1–2: a night raid in 315 by Apollonodes, a general under Kassander, is reported but no information given on whether the city was taken at the time.

14 Williams et al. (2002) 153; Williams and Gourley (2005) 249. The suggestion of a Roman attack on the town is based on a range of evidence for a significant change to parts of the city's infrastructure. Of the more interesting and hotly contested pieces of evidence are the large number of projectile points, including ling bullets and catapult bolts, recovered the acropolis sanctuary, and a forged Roman denarius of military issue which has been taken as evidence in support of roman presence: see Williams and Gourley (2005) 257, note 57, for discussion.

daily life appears not to have suffered extensively and archaeological evidence from the lower town and the agora points towards a period of prosperity in the years following the mid-second century.[15] However, by the mid-first century BC parts of the city were abandoned, most likely due to a major earthquake that flattened structures including the excavated houses in the lower town.[16]

By the mid-first century AD there is evidence of some limited resettlement. To what extent this represented a resumption of civic life in the town is also unclear, but it seems that at least some infrastructure was retained into the second century AD. Under Hadrian an aqueduct was constructed across the valley linking the valley's springs with Corinth and the costal plan of the Corinthia.[17] Whether or not the population that serviced and administered this resource lived in the city itself is not known, but Pausanias records an active population in the place (which he refers to as a city) regardless.[18] At some point between the second century and the mid-fifth the city must have been abandoned altogether. Excavation in several of the major monuments in and around the town has revealed extensive cemetery sites with inhumations buried inside the former structures in the city. The burials span the fifth and sixth centuries, and the location of the settlement or settlements from which the dead originated remains a mystery.[19]

The late antique and early modern history of the valley remains largely unstudied. For the centuries following the use of the site as a cemetery in the sixth century, the first evidence of any large-scale use of the former urban landscape is the appearance of the Cistercians in the 13th century. In the years leading up to 1220 a Cistercian abbey was founded on the northern outskirts of the ancient site, and the landscape in the north of the town was significantly impacted by associated agriculture and waterworks.[20] However the abbey was relatively short-lived and appears to have been abandoned by the 1260s, although the abbey and its precinct remain one of the most conspicuous remains in the valley, and much of its fabric was constructed from *spolia* from the ancient site.

15 Weir (2002).

16 Williams et al. (2002) 145.

17 For a comprehensive study of the Hadrianic aqueduct and its context see Lolos (1997) 271–314.

18 Paus. 8.22.3.

19 Williams et al. (2002) 136, 139.

20 For discussions of the monastery see Bon (1969); Panagopoulos (1979); Campbell (1997).

Regional Context

The physical situation of ancient Stymphalos has played a major role in shaping its economic and political outlook though history. Like many other small sites in upland Arkadia, it is generally considered as culturally remote in comparison to its coastal neighbours or to the major states of the Greek world.[21] Geographically it is located approximately 40 km inland from the Gulf of Corinth in a mountain pass 600 m above sea level, under the shadow of the peaks of Mt Kyllene (which at over 2300 m is the highest mountain in the region). The valley seems to have been historically important for strategic reasons, and provided one of the few accessible land routes from the Gulf of Corinth to the central and southern Peloponnese. Its geographical orientation is predominantly to the northeast and it seems that the major centres of Corinth and particularly Sikyon (located at the head of the pass on the coast) were the most influential neighbouring states through the period of organised settlement in the small town.

Results from excavation—mostly from deposits of Roman and Hellenistic dates—indicate a close relationship with Sikyon. Numismatic evidence suggests wider chronological range from the fifth century BC through the first century AD. In many ways it most graphically reflects the relationship with over 40 per cent of the 350 coins recovered at Stymphalos originating at the Sikyonian mint (mostly bronze and silver).[22] A similar trend appears in the ceramic data. Comparison between material recovered during surface survey at Hellenistic Sikyon and excavation at Stymphalos show a strong representation of Sikyonian ceramic fabrics evident at least from the beginning of the third century BC at Stymphalos.[23] (We do not know as much about earlier fabrics from Sikyon, nor has excavation at Stymphalos yielded a significant sample of pre-Hellenistic fabrics.)[24]

21 For an overview of perceptions of Arkadia in antiquity see Nielsen (1999) 16–79.

22 Weir (2002).

23 Thanks to Peter Stone and Elli Tzavella as well as Matt Maher all of whom have devoted much time to the study of Sikyonian pottery as part of the Sikyon Survey Project, and have also studied the ceramics at Stypmphalos in differing degrees. Peter Stone's MA dissertation on the ceramic assemblages from the Acropolis Sanctuary has been particularly useful in observing parallels between the sites.

24 Williams et al. (2002) 135–187. The great majority of ceramics at Stymphalos were recovered from the Acropolis Sanctuary, the excavations in houses in the lower town, various locations on the fortifications and the theatre. The chronological range spans the fifth century BC through the first century AD, and various areas of the site have stronger representations for specific periods.

Environmental History and Context

The valley takes its shape from its position at the intersection of a series of minor geological faults that define its linear northeast/southwest alignment and its geological composition.[25] This region of northeast Arkadia is markedly distinct from the more familiar landscapes of the coast just a few kilometres to the north, a coastal plain between the sea and the mountains, where major sites are closely connected to the sea and there is inter-visibility between neighbouring cities: a shared landscape with few natural barriers.

Travelling inland from the coast one traverses a series of marine terraces, formed by the gradual uplift of the north-eastern Peloponnese and subsequent earthquake activity in the region. Approximately 20 km from the coast these marine terraces change to a mountainous landscape dominated by bands of poros limestone and conglomerate. Unlike the coastal plain, this landscape is dramatically enclosed and the relationship between regions and local communities is, to this day, defined by a limited number of winding routes through valleys and over passes (see figure 1). In terms of settlement and land use this landscape could be characterised as comprising two zones—valley bottoms and uplands. Both have been exploited for a wide range of subsistence practices from the production of the "Mediterranean triad" to pastoralism and bee-keeping for many thousands of years, with the valley bottoms particularly suited to cereal crops.

The hostility of the terrain introduces a series of advantages over more exposed lowland or island-based settlements for defensive positioning. Limited communication routes, abundance of high defendable ground for inter-visibility and isolated outpost defences, impassable natural barriers such as rivers and mountain ranges, all played a part in ancient defensive strategy. At Stymphalos, it is very likely that the natural landforms not only influenced the location of the city itself, but communication routes into and out of the city, as well as the location of known satellite defences around the regional territory of the ancient town (see figure 2).[26] One peculiarity of the region that seems to have been influential in defensive strategy is the behaviour of natural spring water and surface runoff from rain and snow on the mountain slopes. The Stymphalian valley, along with the neighbouring valleys of Pheneos and Orchomenos/ Kaphyai is one in a series of *karst poljes*, a class of limestone

25 Knauss (1991) 364.

26 Although a systematic study of the hinterland of the valley has not yet been undertaken, a series of satellite towers and sites have been identified at locations around the valley and on hilltops. Additionally, in his comprehensive study of the territory of Sikyon, Yannis

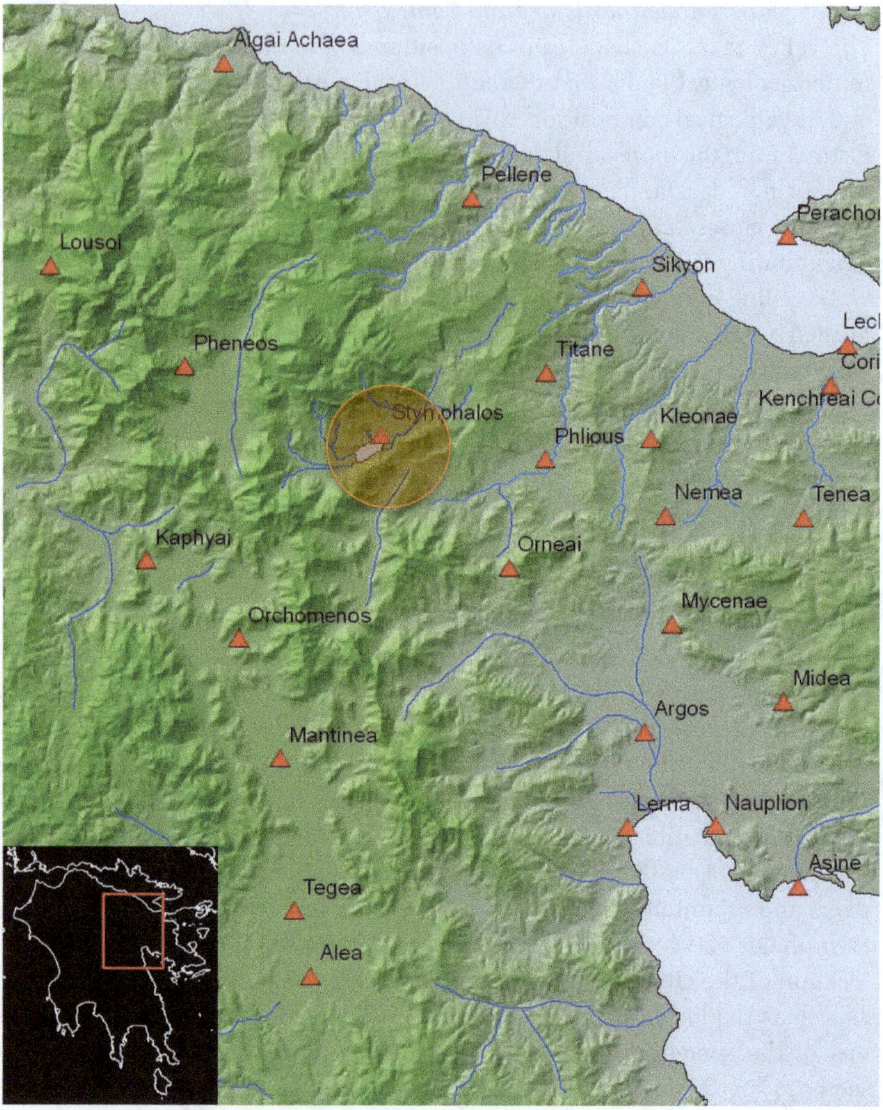

FIGURE 13.1 *Regional map showing Stymphalos in relation to other ancient sites with major centres highlighted*

landscape formation characterised by flat bottomed internally draining basins dependent on natural sink holes for the management of spring water and sur-

---

Lolos has located many such sites on the border between Sikyonian and Stymphlian territory: Lolos (2007).

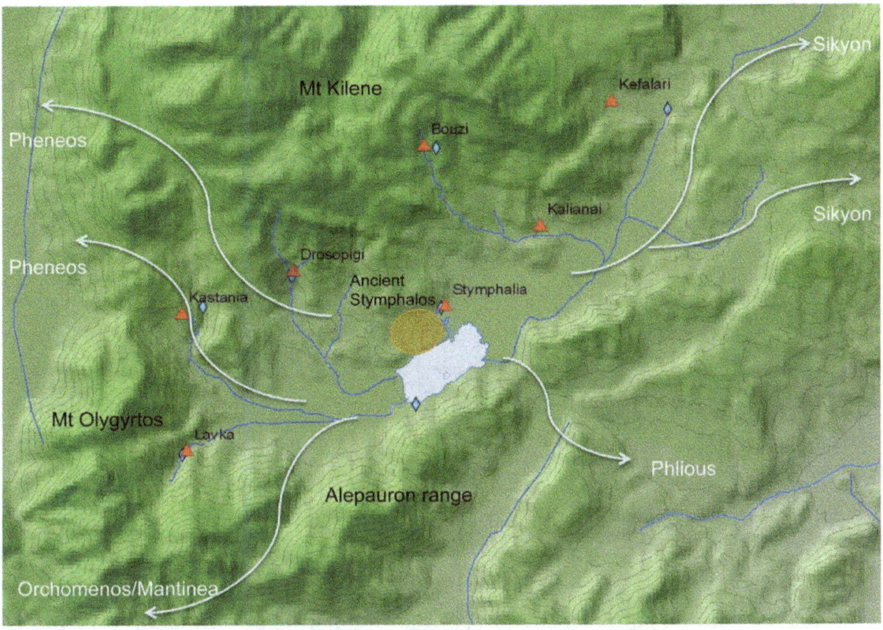

FIGURE 13.2 *Valley map showing position of the ancient site with neighbouring modern villages and ancient transportation routes out of the valley to other ancient centres*

face runoff.[27] As a point of definition therefore, the valley should be considered more a *basin* than an open valley, and the famous lake at its heart is a direct product of its internal drainages.

The internal focus of a basin system—specifically the fact that key essential resources such as water are controllable within one's own territory—might be considered to yield a particular defensive advantage over states that shared resources across territorial borders, such as rivers or coastlines. However at Stymphalos advantages of internal water supply are partially offset by some of the problems posed by internal drainage in a mountain environment. The present lake owes shape and depth only to the systematic management of drainage out of the valley. It has been demonstrated in recent geo-archaeological research that the lake has a complex history that was subject to extreme changeability.[28]

27 Knauss (1991) 363–381.

28 My thanks to Dr Kevin Walsh (University of York) and Prof. Antony Brown (University of Southampton) for sharing initial results of fieldwork seasons in 2008 and 2009. Their

Analysis of auger data from sediments in and around the lake have outlined a rough chronology of the form of the lake over the past three millennia, and most interestingly they have shown how protean a water body it is, due especially to the flat valley floor. Results indicate that over the past several thousand years the lake was a shallow, almost ephemeral, water-body. It seems that even until as recently as the mid-nineteenth century it was very shallow if not completely dry apart from the creeks issuing from the springs at the north of the valley.[29] Coring strongly suggests that this was the case throughout the period of settlement at the Classical and Hellenistic city. It is likely that given the unpredictable nature of the lake from season to season and year to year, the city would have depended on springs for its water supply. Indeed at least since the second century AD (and especially in recent years) the spring waters that help feed the lake have been heavily exploited both inside and outside the valley for irrigation.[30]

The city's location at the heart of this system on the northern lakeshore took advantage of the peculiar natural topography of the valley: a combination of spring sites on the lake's edge, springs to the north east which from a creek on the eastern boundary of the site, and a defensible acropolis on an isolated limestone ridge on the lake's northern shore.[31] In contrast these obvious natural advantages, the city was also particularly vulnerable to breakdown in the natural cycle of drainage from the valley and its situation on what is effectively a flood plain impacted longevity and use of the site. The tenuous relationship between site and lake was clearly a significant enough factor throughout the valley's history for various flood events—both mythological and historical—to be recorded by ancient authors in various contexts.[32] The modern blocking

---

analysis of lake sediments has greatly increased out understanding of the environmental context of the city and its tenuous relationship with the lake.

29 Pausanias records that in summer in his day there was no lake and the waters of the spring formed a river crossing the plain: Paus. 8.22.3; Frazer (1898) 274.

30 The earliest direct evidence of external exploitation of the lake waters is the construction of the Hadrianic aqueduct: see Lolos (1997). In 1881 the tunnel used by the Hadrianic aqueduct was again used by engineers charged with draining the valley (Frazer (1898) 269) and within the last few decades, pump houses have been built to carry spring water further up the valley to the west.

31 The benefits of the topography were also recognised by the Cistercians in the 13th century AD and they established the abbey between the northern outskirts of the ancient site and the southern slopes of Mt Kyllene directly on the course of the outflow of the spring waters from the northern spring.

32 Ancient references to the blocking of the sinkhole take both a mythological and a historical stance. Pausanias 8.22.9 recounts the story of a hunter in pursuit of a deer blocking

of the sinkhole has demonstrated the effects of restricted drainage on the lake level, and much of the ancient site and the fields that surround it are completely inundated during the winter months. Even in its 'dry' months during the summer, the lake in recent years has covered large portions of the southern city.

Previous and Current Work

The fortifications have always been amongst the most conspicuous remnants of the ancient town. They are traceable in occasional sections of standing stone foundations (the mudbrick walls themselves having collapsed) and in the alignment of field boundaries or earthworks where lake sediments have masked surface remains. Early travellers to Stymphalos recorded the fortifications in varying degrees of detail and accuracy. Frazer documented the acropolis and southern fortifications in some detail, although he mistakenly took the later for a dam to hold the lake water from the site, and also attributed the western wall as a precinct wall for the abbey remains.[33] Twenty-five years later, the earliest systematic investigations of the defences were carried out (between 1924 and 1930) under the direction of A. Orlandos who cleared and recorded at least two of the city gates and associated towers. His reports are useful for discerning the plan and layout of these structures but detailed analysis of the chronological developments of the fortifications and their relation to the town they enclosed is absent from his analysis. Notwithstanding, his plan of the Phlious Gate received some attention in subsequent studies of fortifications and therefore, served to highlight the potential of the site for further study.[34]

Recent work on the fortifications began in 1983 with geophysical survey of sections of the defences and internal arrangements of the lower town. At that time it was noted that their layout, which incorporates overlapping sections of wall enclosing a roughly circular area, bears close resemblance to the fortifica-

---

the hole by jumping down it, and also a log falling into the hole and blocking it as a result of careless celebration of the festival of Artemis (8.22.8); Strabo on the other hand documents an attempted siege of the city using sponges to block the hole (8.8.4). In 1896 Frazer describes the lake as some 4 miles long, which—accurate or not—he attributes to the modern blockage of the sinkhole: Frazer (1898) 269.

33 Frazer (1896) 272–273. For other nineteenth-century travelers' accounts see Dodwell (1818) 433–436, and Leake (1830) 108–114.

34 For reports of Orlandos' work see *Praktika* (1922–1924) 117–123; *Praktika* (1925–1926) 51–55, 131–139; *Praktika* (1927) 53–56; *Praktika* (1928) 120–123; *Praktika* (1929) 92; *Praktika* (1930) 88–91. For inclusion of plan of the Phlious Gate in fortification studies see: Winter (1971) 217; Lawrence (1979) 334; Adam (1992) 39–40.

tions of Mantinea, and a provisional date of the mid-fourth century circuit suggested for the present remains.[35] Subsequent architectural study, survey, and excavation has demonstrated that the majority of the archaeological remains at the site do in fact date to the early fourth century or to later modifications. Between 1996 and 2002 sections of three of the larger towers on the acropolis and the west wall, two small towers and two gates have been excavated and provided dating information for each of these structures and for the system as a whole. Additional geophysical and topographical survey was carried out at the same time and it is now possible to trace almost the full length of the walls, fourty-seven towers and ten gates in detail.

The scarcity of material predating the fourth century has reinforced the hypothesis that the city was re-founded in the early fourth century, and its construction in its present form may have been associated with new cities at Mantinea, Megalopolis and Messene built to control Sparta; the city controlled a route north from Arkadia and thus from Lakonia to the Corinthian Gulf.[36] However, the question of the location of Archaic and Classical Stymphalos has never been satisfactorily answered and is of considerable importance for unravelling the construction sequence of the fortifications and for understanding the full significance of their fourth-century construction. The issue is complicated by a brief passage in Pausanias which recounts that the settlement he visited in the second century had been relocated from a site nearby.[37] It is a statement that has caused more frustration than necessary and the events surrounding the birth of the new city are speculative at best.

One possibility is that the Archaic/Classical town was located much further to the west in the valley in the direction of the modern village of Lavka, although no convincing candidate has been identified. What seems more likely is that Pausanias was mistaken in his account, or that he was recording a local history that had gradually changed over centuries—perhaps transforming the act of re-foundation into migration.[38] Whatever the case, archaeological evidence from different parts of the site—though relatively scarce for these

---

35 Williams (1983–1985).

36 Williams and Gourley (2005) 219 note 10. The city's strategic location on one of the primary North-South route from the Corinthian Gulf to the central Peloponnese would have necessitated its importance in such a defensive system. The similarity of the defences here with those at Mantinea and Messene also supports the hypothesis. For the fortifications of cities founded under Theban influence see: Cooper (2000) 155–191.

37 Paus. 8.22.1.

38 On the historiography of 'accuracy' in Pausanias' descriptions of local landscapes, see Hutton (2005) 20–29, 122–125; Pretzler (2007) 12–13, 40–41.

periods—suggests that the current town was rebuilt on a site that had been in use in previous centuries.[39]

The question of the chronological timeframe for the construction of the new town is also difficult. If one draws the comparison with Mantinea alone it is reasonable to see the two built as part of the same strategic initiative. In this case we could expect a date following 365, in which case the new city would have been under construction in Aineias' day. Of interest are the few ancient sources beyond that of Pausanias, which though brief, do yield a little information. In particular Strabo records an unsuccessful siege at Stymphalos by Iphikrates.[40] Xenophon also records that Iphikrates was campaigning in the region in 392 and again in 371/69,[41] but does not relate the details of these. Strabo states that Iphikrates attempted to flood the city by blocking the *katabothros* (the natural drainage sinkhole on the southern side of the lake) with sponges, which would indicate that the Stymphalos he was besieging was in its present location or somewhere nearby. It also means that his target was defended by walls that could be threatened by a rise in the lake level; i.e. of mudbrick construction on the valley floor.[42] It is feasible that if Strabo's story is accurate, Iphikrates may have besieged the walls of the earlier city, although it is equally possible that his attack was directed towards the newly constructed fortifications.[43]

The above illustrates how foggy the events surrounding the fourth-century building/ foundation are. It is possible only to speak in general terms about the construction, but it is clear that it took place in the first part of the fourth century and that the material remains at the site date almost exclusively to this period, but also that the fourth-century construction took place at the site of a pre-existing settlement. Equally problematic are the few ancient sources,

39 In deposits in the lower town black figure pottery of the fifth century has been recovered from isolated sondages, as have possible Bronze Age ceramics. See Williams et al. (2002) 141. On the acropolis.

40 Strab. 8.8.4.

41 Xen. *Hell* 4.4.16 and 6.5.49–51 respectively.

42 In the event this attempt was unsuccessful, but the formula is a familiar one and the parallel between this strategy and the Spartan flooding of the walls of Mantinea in 385 has been noted elsewhere: Williams and Gourley (2005) 219.

43 In recent years at least one third of the site is submerged annually lake during winter months, and the proximity of the spring-fed creek to the east wall could easily inundate the city if left unmanaged. In fact, it may be that this issue of rising lake was the reason the site appears to have been abandoned by the first century AD. Without organized maintenance of the sinkhole the lake quickly grows.

FIGURE 13.3 *Oblique elevated view of valley with area of ancient site delineated with dotted line*

although if Strabo is to be taken seriously, the city was likely founded at a time of instability, if not as the result of a direct destruction. Perhaps ironically, it is for this time frame that the connection with Aineias becomes most vivid, simply in the fact that it reflects the world about which he was writing. And yet the *direct* material connection with this world in terms of the correspondence of archaeology with historic event, remains frustratingly faint.

The Ancient Site and the Fourth-Century Walls

Orientation (*Figure 3*)

Approximately two thirds of the town is situated on gently sloping terrain to the north of the lake. The area was terraced in its original construction in order to provide level ground for the orthogonal grid that expands over most of the urban area to the north of the acropolis.[44] Most of the public buildings and

44 See T. Boy in Williams (1984) 183–185, for discussion of the town plan.

monuments identified thus far are located either on the acropolis or in the small rectangular space between the southern defensive wall and the acropolis to its north. The area is now heavily waterlogged and covered by reeds and has not been investigated in great detail, but it seems that it functioned as the agora for the original city. To the west end of this space lies a fountain house which collects water from the lowest of the major springs in the valley. A number of other minor springs issue from the southern acropolis within the city to the east of the fountain house. This system of springs was intentionally incorporated in the urban layout of the town and the line of the southern fortifications appears to have been designed specifically to enclose and protect these assets.

The fortification rings the entire urban area in a gently curving circuit of overlapping sections of curtain wall, totalling approximately 2.5 km in length. It is possible to trace this almost in its entirety with the exception of a short section to the north which has been obscured by later medieval or early modern terracing (figure 4: site plan). It is feasible that much of the building material from the northern part of the town—both fortifications and buildings within—was incorporated in the construction of the nearby abandoned village of Kionia and possibly the Cistercian abbey to the north of the site. In total the wall is defended by forty-five regularly spaced towers and seven gates all situated between the overlapping sections of wall. At least three of the towers appear to have been modified at a later date in the third century as a result of the changing demands of weaponry in this period (A, B and C on plan).

The Fortifications

So what exactly *can* we say about the Stymphalian walls in the context of this fourth-century defensive environment? It is clear from the style of defences that they were designed in a time of rapid advances in siege warfare, particularly with the use of early catapult technology in defensive positions in mind. The close spacing of towers, the form of the overlap gates, and the size of the tower chambers all point to a transitional period where close range fighting was still the norm and the heavy machinery of the late fourth century had not yet influenced wall design.[45] We are somewhat restricted in detailed interrogation by the construction method in mudbrick, long since collapsed into the surrounding fields. What remains are the foundations of towers, gates and walls

45 The similarity with the walls of Mantinea has already been noted. For further detailed discussion of similarities with Mantinea see Williams and Gourley (2005) 225, and, more generally, see Lawrence (1979) 334; Williams (1983) 201–202; Williams (1984) 175–176. For the characteristics of early artillery towers and the towers of Messene see Ober (1987) 570–604 and (1992) 147–169.

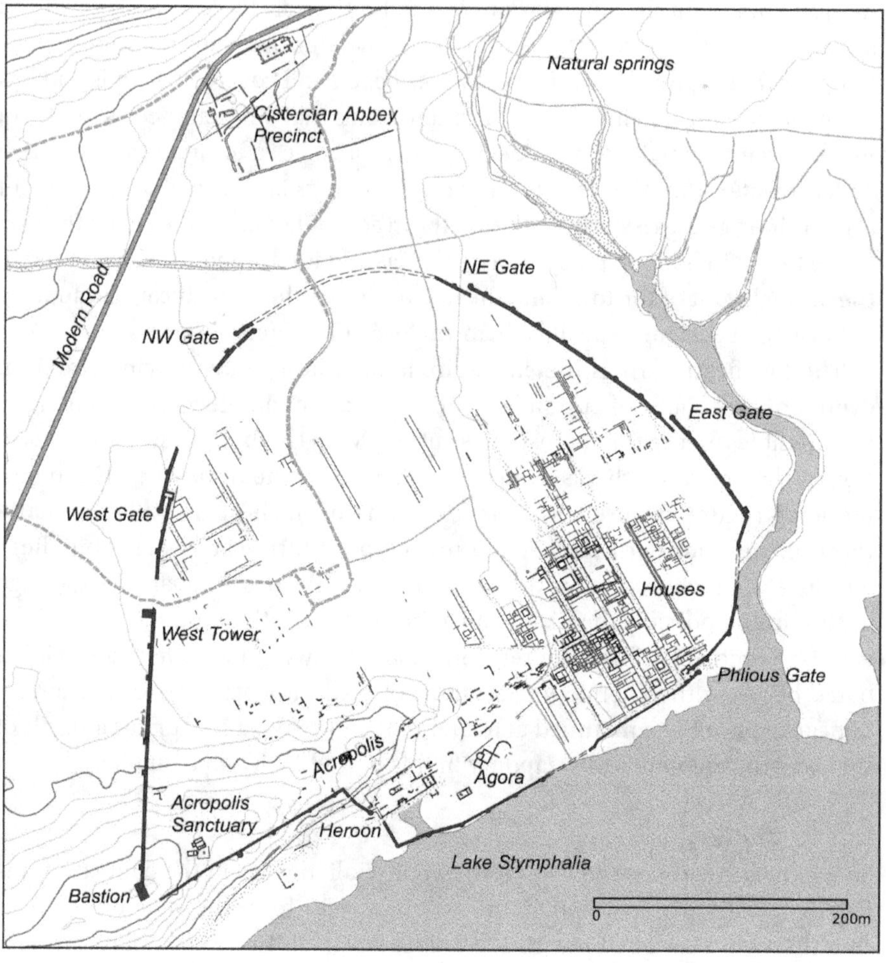

FIGURE 13.4 *Site plan*

from which we can only extrapolate points of general defensive planning, and the approximate form of elements such as curtain, towers and gates. Scope for getting at the 'interesting bits'—the wall walks and tower platforms, for example—is limited, and one must look to comparanda from stone-built sites such as Messene or Aigosthena for many of these details.

Considerations of the Site

Something of the defensible nature of the landscape has been explained above, but in the context of the site itself its influence is immediately clear. The city is nestled in a naturally enclosed space between the lake and Mt. Kyllene to the north and ringed by marshy ground and river on the east. The inclusion of the acropolis to the west leaves only the west wall exposed in any significant way and even that was buffered from the rest of the open ground in the valley by a ridge of high ground. The southern and eastern defences stand entirely on the lakebed just north of the modern lakeshore. They are comprised of a series of overlapping sections of curtain wall with a gate located at the intersection of each overlapping stretch. Regular semicircular towers break the line of the curtain at 36m intervals (see eastern wall on site plan and aerial photograph figure 5). The western wall and the acropolis defences follow a less regular course and enclose the high ground on the acropolis in a triangular configuration with a major tower at its apex. Here too we find regular towers but in the case of the west wall they are rectangular in plan, though spaced at the same interval as those on the east wall. The southern wall of the acropolis follows the cliff edge above the lakebed and is equipped with much smaller towers.

At 2.5km the wall encloses a relatively small area when compared to contemporary fortification at Messene or Mantinea, and while it is tempting to ascribe a strategic approach to restricting the defendable area to the bare minimum here, it is more likely a simple response to the size and population of the town.[46] Notwithstanding, it is notable that the curving route of the wall bears little relation to the arrangement of the urban layout within beyond the positioning of gates on the major thoroughfares, and it is in this (and the curving form) the strongest parallel with Mantinea lies.

The inclusion of the acropolis marks an obvious departure from the Mantinean comparison and the triangular arrangement of walls and major tower at the apex hints at other contemporary sites in the region. The unstudied site of Alea 5km to the southwest provides a remarkable parallel of uncanny similarity. Here too the main fortification system encloses the better part of the site on the valley floor, but includes a small triangular defended area on a nearby acropolis.[47] At neither site do the acropolis defences enclose all of the high ground, but instead appear to be built with the intention of restricting the defensive area to an easily manageable regular zone. At Alea a very well preserved cross-wall closes the triangular area midway up the acropolis and is

46 Winter (1971) 111–114. In observing general trends concerning the size of defensive circuits from the Dark Age through Hellenistic times, Winter notes that the strained financial state of many mainland cities, and the general need for strengthened defenses in the late fourth century and Hellenistic periods, led many fortified sites to restrict the area enclosed within the walls.

47 Lawrence (1979) 396–397.

FIGURE 13.5 *Towers on east wall, aerial view*

equipped with its own gate giving access to the city proper.[48] At Symphalos there is not enough evidence to argue conclusively the case for such a wall but occasional cuttings and natural topographic features make this a strong possibility. Indeed deposits from the sanctuary at the acropolis tower have yielded substantial numismatic evidence from the third century which has been cited as evidence of a garrison presence at that time and it and it is reasonable to expect that this could have been the case through much of the city's existence.[49]

The triangular plan of the acropolis defences is likely to have been shaped as much by a need to maintain lines of visual communication as by the necessity of defensive strength. The artillery tower that now stands at the apex of the acropolis dates to the third century; built to house the heavy artillery of that period, evidence suggests that it replaced a similar tower dating to the fourth century.[50] From such a position it is possible to maintain visual contact both with other sites in the valley and with the entirety of the fortification circuit that rings the town. One assumes that it must have played a principal role in both internal and external signalling and command of personnel.

Entry and Exit

The location of city gates was clearly an important factor in maintaining the balance between daily civic activity and defence. Movement of people in and out of the town was a necessity of daily life in time of peace, and the attention Aineias devotes to means of defending gates shows the importance of the gate in overall defense.[51] The conflicting relationship between strength of defence and practical movement is evident in the range of gate plans that one encounters in Greek fortifications in the period and the level of sophistication of gate arrangements provide a good measure of the defensive capabilities of the defenders.[52]

---

48 Lawrence felt that the fortification of Alea was perhaps undertaken in the late fourth century in order to house a garrison of either Demetrios Poliorketes or Kassander, but inspection of the site in recent years has shown that the tower at the apex belongs to a later phase of the walls, and the original circuit likely dates to the early or mid fourth century.

49 Weir (2002).

50 Williams and Gourley (2005) 249–250. For a discussion of Hellenistic artillery towers in the region see also Winter (1997) 247–292.

51 Ain. Tac. 28.6–7, 33.3–4 by way of example, on defense of gates from fires etc.

52 For a survey of gate technology in Greek fortifications see Adam (1986) 5–46.

At Stymphalos the placement of gates on the level terrain of the valley seems to have been determined more by the interior arrangements and external communication routes than by defensive strategy. Two major thoroughfares dissected the city linking the Northwest Gate to Northeast and West Gate to the East Gate have been traced inside the town and it is likely that there are more on the same axis. This configuration divides the regularly spaced insulae of the orthogonal plan into at least thee main blocks.[53] On the opposite axis the Phlious Gate in the southeastern corner of the city seems to have fed into a north-south avenue and an inner ring road located on the inside of the wall (evident from resistivity results on the South Wall: figure 6). This arrangement results in a wall with a gate every 250 to 300 m around the full length of the defenses, with the exception of the acropolis where gates are positioned less regularly. In such a configuration the strength of the gates relies on the defensibility of the structures themselves, or on those defending them, and little strategic attention to natural defensive features. The correspondence between gates and the position of arterial internal roads suggests that the openings in the walls were planned using the internal layout as the starting point and therefore the length of the overlapping sections reflected these spatial decisions.

As at Mantinea, the majority of the Stymphalian gates are of the overlap type.[54] The best surviving of these is the recently excavated Phlious Gate located on the southern wall. The plans of several other gates are known through geophysical and surface survey and individual details of their construction discernable in more or less detail in each case. The basic form of the overlap gate is generally considered to be a precursor to the frontal gate used in many Hellenistic fortifications though at many sites gates of both type were in use at the same time. The likelihood is that the choice of gate type was influenced by a wide of factors, ranging from technological capabilities, population size, ease of traffic flow, terrain etc.[55]

---

53 See T. Boyd in Williams (1984) 185 for a discussion of the orthogonal grid and the position of main thoroughfares.

54 Fougeres (1898). There has been no systematic study of Mantinea since the French work in the nineteenth century. Plans of gates are reproduced in various publications including Marsden (1969) and Adam (1989). Their accuracy is based on the original French publications.

55 Winter (1971) 215–224. In his comprehensive study Winter assigns gates to categories: Overlap gates (which he calls Type 2) and frontal gates (Type 1). The advantage of the overlap type is the protection provided against rams, mines, and close range weaponry which could not be brought into direct contact with the gate without significant risk to the

FIGURE 13.6 *Resistivity Phlious gate with survey of cropmark features*

In plan, the overlap is comprised of two overlapping sections of curtain wall with a more or less simple corridor between them inside which sits the gate. The length of the gate corridor varies at Stymphalos, though on the East Gate and West Gates surviving traces of internal curtain suggests a length of over 20 m. The width of the corridor also varies, but in most cases is approximately 6 m between the two walls, and narrows to as little as 3 m at the gate threshold itself. Most of the gates at Stymphalos are equipped with a gate court within the overlap corridor. This is achieved through the uses of an outer and an inner entrance, and an excavation of the Phlious gate has demonstrated that the gates themselves were mounted in this inner opening. The shape of the gate court is

attackers. The recessing of the gate at the back of the corridor allowed for the defenders to control the traffic and accordingly gates of this type were preferred on level terrain where the funneling of traffic was necessary.

not consistent from one gate to the next, but the concept remains the same and the shape and size of the gate court may have been determined by the extent of traffic flow. The final element in the typical overlap form is the position of towers at the gate opening. In all the gates at Stymphalos a round tower caps the external leaf of the wall and a corresponding tower guards the entrance to the overlap on the inner curtain. This arrangement effectively conceals the gate from the view of those outside the wall but also provided extra strength to the wall at the gate opening. Such gates provided effective means of defence against typical threats of the day, and also provided a means of careful control of traffic. Of particular advantage in the overlap gate was the concealing of the gate itself behind a defensive 'breakwater' of the outer projecting section of curtain. That, and the fact that the gate opening was aligned at a perpendicular angle to the line of the wall, means that rams and light infantry could not gain easy access to the front of the gate. It has been suggested that the overlap gate was in most common use on fourth-century defences, and as defensive practices changed to respond to the greater threats of the late fourth and third centuries, that it was gradually abandoned in favour of frontal gate which facilitated greater traffic flow.[56]

Excavation of the Phlious gate in 1999 has shown that the basic plan form of the overlap was flexible enough to be modified with changing trends in defense. This gate seems to have been adapted in the third century to incorporate a large circular gate court in the corridor.[57] It has been noted for its similarity with the Arkadian gate at Messene,[58] but the two share a very different defensive premise, the former essentially an elaborate overlap corridor, and the latter a 'Type 1' courtyard gate in Winter's classification.[59] It is possible the West Gate at Stymphalos (known only through resistivity survey shown in figure 7) was also modified at a later date. In this gate the basic overlap is enhanced by the inclusion of a large rectangular court which takes up the width of the gate corridor and the thickness of the internal wall. Such a court suggests a higher level of traffic though this gate, but also a more sophisticated approach to controlling this traffic than in simpler gates elsewhere at the site. Indeed

56 Winter (1971) 222–223. Winter suggests that the advantages of overlap plans were outweighed by practical concerns of traffic after the Hellenistic period. In the following centuries Type I frontal openings protected by gatecourts were favoured over the more awkward lateral entrances, and the Type II gate seems to have been used only for entrances of secondary importance.

57 Williams et al. (2002); Williams and Gourley (2005) 233.

58 Winter (1971) 217; Lawrence (1979) 334; Adam (1992) 39–40.

59 Despite the distinction it is highly likely that both gates are products of the third cen-

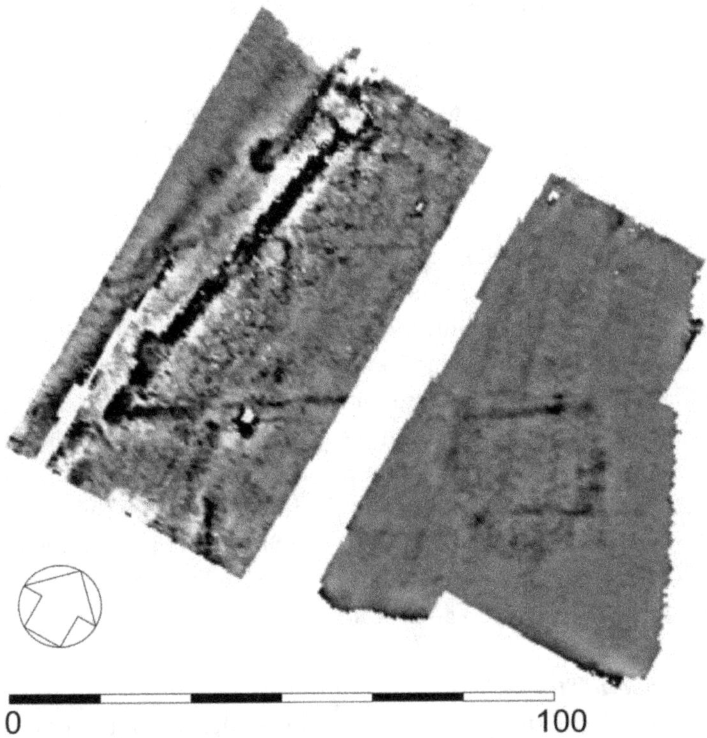

FIGURE 13.7 *West gate resistivity plot showing wall and gate on left of image and internal city divisions to the right. Scale is in metres.*

it may be that in the third century some of the lesser gates were blocked up, thus reducing the access points to the town, and those that were left open were strengthened and adapted for higher traffic levels.

That the walls were substantially overhauled in the third century is made clear from the construction of a series of artillery towers whose excavated deposits show a construction in the early part of this century (discussed below). It might be expected then that the older gates of the original fourth-century defenses would be modified at this time also, but to what extent such activity took place is unclear. Detailed investigation of the gates at Stymphalos has been hampered by high water levels which has prevented testing of original

tury and not the fourth and in the case of the Phlious gate there is direct evidence for modification This suggestion contradicts the common belief that both are of an early date.

construction deposits. Modifications and changes to gates have been suggested therefore only on the basis of deviations in their plan from the standard examples found in other parts of the site.

Evidence from the Pheneos Gate to the south of the acropolis on the west of the agora seems to indicate a later modification of a different sort. It is one of two gates that were not built in the overlap style (the other is at the top of a ramp which runs up the south side of the acropolis) and was much simpler in its construction, comprising a simple opening in the wall (see figure 8). At some point in its history this gate seems to have been converted from a defensive structure into a more monumental entrance into the agora area, embellished with columns inclosing a porch and with benches in the gate passage itself.[60] Stratigraphic evidence suggests a late date for this modification (late second, early first century BC), though it may be that the gate always held some kind of special significance given its position in relation to the agora. This seems all the more feasible since its defensive strength arises from its position on the circuit directly below the acropolis defences. Approaches to the opening are funnelled between the shore of the modern lake and the acropolis and thus the acropolis defences served a function similar to an overlap gate.

Construction and Strength

When compared with the better preserved stone examples of contemporary fortification circuits, the surviving remains of the Stymphalian circuit might seem rather unimpressive. It is apparent however, that no matter how aggressive the passage of time has been on the fabric of the walls themselves, the surviving evidence betrays a close parallel in mudbrick to stone contemporaries and there is no reason to expect technological differences between the two.[61] The curtain at Stymphalos was bedded on a shallow stone foundation built in varying masonry styles and in varying heights in different parts of the circuit. The walls themselves were built using sun dried brick as a facing with an earthen core. Their height can only be extrapolated from the width of the foundation and though it is at best a coarse approximation it bears up in structural principle. For the majority of the fortified circuit on the valley floor the width of foundation is between 3.5 and 4 m. On the south wall of the acropolis this is reduced to 2.5 m. Between the foundation and the earthen wall, a levelling course of broken roof tile seems to have acted as a bonding interface between the two. The total height of the wall would likely have been in the range of 4 to 8 m depending on location and situation. Excavated deposits on the inte-

---

60 Williams and Gourley (2005) 237.

61 For a well-studied contemporary site from northern Greece—New Halos—where the

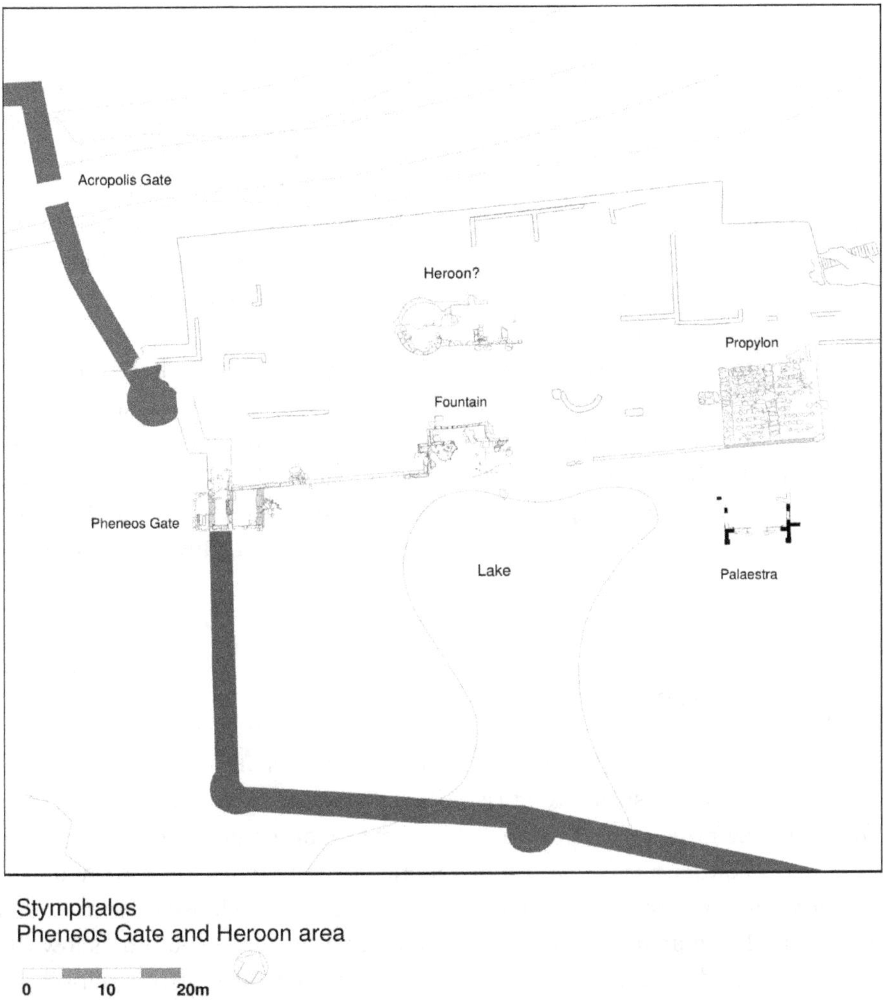

FIGURE 13.8 *Location of Pheneos Gate with respect to internal hero shrine and fountain house*

rior of the wall on the acropolis yielded large quantities of roof tile, some in more or less organised deposits suggesting that the wall was equipped with a covered wall walk as attested from epigraphical and archaeological evidence elsewhere.[62]

mudbrick fortifications have been conjecturally reconstructed on the basis of stone examples, see Reinders (1989) 40.

62 The construction of brick walls is recounted in occasional inscriptional evidence such as the survival from Athens of IG.II.2.463 which outlines the requirements for the renovation

Earthen construction is not unusual in fortifications and evidence from excavation shows the bricks were produced locally. The bricks themselves are formed from the clay-rich alluvial soils of the lakebed, whereas the mortar used to bond the bricks seems to have been produced much closer to source at the time of assembly. Examples from the acropolis tower, rebuilt in the third century, show that bricks were of regular dimension (c. 45×10 cm). Here the bricks have been brought up from the valley floor, and mortar mixed on the acropolis from characteristic terra-rossa soils found overlying the limestone of the acropolis hill. It is highly likely that the finished wall was externally rendered and possibly painted in order to protect the core from weather. The construction of a defensive system using locally available building materials that were easily produced would no doubt have been a major advantage in such defences. It facilitated rapid repairs and easy modification. Easy repair of damaged or breeched sections of wall is possible when one's raw materials are close at hand and though there are no direct examples of such action from Stymphalos, the reality of such repairs can be realised locally, within the city (itself largely built in mudbrick).

Manning the Walls

Recent studies have demonstrated a gradual progression in tower design and placement from Classical through late Hellenistic times: the ancient equivalent of a creeping arms race as new modes of attack and defence came into play in practical situations.[63] The characteristic development of the early fourth century was the use of the small tension catapult in defensive positions on walls—and principally from towers—and fortifications of the time show the effects of the dissemination of this technology. Defences of this period are usually equipped with towers of c. 24 m square, of either rectangular or semicircular plan. Standing examples from very well preserved sites such as Messene and Aigosthena employ one or more chambers above the level of the wall walk to elevate weaponry even higher than the line of the wall. In some cases, as in tower N at Messene, provision was made for a fighting platform on the room of a single chamber tower. The internal chambers were large enough to hold a

of the brick wall at Athens between the years 307–304. For discussion and translation of this inscription see Caskey (1910); Holland (1950); Winter (1959).

63 For a detailed study of developments in tower design with the introduction of so-called first generation catapults in the years after 400 see Ober (1992). On pre-artillery towers see Marsden (1969) 138; Winter (1971) 153–161; Lawrence (1969) 376–382.

range of weapons, and in most instances were spaced at frequent regular intervals around the curtain.[64]

The fourth-century towers at Stymphalos are typical of the type and likely very similar to stone examples (see figure 9: reconstruction). In plan they are semicircular, circular or rectangular and were positioned at approximately 30 m intervals around the entire circuit (the measurement does not include the width of the tower). The close spacing provided ample enfilading cover from archers or slingers on the wall from tower to tower and thus provided an active barrier from reaching the wall, provided that the wall could be defended by enough men. The close spacing also had the advantage of strengthening the line of the wall so that if sections were compromised then the towers would continue standing even if sections of curtain had completely collapsed or were being repaired.[65]

In their foundations they appear as simple projections 2.5 m in length from the face of the curtain, but it is likely that as in stone examples the upper chambers comprised the entire width of the curtain including the projection. The 4 m width of the foundations also suggests that the wall walk/gallery at the top of the curtain was comfortably wide. At Messene the wall is substantially thinner and the width of the wall walk only enough to comfortably accommodate two men passing. Though at Stymphalos, one should allow for a sloped wall face, we might reconstruct a fighting platform of 3 m deep on at the top of the wall. The obvious question of how tower chambers and wall walk were accessed internally for the soldiers manning the wall has never been successfully answered at the site. It is notable that no foundations of stair platforms have been identified apart from one of unknown location featured in a photograph in Orlandos's publication.[66] It seems most reasonable to reconstruct a system of ladders for internal access though as in the case of the tower chambers we rely entirely on other surviving examples for such reconstruction.

---

64 The towers at Stymphalos belong to Ober's 'First Generation' which were designed specifically for the use of small scale catapult technology housed within tower chambers. In stone examples such towers most commonly only preserve chamber walls of 0.60 to 0.75 m in thickness, and were distinct in size and purpose from towers of the Second Generation (characterised by thicker chamber walls over 1.00 m thick, amongst other things).

65 On the spacing of towers, it is worth noting that the average curtain lengths between towers at Mantinea is 33 m (Reinders 1989, 81), at nearby Kleitor 35 to 30 m (Winter 1989, 198), and at Stymphalos 32 to 29 m.

66 Orlandos (1926) 133.

FIGURE 13.9 *Reconstruction of west wall tower of 4th century date*

Towers of differing form in different parts of a city are not uncommon in defences of all periods,[67] and at Stymphalos the west wall and east/ south-east wall feature different tower types. The reasons behind this are not clear, although it has been suggested that they differ in constructional date.[68] It is notable that a whole section of wall is defended only by rectangular towers, while the majority of the circuit has semicircular examples. The tower size is comparable and the likelihood is that the difference lies in the nature of the terrain being defended and the suitability of one type of tower for different terrains. For example on the acropolis the semi-circular towers are markedly smaller than in the rest of the site and the curtain wall is only half the width of corresponding sections on the valley floor. There is no doubt that the height advantage that these towers held, perched above the lake plain, influenced their design and their use. It would have proven impossible to defend the south acropolis wall with artillery, since there was only room for personnel with small-scale weaponry (average tower dimensions here 3×3m square).

67 Winter (1971) 192–193.

68 Williams and Gourley (2005) 241.

These different 'zones' marked by the use of differing tower types are interesting to consider in light of the defensive demands on different parts of the city. No doubt certain areas were considered more secure than others, while some sections might have been considered quite vulnerable. The west wall, with its exclusive use of rectangular towers, is the most exposed and has little in the way of natural defences and therefore it makes sense that this tower type and the 4.5 m width of the curtain was a response to these terrain deficiencies. On the other hand, the semi-circular towers are used only on curving sections of curtain and on low ground near the natural defences of the marshy ground. It would make sense to associate their plan with the demands of defending a curving curtain on low ground then. The characterization of defensive zones becomes much more apparent in subsequent centuries, when the inadequacies of the outdated fourth-century defenses were improved in the face of demands of third-century warfare.[69]

Later Developments

Viewed in the context of the later fourth century advances in siege warfare the fortifications of Stymphalos could be considered to rely on relatively simple provisions. A strong and cohesive circuit capable of easy communications from key locations, either from tower to tower or by means of signalling stations, could successfully hold an enemy without heavy artillery from the walls for a time. The ability of gates to control access in an out of the town was no doubt the weakest point in the walls and their strength or strategic configuration of the utmost importance. Considering the events that followed the construction of such defences in the later fourth century, it is clear how short-lived the technology of the day could be. Aspects of the daily routine, such as monitoring activity through openings in the wall, no doubt remained the same and the recommendations found in the text of Aineias probably changed little, but the threat was ever greater.

How should or could such systems adapt? The Stymphalian response to the developments of the late fourth century took the form of strengthening various positions around the walls, but beyond a few specific structures, the fourth-century characteristics of the fortification have changed very little. The most conspicuous of the later developments is the construction of three (and possibly four) artillery towers on the acropolis (one hexagonal and two rectangular) and the west wall, in which existing smaller towers were entirely rebuilt on a

69 See Williams and Gourley (2005) 243–251 for the upgrading of west wall and acropolis defenses.

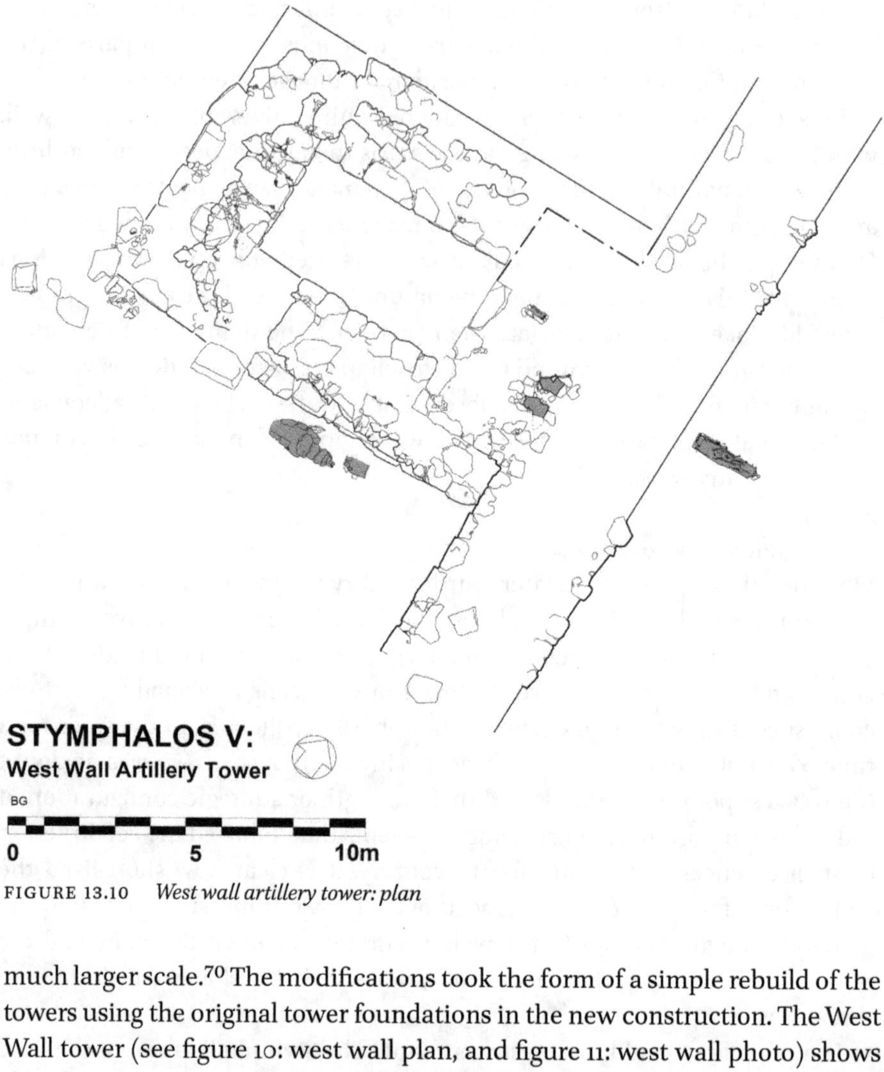

FIGURE 13.10 *West wall artillery tower: plan*

much larger scale.[70] The modifications took the form of a simple rebuild of the towers using the original tower foundations in the new construction. The West Wall tower (see figure 10: west wall plan, and figure 11: west wall photo) shows the clearest evidence for this with a small rectangular tower incorporated into the later tower foundations. Coins from the construction debris date this renovation to the early 3rd century and it is likely that this was part of a specific program of upgrades to the fortifications at this time.[71] The new towers have been interpreted as emplacements for large-scale defensive stone throwing

70 For detailed discussion of the context and construction of these towers see Williams et al. (2002) 155–160 and figs. 6, 7; Williams and Gourley (2005) 243–252 and figs. 12, 13, 14.

71 A small hoard of bronze and silver coinage from the fourth and third centuries (latest

FIGURE 13.11 *West wall aerial photo showing large artillery tower surrounding smaller rectangular tower*

machines which came into use in the third century.[72] Their position on the wall is of interest since it seems to suggest a western focus for such defences and no similar towers were built on the eastern and southern sides of the city. They are spaced at roughly the same intervals of the gates (c. 250–300 m), and the towers chosen for modification were clearly of strategic significance on the original circuit.

On the acropolis south wall a hexagonal tower was constructed directly to the south of the Acropolis Sanctuary (see figure 12),[73] while on the top of the

---

datable to Demetrios Poliorketes) provided the most accurate construction date for the larger artillery towers at Stymphalos. See Williams and Gourley (2005) 250.

72 All of these towers are large enough to have housed large-scale torsion catapults in differing configurations. See Winter (1997) 247–292 for a study of regional artillery towers of the Hellenistic period, and for more general studies see McNicoll (1986) 305–309 and Winter (1989) 191.

73 The hexagonal tower is of interest for the relative infrequent use of this plan type. It seems that it was exclusive to the age of artillery. See Marsden (1969) 147–148; Winter (1971) 194–196; Lawrence (1979) 387. Unlike circular and rectangular towers which existed in the

FIGURE 13.12 *Aerial view of south acropolis wall with hexagonal tower and nearby sanctuary on acropolis terrace*

hill a 21 × 11 m tower was built at the apex of the triangular defence. On the west wall, the West tower was built on a slight rise in the limestone topography, and further up the wall and earthwork on the inner face of the wall may represent a fourth tower (but this is not certain). The absence of any changes to towers on the eastern defences may indicate that the natural defence of the marshy ground that extends far to the east of the wall itself (thanks to the series of springs to the north) was enough to ensure any enemy siege machines could be kept at bay.

The suggestion that at least two of the original gates were adapted in the third century has already been touched upon above. Given the changes deemed necessary to modernise towers, it stands to reason that the gates should also be

pre-artillery age, the hexagonal design seems to have been developed specifically to meet the demand for maximum range of movement of the artillery pieces housed within the tower chambers.

renovated, although to test this more systematically we need more evidence from excavation. It is in discussing the later changes to the site that the very short chronological timeframe of fortified defence at the site is made apparent. Evidence from the acropolis tower and the west tower seem to indicate that these large towers went out of use by the early to mid-second century although apparently not as a result of military activity, but because of general abandonment.[74] Deposits from three sites around the walls would also seem to indicate the fortification circuit in general went out of use at approximately the same time. The picture is different for the city itself. The period between the mid-second and the mid-first centuries seems to have been one of prosperity[75] and excavation of the early Roman houses in the lower town shows that the city continued to thrive. It seems that in the years after the destruction of Corinth, either Stymphalos could not, or did not need to, maintain its defences in any sort of organized way.[76]

Conclusion

In historical terms the fortifications of Stymphalos were relatively short-lived. Archaeological evidence points to a working life of no more than 250 years. They were built in a time of rapid changes from fifth-century to fourth-century defensive warfare and display many of the most fundamental aspects of defence of the time. Nonetheless, in less than one or two generations the defences needed considerable modification, and although this was done only in specific areas the situation is indicative of the progression of technologies of offence and defence. Still, it seems incongruous that in the text of Aineias there is a marked absence of discussion of these issues. One of the challenges in using the archaeology of fortifications as a vehicle for contextualizing Aineias' world is defining the division between hopeful interpretation and the hard fact of material remains. No doubt, to the population of fourth-century Stympha-

74 Ceramics from both towers indicates nothing later than the mid-second century BC, with the majority of the assemblage dating to the late third and early second century. Thanks to Peter Stone for his work on this.

75 Weir (2002).

76 It has been suggested that Stymphalos suffered a similar fate to Corinth at the hands of Mummius in 146. See Williams and Gourley (2005) 257. Whether or not such a violent act was forced upon the city, the extensive fortifications which began their life in the fourth century, acted as no more than a boundary wall enclosing the city of the late second/early first centuries.

los, the newly built fortifications were a simple necessity, part of the basic civic infrastructure, and only that. You either had them or you did not: If they were not defendable due to being outdated or weak in some way then the town was in danger or as good as defeated. Of course it is impossible to know how the occupants of Stymphalos viewed their walls. One suspects from the fact that Aineias puts very little effort into specific prescriptions for the form of fortifications, that defence was more about how you used them than their physical form.

Bibliography

Adam, J.P. (1982) *L'architecture militaire grecque*. Picard, Paris.

Adam, J.P. (1986) 'Approche et defense des portes dans le monde hellenisé', in Leriche, P. and Treziny, H. (eds) *La fortification dans l'histoire du monde grec*. Editions du Centre national de la recherche scientifique, Paris. 5–43.

Bon, A. (1969) *La Morée franque*. E. de Boccard, Paris.

Bakhuizen, S.C. (1986) 'La grande batterie de Goritsa et l' artillerie defensive', in Leriche, P. and Treziny, H. (eds) *La fortification dans l'histoire du monde grec*. Editions du Centre national de la recherche scientifique, Paris. 315–321.

Campbell, S. (1997) 'The Cistercian Monastery of Zaraka', *Échos du monde classique/Classical Views* n.s. 16: 177–196.

Caskey, L.D. (1910) 'The Roofed Gallery on the Walls of Athens', *American Journal of Archaeology* 14: 298–309.

Cooper, F.A. (2000) 'The fortifications of Epaminondas and the Rise of the Monumental Greek City', in Tracy, J.D. (ed.) *City Walls: the Urban Enceinte in Global Perspective*. Cambridge University Press, Cambridge. 155–191.

Fossey, J., van de Maele, S. (eds) (1992) *Fortificationes Antiquae*. J.C. Gieben, Amsterdam.

Fougères, G. (1898) *Mantinée et l'Arcadie orientale*. Fontemoing, Paris.

Fraser, J. (1898) *Pausanias' Description of Greece*, vol. IV. Macmillan, London.

Habicht, C. (1998) *Pausanias' Guide to Ancient Greece*. University of California Press, Berkeley.

Holland, L.B. (1950) 'The Katastegasma of the Walls of Athens', *American Journal of Archaeology* 54: 337–356.

Hope Simpson, R. and Dickenson, O. (1979) *A Gazetteer of Aegean Civilization in the Bronze Age*. Åström, Goteborg.

Howell, R. (1970) 'A survey of Eastern Arcadia in Prehistory', *The Annual of the British School at Athens* 65: 97–98.

Hutton, W. (2005) *Describing Greece: Landscape and Literature in the Periegesis of Pausanias*. Cambridge University Press, Cambridge.

Garlan, Y. (1974) *Recherches de poliocétique grecque*. Bibliothèques de l'Ecole française d'Athènes et de Rome—Série Athènes 223, Paris.

Knauss, J. (1990) 'Der Graben des Herakles im Becken von Pheneos', *Mitteilungen des Deutschen Archaologischen Instituts (Athen)* 105: 1–52.

Lawrence, A.W. (1979) *Greek Aims in Fortification*. Oxford University Press, Oxford.

Leriche, P., Treziny, H. (eds) (1986) *La fortification dans l'histoire du monde grec*. Editions du Centre national de la recherche scientifique, Paris.

Lolos, Y. (2007) *Land of Sikyon* (*Hesperia* Supplement 39), American School of Classical Studies at Athens, Princeton.

McNicoll, A. and Milner, N.P. (1997) *Hellenistic Fortifications from the Aegean to the Euphrates*. Oxford University Press, Oxford.

McNicoll, A. (1986) 'Developments in Techniques of Siegecraft and Fortification in the Greek World', in Leriche, P. and Treziny, H. (eds) *La fortification dans l'histoire du monde grec*. Editions du Centre national de la recherche scientifique, Paris. 305–424.

Marsden, E.W. (1969) *Greek and Roman Artillery*. Oxford University Press, Oxford.

Munn, M. (1993) *The Defence of Attica*. The University of California Press, Berkely.

Nielsen, T.H. (1999) 'The Concept of Arkadia: the People, their Land, and their Organisation', in T.H. Nielsen, T.H. and Roy, J. (eds), *Defining Ancient Arkadia* (Acts of the Copenhagen Polis Centre, 6; Historisk-filosofiske meddelelser, 78). Kongelige Danske Videnskabernes Selskab, Copenhagen. 16–79.

Orlandos, A. (1926) 'Anaskaphai en Stymphalo', *Praktika*: 131–138.

Ober, J. (1987) 'Early Artillery Towers: Messenia, Boiotia, Attica, Megarid', *American Journal of Archaeology* 91: 570–604.

Ober, J. (1992) 'Towards a Typology of Greek Artillery Towers: the first and second generations (c. 375–275 BC)', in Fossey, J. and van de Maele, S. (eds) *Fortificationes Antiquae*. Brill, Leiden. 147–169.

Panagopoulos, B. (1979) *Cistercian and Mendicant Monasteries in Medieval Greece*. University of Chicago Press, Chicago.

Pretzler, M. (2007) *Pausanias: Travel Writing in Ancient Greece*. Duckworth, London.

Scranton, R.L. (1941) *Greek Walls*. Harvard University Press, Cambridge, MA.

Reinders, R. (1989) *New Halos: a Hellenistic Town in Thessalia*. Brill, Leiden.

Tracy, J.D. (2000) *City Walls: The Urban Enceinte in Global Perspective*. Cambridge University Press, Cambridge.

Whitehead, D. (1990) *Aineias the Tactician, How to Survive under Siege. Translated with Introduction and Commentary*. Clarendon Ancient History Series, Oxford.

Weir, R. (2002) 'The story of an Arcadian Polis: Coins from the Excavations at Stymphalos', *The Labyrinth (online journal, Classics Department University of Waterloo)* No. 79, at http://artsonline.uwaterloo.ca/labyrinth/

Williams, H. (1983) 'Stymphalos, A Planned City of Arcadia', *Échos du monde classique/Classical Views* n.s. 2: 194–205.

Williams, H. (1984) 'Investigations at Ancient Stymphalos', *Échos du monde classique/ Classical Views* n.s. 3: 174–186.

Williams, H. (1985) 'Investigations at Ancient Stymphalos', *Échos du monde classique/ Classical Views* n/s. 4: 194–205.

Williams, H. and Cronkite, S. (1995) 'Excavations at Stymphalos, 1994', *Échos du monde classique/Classical Views* n.s. 14: 1–22.

Williams, H., Schaus, G., Cronkite, S. (1996) 'Excavations at Stymphalos, 1995', *Échos du monde classique/Classical Views* n.s. 15: 75–98.

Williams, H. et al. (1997) 'Excavations at Stymphalos, 1996', *Échos du monde classique/ Classical Views* n.s. 16: 23–73.

Williams, H. et al. (1998) 'Excavations at Ancient Stymphalos, 1997', *Échos du monde classique/Classical Views* n.s. 17: 261–319.

Williams, H. et al. (2002) 'Excavations at Ancient Stymphalos, 1999–2002', *Mouseion* 2: 135–187.

Williams, H. and Gourley, B. (2005) 'The Fortifications of Ancient Stymphalos', *Mouseion* 5: 213–259

Winter, F.E. (1959) 'Ikria and Katastegasma in the Walls of Athens', *Phoenix* 13: 161–200.

Winter, F.E. (1972) *Greek Fortifications*. University of Toronto Press, Toronto.

Winter, F.E. (1986) 'A summary of recent work on Greek fortifications', in Leriche, P. and Treziny, H. (eds), *La fortification dans l'histoire du monde grec*. Editions du Centre national de la recherche scientifique, Paris. 23–29.

Winter, F.E. (1997) 'The Use of Artillery in Fourth Century and Hellenistic Towers,' *Échos du monde classique/Classical Views* 16: 247–292.

CHAPTER 14

Technology in Aineias Tacticus: Simple and Complex

*Tracey E. Rihll**

Introduction

Aineias the Tactician has little to say in his *Poliorketika* about specific siege technologies such as ramps, mantlets, and siege towers, and makes only passing reference to a number of other military technologies, including one of the earliest mentions of the catapult. Occupying a much larger and more significant place in his text are civilian technologies that can be pressed into military service. In the course of giving his advice on how to survive under siege, Aineias refers to two- and four-wheeled vehicles, a variety of types of arms and armour, baskets, cautery, chains, chests and other containers, the crane, curtains and drapes, felt, fortifications and defences of various types, incendiaries, ink, ladders, lanterns and lighting, leather, locks and bolts, nets, papyrus, roads, rushes, sewing, sponges, sulphur, tattoos, thatched roofs, tin, transport and distribution, vinegar, water clocks, wax tablets, and wound dressings. His text thus offers us a special opportunity to examine the material culture of his place and time, Greece c. 360 BC, and in the process challenges our assumptions about the quantity and quality of technology in ancient Greece, and about what constitutes 'military technology' in particular. Improvisation transforms a pitchfork into a weapon, as a modern IED (improvised explosive device) transforms fertilizer or mothballs into one.

* I would like to thank Maria Pretzler for organizing the excellent conference at which the ideas in this paper first saw the light of day, and RIAH for financial support. I am also grateful to the participants at Gregynog in September 2010 and those at an EMAET seminar in Athens in May 2011, where this paper received its second scrutiny, for their useful comments. J.V. Tucker tackled it in written form and offered his usual useful advice.

 | DOI: 10.1163/9789004352858_015

Methodology

Anachronism on technological matters in ancient history is a largely unrecognised problem. We are often reconstructing the past from written sources alone, but words can disguise huge technical changes because technologies are replaced faster and more often than the nomenclature by which they are known. For example—and this point is best made with modern, clear and obvious examples—goods are now *ship*ped by train or lorry; we watch digital *films*; and we now *write* without scratching or marking a physical surface. A future historian reliant on written text might suppose that an invoice from 2010 for shipping was still evidence of transport by water, that audio-visual entertainment was still delivered on a thin celluloid ribbon, and that writing was still done with a small stick of one variety or another rather than by touching a keyboard of one variety or another. So the existence and occurrence of a specific word for a given technology should not be assumed to represent the same thing whenever and wherever it occurs: each case must be examined individually.

As historians we need to be alert to the facts that (i) technologies are functional and (ii) they usually change over time, reflecting changing needs and applications in the society that produces and uses them. Technologies changed dramatically over the course of antiquity, just as society changed dramatically during that time. Many really fundamental technologies are not known to predate the Greeks. Some of them probably go back to the Stone Age, but not all.

Another preparatory point must be made. It seems to be human nature to ignore the ordinary in favour of the exotic. However, technology-in-use is what makes life what it is for most of the people most of the time.[1] Exotic technologies have little or no impact, at least as long as they remain exotic.[2] Bizarrely, once a technology is commonplace and taken for granted, it is not usually perceived as 'technology' anymore; in contemporary life the term is used to refer to 'complex systems of machines and techniques'.[3] Yet somebody invented the chair, for example—cavemen did not sit on chairs—and probably many different people invented that particular technology in many different varieties in many different places and times. By ignoring technologies-in-use we ancient historians are overlooking a rich source of information: much social and economic history hides in the everyday technologies that the ancients made and used, as I hope to demonstrate below.

1 Lienhard (2000) 55–69; Cuomo (2007) 1.

2 Edgerton (2006) 1–27; Lienhard (2006).

3 Nye (2006) 15.

It also seems to be natural to assume single linear development of technologies and a succession of one item over another along that path, rather than to contemplate parallel paths and separate but interlocking weaves. There are usually several ways by which a given task can be achieved, and very few technologies have no alternative. Which technology is selected by a particular user on a particular occasion may be the result of a compromise between factors, or the overriding importance in that case of one factor over all others. A systematic comparison of such technologies reveals their strengths and weaknesses relative to each other, and indicates the contexts in which each is likely to be selected for use in preference to the other. Studying the technological choices on any given occasion can therefore be used as a historical tool better to understand the people who made the decision to use one rather than its alternatives in antiquity.

With these methodological points in mind, I shall begin my analysis of Aineias' *Poliorketika* with a review of the technologies Aineias experienced in daily life. I shall then turn to analyse in detail specific examples from each end of the complexity spectrum: ladders, which Aineias refers to twelve times (in comparison to the catapult's once), at the simple end, and his telegraph system, at the complex end. His text offers us a special opportunity to examine the material culture of his place and time, Greece c. 360 BC, and in the process challenges our assumptions about technology in ancient Greek society.

Technologies in an Ordinary Greek Town in the Fourth Century BC

Aineias refers explicitly to a range of technologies. Besides a variety of types of arms and armour, fortifications and defences of different sorts, and missiles, incendiaries and wound dressings, all of which are to be expected given the nature of his treatise, he offers a snapshot of the material culture he knew and expected to be available in any town in his times. There are technologies to do with communication, construction, consumption, entertainment, health, production, and transport.

Two- and four-wheeled vehicles are used as troop carriers where there are roads (16.14). Two-wheeled carts are upended and tied together to serve as a mobile siege shed (37.9). The special quality of two-wheeled carts is their ease of tipping, in contrast to a four wheeled cart which is more stable and can carry a larger, heavier load, but is slower, and in particular cannot be tipped. Therefore if the load can be dumped in a heap—sand, for example—then a two-wheeled cart is preferential as it is quicker and easier to unload. For military application, two carts are set in the tipping position to make a mobile triangular shed very quickly, in which the shafts form the apex and would be covered with material, ideally something fire-resistant like felt, rushes or rawhide, whose fire-retardant properties are emphasised (33.1–3). (This might require the slaughter and skinning of two or more bovines if no raw hide was available at the time.) The fighters would take shelter beneath this cart-shed and propel it on its own wheels.

Artificial lighting is an important technology. Lighting is what enables people to live other than according to the rhythms of natural time, governed by daylight. In the late nineteenth century, lighting was the main use of electricity; at the start of the second World War a third of British homes still had no electricity, and in 1955, although nearly 90% of houses had by then been connected to the electricity supply, only 58% had power sockets, averaging three per house. The other homes that were connected were still wired up *only* for lighting, and any appliances they had were designed to be plugged into a light socket when needed.[4] (By 1960, over 50% of British homes still had no electric kettle, fridge, washing machine or TV. What we take for granted is really *very* recent.) The bare flame lighting technologies of Aineias' world would have been familiar to most people in most places at most times in history.

Besides lamps, Aineias assumed that lanterns are available to all: to guards on the wall, to anyone walking the streets at night, to artisans and craftspeople working at night (22.21–24). Artisans and craftspeople working with lanterns are thus distinguished from domestic indoor environments where plain lamps would be used, presumably because artisans were working outdoors or under cover only, as is common for people working with fire at a hearth, oven, kiln or furnace (moreover, as Pliny pointed out *Natural History* 36.200–201, most manufacturing processes require fire at one or more stages of production). Lanterns are housings for lamps that stop the wind blowing out the flame, for use wherever this is likely. Translucent materials used for this purpose included parchment and horn, alternatively a ceramic lattice or bronze mesh could be used to stop gusts while letting light out, the last attested as early as Aineias' contemporary Theophrastos.[5] Lanterns could be designed to make their light shine in just one direction, or permanently or temporarily as necessary; in Aineias' case lanterns are covered to cast their light down to the ground only (26.3). This was a technology that developed through antiquity. By Roman imperial times we find Antioch and Byzantion at least had street lighting.[6] But this was not an issue of technological development as much as one of

4 See the displays in the Museum of Science and Industry (MOSI), Manchester, Electricity gallery.

5 Theophr. *Fire* 22. Surviving examples include BM 1856,1226.670 and 1996,0712.1.

6 Amm. 14.1.9. Bouras and Parani (2008).

political development. The problem of keeping a light burning outdoors had been definitively solved already by the lanterns of Aineias' time; it was the security problem which had not. Light is an asset in peaceful world. In a hostile one, it makes its bearer an easy target or a potential traitor.

Security was also the driver for lock technology, on which significant intellectual effort had been expended by Aineias' time, both by those trying to secure gates and by those trying to open them (18–20). Domestic doors had been lockable since Homer's time if not before (e.g. *Odyssey* book 22), and Aineias' focus is on explaining how a smooth cylindrical bolt to keep gates shut, which was presumably designed to act as a tenon securing the crossbar from being slid sideways (and which was ordinarily kept, along with its associated bolt-remover, in the possession of the general, when not actually in use in the gate), can be and in the past had been surreptitiously removed to throw open the gates to the enemy. The independent removable bolt and bolt-catcher, which were both inserted into holes in the wood, were in domestic contexts replaced by a design in which the locking bolt was integral or fixed to the door, and a more or less complex bolt-lifter (key) was the only removable part. The story of the Gordian knot preserves an alternative technology for the same purpose: securing by tying knots. This was a deterrent with an obvious shortcoming, as shown by Alexander, but we should remember that any lock, especially a padlock, is only a deterrent too. Some fibres used in antiquity were very tough, and it was no mean feat to cut through a thick knot of such type.

The mass production of baskets was not suspicious (29.11–12). Baskets were strong and light containers that barely survive in the archaeology but were used in those conditions in which they had advantages over ceramic, and are still used today in a variety of applications from baby beds (Moses baskets) to laundry baskets, cheese formers to fruit baskets. The world Aineias knew consumed baskets in quantity, including as shields and helmets when necessary. It did not, however, assume a dock; unspecified technologies for beaching and relaunching ships feature instead (10.12). Ladders are one such, and we will now undertake detailed analysis of this to show that any technology, even the simplest, is just as prone to anachronism as, say, concepts, morals, and practices are.

The Ladder

The ladder is a very simple technology. Its purpose is to enable people safely to get up or down a vertical surface. Such surfaces are by and large man-made, consisting of the built environment and ships, where a ladder enabled those needing to fix the roof or rigging to climb up.[7] Ladders can be firm or soft (rope). Firm ladders are required if the aim is to ascend; if and only if it can be suspended from above before use, a soft ladder is an alternative. In antiquity firm ladders were incorporated into a variety of more complex military technologies, such as the stock of a palintone catapult[8] and a boiling-oil or water bucket-chain device.[9] The name (κλῖμαξ) clearly derives from the firm version, which would suggest that it preceded the soft ladder in ancient Greece.

The ladder's chief functional competitor is the plank.[10] A comparison reveals their respective strengths and weaknesses.

The plank (ἀποβάθρα, pons)	**The ladder (κλῖμαξ, scala)**
Easy to make	Much harder to make
Continuous	Discontinuous
Relatively heavy	Much lighter
Easy to use	Need to be self-aware when using it (cf. steps)
Must be used at low incline	Can be used at steep incline
Many others uses (e.g. as see-saw, lever, barrier, float, floorboard etc.)	Limited alternative uses
Hands-free use	Need/temptation to hold on if it is set at a steep incline

Continuity trumps all other aspects if the problem is to embark an animal onto a ship: thus animals board via a gangplank while the humans use a ladder in

7 Rope ladders on a ship of c. 50 AD are depicted in Casson (1971) fig. 151.

8 Ktesibios/Heron *Belopoiika* 100.5–101.1.

9 Apollod. *Pol.* 182–184.

10 Another alternative is a mast or tree trunk with footholds (cleats) attached in the manner of a telephone pole, as depicted in a C3 AD relief of a Roman boat illustrated in Casson (1965) plate V 1; I am indebted to Mark Humphries for this reference. Another alternative apparently existed in Egypt; this was a wooden beam cut into a zig-zag shape on one face, and flat on the reverse. These have been found in a variety of sizes in various tombs, and probably belonged to the model ships found in the same (Creasman 2010). When set at a low angle, the zig-zags become steps to embark ship. It thus forms a narrow flight of steps.

Euripides *Helen* (1555–1572). The ladder also appears as the means to embark ship in Euripides *Iphigeneia at Taurus* and *Rhesus*, on a black-figure vase from the mid-fifth century, and on a red-figure vase from the late fifth or early fourth century, for example.[11]

Ladders before Aineias

Ladders existed a long time before Aineias; the earliest evidence known to me is a depiction in an Egyptian fifth dynasty tomb, dating to the 2400s BC.[12] It would appear to be an up-ended sledge,[13] and was here being employed apparently in a siege situation. Ladders are also depicted in an Assyrian relief now in the British Museum of siege of a town by Tiglath-Pileser III in c. 730 BC; this ladder is long and is shown crossing a ditch as well as scaling the wall. The earliest ladders in Greek material are mentioned in a domestic context, for gaining access to the upper floor and the roof, in Homer's *Odyssey* (1.330, 10.558, 11.63, 21.5) but they do not appear at all in the poem about a siege, to wit the *Iliad*. Homer's Achaians did not even entertain the idea of trying to take Troy by escalade.

If siege was not common in classical Greek warfare, the typical ancient Greek fighter would have known very little about siege techniques or about the defence of a city under siege. That the Athenians were reputedly good at siege warfare suggests that knowledge of it was uncommon. For the average Greek, if not for most Athenians most of the time too, escalade or any defence against it would be unfamiliar technologies. Common experience shows generally that unfamiliar technologies usually involve a struggle, as we fumble and fail in one way or another until we gain the requisite familiarity to use new technologies properly. That is why we have training, not only in vocational skills such as brick-laying (where the objects being manipulated are about as simple as a ladder) but also in the use of supposedly 'user-friendly' devices whose very name indicates the existence of its opposite. Therefore we should allow for a certain amount of ineptness in the application of any technology that we have reason to believe is new in the society in which we are interested, or new to the users in the case in question.

In Thucydides 3.20–22 the Plataians and Athenians besieged with them made ladders to scale the Peloponnesians' circumvallation walls. Thucydides says (3.20.3) that "they made the ladders equal to the enemies' wall."[14]

---

11 Eur. *Iph.Taur.* 1351; *Rh.* 73; Beazley 214678, 217518 respectively.

12 Clarke and Englebach (1990) fig. 83; thanks to my colleague Troy Sagrillo for this reference.

13 Clarke and Englebach (1990) fig. 85.

14 Thuc. 3.20.3 "κλίμακας ἐποιήσαντο ἴσας τῷ τείχει τῶν πολεμίων".

This could mean the ladders were as long as the wall was high, in which case the ladders would have been too short to reach the top when leant against them at an angle (unless the wall was so low it would not matter). Alternatively, it could mean they were equal to the height of the wall when they were leant against it. The fuss Thucydides records about seeing and counting courses of brick to get the height estimate correct suggests that the latter interpretation is the right one.[15]

Brasidas put a ladder against the walls of Poteidaia but was spotted before mounting it and ran away.[16] Escalade this was not, apparently, even in intention. An attempt at a proper escalade may be indicated by Thucydides 5.56.5, where he tells us that the Argives took ladders to Epidauros but failed to take the city and withdrew (419/8 BC). Long ladders for scaling walls appear in Aischylos' *Seven against Thebes*,[17] Euripides' *Phoenissai*,[18] and in Xenophon's *Kyroupaideia*.[19] Ladders were used successfully in what was ultimately (and for other reasons) an unsuccessful attack on Phleious in c. 369 BC.[20] Of approximately the same date, ordinary ladders are depicted on a Paestan bell-krater and an Athenian hydria.[21]

The Nereid monument (see photo 1),[22] carved c. 390–80 BC, shows ladders in use in an escalade. Note the incline of the ladder. This draws our attention to another area of potential misinterpretation of ancient technologies. Because ancient ladders looks like modern ladders, we naturally assume that they were used in the same way that we use them today. But the angle at which a ladder is set against a wall is not self-evident, nor intrinsic to the ladder. Modern health and safety advice is to set a ladder quite close to the vertical, at an angle of 75 degrees. We happen to have an ancient analogue to this advice which makes a striking contrast with it: Polybios recommends a ladder be set half its length away from the wall.[23] This gives an angle between the ground and the ladder of

---

15 There is a summary version of this event in Polyain. *Strat.* 6.19.3.

16 423/2 BC, Thuc. 4.135.1.

17 Aesch. *Seven.* 466; the play was produced in 476 BC.

18 Eur. *Phoen.* 1173 and elsewhere, of c. 410 BC.

19 Xen. *Cyr.* 5.3.12 and 16, probably written in the 370s BC.

20 Xen. *Hell.* 7.2.5–9.

21 BM GR 1865, 0103.27 and BM GR 1856, 1001.16 respectively.

22 BM GR 1880, 0130.7.

23 Polyb. 9.19.7: "To suit the convenience of those mounting it, the ladder must be planted half the ladder's length from the wall. If they are planted further away they are liable to break when crowded, and if planted closer to the perpendicular they are very insecure for the people using them" ("τὴν δ' ἀπόβασιν τῆς κλίμακος πρὸς τὴν τῶν ἀναβαινόντων

PHOTO 14.1 *Ladder depicted on the Nereid monument, Xantos 390–380 BC (British Museum)*
PHOTOGRAPH M. PRETZLER

60 degrees. Ladders at this sort of incline would need either to rest on the top of the wall, as the embarkation ladder rests on the gunwale of Beazley 217518, or the feet of the stiles would need to have been strongly fixed to the ground so that they could not slide backwards causing the top of the ladder to slide down the wall. At an angle of c. 60 degrees they would also need to be very short or very strong not to bend and break in use, as noticed by Polybios in the same place. This must have militated against their use to surmount high walls; short ladders would not reach the top, and a ladder with long and sufficiently strong wooden stiles to span the gap would weigh a lot, making deploying it especially challenging.

On the positive side, ladders used at this sort of angle would be very difficult to push away backwards; both gravity and the distance required would work against the defenders on the wall. In the Nereid monument depiction, the ladder appears to be secured by two men kneeling under it; their pose suggests

συμμετρίαν ἡμίσειαν εἶναι δεήσει τῆς κλίμακος, ἵνα μήτε πλεῖον ἀφιστάμεναι διὰ τὸ πλῆθος τῶν ἐπιβαινόντων εὐσύντριπτοι γίνωνται μήτε πάλιν ὀρθότεραι προσερειδόμεναι λίαν ἀκροσφαλεῖς ὦσι τοῖς προσβαίνουσιν.").

This advice is coming from a man who has experienced military command and who claims expert knowledge, and is here lecturing his reader on the correct use of ladders. Such digressions to provide explicit useful knowledge were intentional in his *History*, 10.47.12–13. Escalade might seem to be such an obvious method of attacking a walled city that we might expect it to be used automatically and wonder that people needed instruction in how to do it. However, the fact that Polybios does spell it out suggests otherwise, and a story in Polyainos 7.22 suggests that the Thracians were completely unfamiliar with ladders at the time of the stratagem involved.

that they were pulling down on ropes attached to the top of the stiles, presumably to try to prevent the defenders shifting it. Alternatively, but I think less likely, they may have been pulling ropes that extended the ladder, like a modern extendable ladder, which technology is known from literary sources of a slightly later period.[24]

To summarise the findings so far, ladders before Aineias' time look like our ladders, and serve some of the purposes of our ladders. Given that people are about the same size, standard ladder rungs were probably about as far apart as they are now (one foot or c. 30 centimetres). But ladders were apparently erected at a lower incline than we are advised to. Standard ladders would give access to standard heights. An attack mounted with standard ladders reveals an assumption by the attackers (right or wrong) that the defenders' city walls were approximately the height of a house, or of a floor within it: thus in Lysias 1.9 the wife with newborn temporarily moved downstairs because it was dangerous for her to go up and down the ladder with the baby. (This is not the place to discuss the archaeological evidence, such as it is, for stairs in classical Greek houses; I merely flag up here the disproportionate amount of assumption and hypothesis to evidence that one who looks will find in the very limited literature on this topic. The so-called Nekromanteion in Epirus has the best evidence, in my opinion, for a wooden staircase, and it consists of two notches in the long top step of three stone steps leading apparently straight into a wall. The terminus ante quem is 167 BC.) A ladder also appears as a means to go upstairs on a red-figure vase from the late fifth century (Beazley 215618; I know of no ancient Greek image of a staircase), and they served similarly even if the accommodation was, very unusually, underground.[25] If ladders are said to have been custom-made, that indicates they had an unusual height to scale, or they were required in unusual numbers. Let us then turn to Aineias' text.

Ladders in Aineias

Ladders appear in chapters 11, 36 and 38. In 11.3 they are used in a ruse: ladders were set out for access to the roofs of various public buildings which were alleged to be in need of maintenance, but the intention was for them to be used militarily by political conspirators at a later time.

In chapter 36 ladders are used by the enemy for an assault on the city walls; a proper escalade. From 22.29 and 40.1 we know that the walls should be suffi-

24 Apol. Mech. *Poliork.* 175.

25 Xen. *Anab.* 4.5.25. Archaeological evidence for the domestic wooden staircase (as distinct from the ladder, stone steps, and ramps) appears to begin in the Hellenistic period, but that is an argument for another time and place.

ciently high to dissuade people from trying to jump down on either side; that is obviously a vague and variable quantity, but we should be thinking about the perception of a man who was expecting to have to fight for his life as soon as he had landed, and was probably wearing armour and carrying a shield, so the 10–12 foot (c. 3–4 metre) range seems reasonable. Aineias' advice is that if a ladder protrudes above the wall, the defenders should push it away with a pitchfork; if they cannot see or move the ladder, they should push away the man mounting it instead. Clearly if the man mounting the ladder is in pitchfork range then the defender is in *his* spear range. As an alternative to this civilian technology (pitchforks) mobilized for military use (to push away scaling ladders or invaders climbing them), Aineias introduces the famous 'door-like' board made out of wood (οἷον θύραν ἐκ σανίδων, 36.2). There are various reconstructions of this device, e.g. Waschow's, Barends', and Garlan's[26] (see figure 1). Bettalli[27] recognized the chief problem with these reconstructions—they bear scant resemblance to the text—but did not attempt one of his own.

Aineias gives no dimensions, but this wooden structure has to fit the requirements that

(i) it can be slid out easily by a defender[28] and will slide in again just as easily on its own when the ladder leans against it, and
(ii) it apparently needs no visible means of support.[29]

To my mind these conditions rule out any vertical motion, and clearly imply that at least half of its weight remains on the *parodos* (the wall walk), else this board would tumble down from the top of the wall. It is apparent from 32.5 that Aineias envisages the top of the walls to be sufficiently wide and flat that the defenders could park wagon-sized stones there. Standing walls of the period, for example those at Messene or Herakleia on Latmos, are wide and flat-topped and give a clear indication of what he may have had in mind.

My knowledge of Greek technology of the day in general also inclines me to a simple solution (which is not the same thing as a simplistic solution). On my reading, Aineias is talking about a board that is 'like a door' in all

26 Waschow (1938); Barends (1955); Garlan (1974).

27 Bettalli (1990) 326–327, commentary on Ain. Tac. 36.2.

28 This is an alternative to wielding a pitchfork and therefore one should expect it to be operable by a single individual.

29 All elaborate reconstructions assume that something has dropped out of the text here, notably a description of the posited support structures, especially if they involve the door-like board hanging vertically and being moved manually.

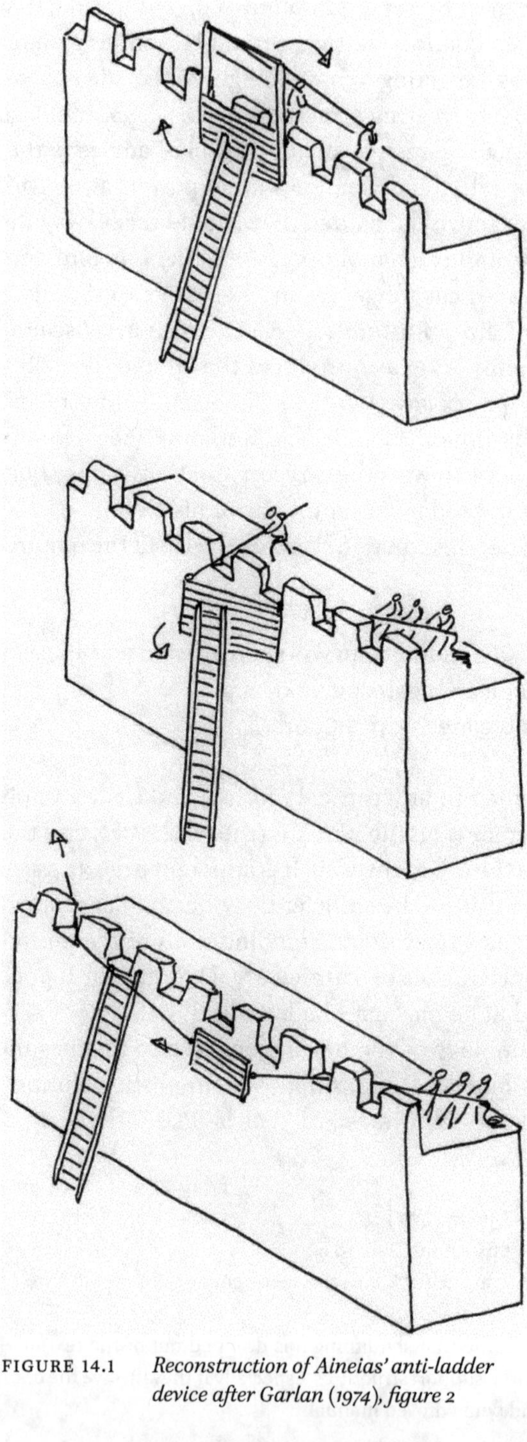

FIGURE 14.1 *Reconstruction of Aineias' anti-ladder device after Garlan (1974), figure 2*

PHOTO 14.2 *Remains of the walls of Herakleia on Latmos*
PHOTOGRAPH T.E. RIHLL

FIGURE 14.2 *Reconstruction of Aineias anti-ladder device*
T.E. RIHLL

respects, and probably *was* a door before the enemy arrived at the walls, and could be a door again after they have left (which would render this another civilian technology pressed into military service, like the pitchfork for which it serves as an alternative). It is laid down flat on the flat top of the wall, with a roller or two (a broom handle or suchlike would do nicely) underneath (see figure 2). Philon later identified threshing sledges (*tribola*) as useful for dislodging ladders placed against the wall,[30] and a threshing sledge is a sort of sturdy door with built-in runners. Aineias says that when the ladder is brought forward (ἔπειτ̓ ἐπὰν προσφέρηται ἡ κλῖμαξ) the door is slid forward under the ladder (προϋποτιθέναι τῇ κλίμακι τῇ προσφερομένῃ) to (discreetly) protrude up to half its length in front of the wall. The ladder would probably be carried vertical to the wall for obvious ergonomic and practical reasons (and as invariably depicted in Medieval siege scenes) and then lowered onto it, so if the assailants' ladders were long enough for the job they would necessarily be higher than the wall on the approach, and any anti-ladder technology on the wall would necessarily be 'under' the top of the ladder. This interpretation does not require great dexterity on the defenders' part, or unlikely compliance on the part of the assailants (*pace* Garlan). I think Hunter and Handford understood Aineias' door-like anti-ladder device in this uncomplicated way too.

30 Philon 1.79. The MSS have triboloi, which may be a variant or mistake for tribola; the context makes absolutely clear Philon meant threshing sledge, οἷς ἀλοῶσι.

Thus interpreted, Aineias' device was much more subtle and smart than are the reconstructions envisaged by Waschow, Barends and Garlan. Thus interpreted, it was intended to mislead the attackers into thinking that they had erected their ladder properly, against the wall, when actually they had not. In my visualisation of Aineias' text, as men started climbing and the ladder started to flex and press against the 'wall'—actually, the top of a door, projecting perhaps two and a half feet in front of the real wall—it would have slid back automatically (a feature specified by Aineias but ignored in other reconstructions) and, since the ladder would have been erected at a significant angle, the top of its stiles would have come to rest not at the top of the wall, where the attackers thought they had put it, but some point below it—possibly at the worst possible place, just too short to allow the climbers to get over the top of the wall, leaving them at the mercy of the defenders. It is also possible that in the act of sliding back, the 'door' could destabilize the ladder, so that the climbers fell off, and the longer the ladder was, the greater would be the danger. A century or so later Philon thought 20 cubits (c. 30 feet) was the maximum possible height a ladder could reach, so recommended building walls higher than that.[31]

We do not need to exaggerate the issues for anti-scaling-ladder technologies to work. According to the UK Health and Safety Executive in May 2010, in contemporary Britain there are on average fourteen deaths and 1,200 injuries per annum caused by accidents with ladders (2011's cache even included someone supposedly demonstrating ladder safety). This is here and now, in a world with a surfeit of 'how to' information, a lively health and safety industry, a compensation-driven legal culture, and most importantly peace, when other people are not deliberately trying to cause injury to those climbing.[32]

As a bonus for the defence, in a post-debacle analysis the offence may have concluded that their ladders were too short, and consequently lost faith in their methods or their engineers. Polybios famously castigated Philip V for this mistake, if such it was, and in due course cited it again to introduce a short lecture on the importance of numeracy for a military commander.[33] An alternative explanation for that event would be that the Melitaians and other defenders mentioned by Polybios[34] had read and understood Aineias Tacticus' text, and it is rather testimony to the success of his door-on-the-wall defence. In

31 Philon *Pol.* 1.12.

32 Accidents happened in antiquity too of course; for example, Trygaios fell off his ladder to Heaven and sustained a head wound: Aristoph. *Peace* 69.

33 Polyb. 5.97.5–6; 9. 18–20.

34 He says this 'mistake' was made frequently, Polyb. 5.98.6.

any event, ancient Greek reticence about escalade[35] could and probably should be seen as another example of the Greeks' uncommon good sense. It is a very hazardous business.

In 38.7 the ladders are made of rush-rope, suspended from above by the defenders to let some of their colleagues climb down to retrieve ammunition and then climb back up.[36] The first technology suggested for this purpose was nets designed to hold large quadrupeds (boar, deer) which would obviously work at least as well and probably better than the rope ladders suggested as an alternative because nets were larger and therefore easier to climb.[37] Soft ladders and nets were tactically shrewd for this application, because they were being used on the enemy's side of the wall, and if they had fallen into the wrong hands, firm ladders would have been useful whereas these soft ladders would not. Aineias' advice is both astute and sophisticated.

Ladders after Aineias

Just because a technology has existed before and been used before, it cannot be assumed that it is on the technology shelf for all those who come after. Technologies and techniques, like knowledge and skills in general, can be and frequently are lost. We have seen that Aineias' door-like anti-ladder device may have been still working effectively in the second century BC, so we look ahead very briefly to discover how ladders were used after Aineias.

In Polyainos, writing in the second century AD, the use of ladders in sieges is overwhelmingly associated with stratagems and sneak attacks, as might be expected given his mission.[38] It usually involves a relatively small number of ladders being brought up to one part of the circuit and mounted by a small number of specially selected men, while the rest of the army hangs back or launches a conventional attack on another part of the circuit. No ladders are depicted in most of the scenes of siege or construction of forts on Trajan's column (spirals 2, 3, 5, 9, 10), but three are shown in spiral 17, which depicts the siege of the Dacian capital in the Second War. In this attack the Romans appear to be using ladders to scale a wall not much more than eight or nine feet high. The incline of the ladder as depicted there is still low by modern standards, but not as low as it had been in earlier Greek depictions, and in fact it fits Polybian recommendations of c. 60 degrees. It is slightly steeper

35 Polyb 5.98.7–9.

36 The baskets' function is probably to bring up the stones not the men.

37 Compare Ain. Tac. 11.6, where 'sails with their ropes outwards' also served to allow people to climb the walls.

38 E.g. Polyain. *Strat.* 4.6.18, 6.5, 7.6.2, 7.22, 8.11, 8.14.3.

again in the construction scene in the tomb of Trebius Justus, from the fourth century AD.[39] Despite the staircase's advantages over the ladder, particularly for someone carrying a load up or down it, as is the norm on a construction site, builders in Britain only began using staircases instead of ladders as part of the scaffolding in the present millennium.

We turn now from consideration of the oldest and simplest of technologies, ladders, to a potentially complex and apparently quite new technology, to the development of which Aineias himself contributed: telegraphy.

Fire Signals, πυρσοί

Fire signals are a simple form of telegraphy, of communicating at a distance. Obvious alternatives that were sometimes employed in antiquity are smoke signals and light signals. Fire signals use the light given by a flaming torch to send a message; smoke signals use the smoke from a controlled fire; light signals use a polished surface to reflect sunlight. Their functionality, range and ease of use vary considerably with the weather and the time of day: light is not an option at night but fire works best then, for example. When the Greeks built lookout towers they appear to have located them in direct sight of each other,[40] which would have allowed any of these alternative systems to be used.

Fire Signals before Aineias

A fire signal may be anything from a hand-held flaming torch to a huge raised beacon. Before Aineias fire signals were apparently used to send a single unequivocal message; in Homer 'Help!' and in Aeschylus' *Agamemnon* 'Success!' for example.[41] An alternative and commonly intended message, in a variety of times and places, has been 'Prepare!', usually for an anticipated invasion; the most famous example in British history was to announce the arrival of the Spanish Armada.[42] In the fifth century BC invasions usually involved known and familiar enemies, so the Athenians, for example, could probably guess quite accurately how many and what sort of Lakedaimonian fighters might show up on their border. The simplest fire signal was appropriate in those cir-

39 The tomb is on the Via Latina in Rome and is no longer visible, but there is a reproduction in Adam (1994) 83, fig. 181.

40 Various papers in Van de Maele and Fossey (1992); Woolliscroft (2001) for Roman examples.

41 Hom. *Il.* 18.207–214; Aesch. *Ag.* 281–316.

42 Thanks to J.V. Tucker for this example.

cumstances. Aineias lived in a less predictable world, since he was writing his text in the middle of the fourth century BC.[43] At this time an upstart northern monarch was fighting by other rules, and had a worrying habit of showing up unannounced, to take cities and regions by surprise. He also used some new tools and techniques: catapults, siege machinery, unfamiliar infantry units carrying unfamiliar weapons, and, potentially the most dangerous of all, a donkey carrying gold to persuade your fellow townsmen that submission was a better option than resistance.[44] By these means Philip II subjugated one area after another in the north, and during Aineias' lifetime was working his way down into central Greece.

Fire Signals in Aineias

Fire signals are not much discussed in the surviving *Poliorketika*, but Aineias dealt with them at length in some other book(s) of his on the art of war, perhaps in the then already-written but now lost *Paraskeuastikē*,[45] or perhaps in the then yet-to-written but now also lost *Stratopedeutikē*, as Polybios almost says when he apparently quotes Aineias on a new, improved technology of fire signalling.[46] Polybios is explicit and emphatic that this telegraphy technology had seen significant recent development (10.43.1).

For convenience I give the text here (Loeb trans.):

> (10.44.1) "Aineias, the author of the work on strategy, wishing to find a remedy for the difficulty [of sending information via fire signals] advanced matters a little, but his device still fell far short of our requirements, as can be seen from this description of it. (2) He says that those who are about

43 Whitehead (1990) 10 (and Lane Fox in this volume, 44–45) for the date.

44 Dem. 11.17; Plut. *Reg et Imp. Apophth.* 14 (*Mor.* 178B); Rihll (2007) 78–85.

45 So Hunter and Handford (1927) xiii and others on the basis of a comment in *Poliorketika* 7.4 that fire signals have already been discussed there so Aineias does not repeat it here.

46 Polyb. 10.44.1. Since the *Stratopedeutike* was not yet written when Aineias wrote the *Poliorketike* and it was planned to deal with outpost duties and signalling (21.2), it is possible that he repeated himself there, or even that he invented his new system between writing the *Paraskeuastike/Poliorketike* and the *Stratopedeutike*. Polybios is not a neutral reporter on this device, since he claims credit for an alternative technology, devised by Kleoxenos and Demokleitos, and allegedly perfected by Polybios himself, that he is about to go on to describe (10.45.6 ff.). Therefore we should take care not to dismiss Aineias' system too quickly in favour of Polybios' device, since that was precisely what he was trying to achieve through his choice and deployment of words in describing it.

to communicate urgent news to each other by fire-signal should procure two earthenware vessels of exactly the same width and depth, the depth being some three cubits and the width one. (3) Then they should have corks made a little narrower than the mouths of the vessels and through the middle of each cork should pass a rod graduated in equal sections of three fingerbreadths, each clearly marked off from the next. (4) In each section should be written the most evident and ordinary events that occur in war, (5) e.g. on the first[47] 'Cavalry arrived in the country', and the second 'Heavy infantry', (6) on the third 'Light-armed infantry', next 'Infantry and cavalry', next 'Ships', next 'Corn', and so on until we have entered in all the sections the chief contingencies of which, at the present time, there is a reasonable probability in war time. (7) Next he tells us to bore holes in both vessels of exactly the same size, so that they allow exactly the same escape. Then we are to fill the vessels with water and put on the corks with the rods in them and allow the water to flow through the two apertures. (8) When this is done it is evident that, the conditions being precisely similar, in proportion as the water escapes the two corks will sink and the rods will disappear into the vessels. (9) When by experiment it is seen that the rapidity of escape is in both cases exactly the same, the vessels are to be conveyed to the places in which both parties are to look after the signals deposited there. (10) Now whenever any of the contingencies written on the rods occurs he tells us to raise a torch and to wait until the corresponding party raise another. When both the torches are clearly visible the signaller is to lower his torch and at once allow the water to escape through the aperture. (11) Whenever, as the cork sinks, the contingency you wish to communicate reaches the mouth of the vessel he tells the signaller to raise his torch (12) and the receivers of the signal are to stop the aperture at once and to note which of the messages written on the rods is at the mouth of the vessel. (13) This will be the message delivered, if the apparatus works at the same pace in both cases."[48]

---

47 Hunter and Handford's visualisation (1927, 123) includes an 'armchair' misunderstanding of the technology they are trying to elucidate: by assuming the messages start at the *top* of the rod, they would have the signallers waiting the longest time to send the likeliest message. But since the message is read off when it is level with the rim of the vessel, the reverse order is required, with the likeliest message at the *bottom* of the rod, so we start there.

48 Polyb. 10.44. "Αἰνείας δὲ βουληθεὶς διορθώσασθαι τὴν τοιαύτην ἀπορίαν, ὁ τὰ περὶ τῶν Στρατηγικῶν ὑπομνήματα συντεταγμένος, βραχὺ μέν τι προεβίβασε, τοῦ γε μὴν δέοντος ἀκμὴν πάμπολυ τὸ κατὰ τὴν ἐπίνοιαν ἀπελείφθη. (2) γνοίη δ' ἄν τις ἐκ τούτων. φησὶ γὰρ δεῖν τοὺς μέλλοντας

Aineias' system is a refinement of the simple fire signal insofar as it could send a variety of different messages of the users' own devising and definition, that is, several of any messages the users might anticipate wanting to communicate. Aineias' telegraphic device could have been particularly useful for communication between allies in leagues composed of multiple poleis that had put local rivalries aside to create a larger resistance force and in which one party might be reputed good in its heavy infantry (e.g. Spartans), another in its light infantry (e.g. Akarnanians), another in its cavalry (e.g. Boiotians), another in its fleet (e.g. Corinthians), and so on. They could have prepared various estimates for the size of the invading forces for example, having messages saying 100 invaders, 300 invaders, and 1000 invaders (though Polybios did not apparently realise this when he wrote his critique). It also could have provided reassurance to those poleis which geography placed on the front line that appropriate help could be summoned quickly.[49] Anticipated messages could be sent much more quickly by telegraphic devices such as this than by any other means.

ἀλλήλοις διὰ τῶν πυρσῶν δηλοῦν τὸ κατεπεῖγον ἀγγεῖα κατασκευάσαι κεραμεᾶ, κατά τε τὸ πλάτος καὶ κατὰ τὸ βάθος ἰσομεγέθη πρὸς ἀκρίβειαν: εἶναι δὲ μάλιστα τὸ μὲν βάθος τριῶν πηχῶν, τὸ δὲ πλάτος πήχεος. (3) εἶτα παρασκευάσαι φελλοὺς βραχὺ κατὰ πλάτος ἐνδεεῖς τῶν στομάτων, ἐν δὲ τούτοις μέσοις ἐμπεπηγέναι βακτηρίας διῃρημένας εἰς ἴσα μέρη τριδάκτυλα, καθ' ἕκαστον δὲ μέρος εἶναι περιγραφὴν εὔσημον. (4) ἐν ἑκάστῳ δὲ μέρει γεγράφθαι τὰ προφανέστατα καὶ καθολικώτατα τῶν ἐν τοῖς πολεμικοῖς συμβαινόντων, οἷον εὐθέως ἐν τῷ πρώτῳ, διότι (5) 'πάρεισιν ἱππεῖς εἰς τὴν χώραν,' ἐν δὲ τῷ δευτέρῳ διότι 'πεζοὶ βαρεῖς,' ἐν δὲ τῷ τρίτῳ 'ψιλοί,' τούτων δ' ἑξῆς (6) 'πεζοὶ μεθ' ἱππέων,' εἶτα 'πλοῖα,' μετὰ δὲ ταῦτα 'σῖτος,' καὶ κατὰ τὸ συνεχὲς οὕτω, μέχρις ἂν ἐν πάσαις γραφῇ ταῖς χώραις τὰ μάλιστ' ἂν ἐκ τῶν εὐλόγων προνοίας τυγχάνοντα καὶ συμβαίνοντα κατὰ τοὺς ἐνεστῶτας καιροὺς ἐκ τῶν πολεμικῶν. (7) τούτων δὲ γενομένων ἀμφότερα κελεύει τρῆσαι τὰ ἀγγεῖα πρὸς ἀκρίβειαν, ὥστε τοὺς αὐλίσκους ἴσους εἶναι καὶ κατ' ἴσον ἀπορρεῖν: εἶτα πληρώσαντας ὕδατος ἐπιθεῖναι τοὺς φελλοὺς ἔχοντας τὰς βακτηρίας, κἄπειτα τοὺς αὐλίσκους ἀφεῖναι ῥεῖν ἅμα. (8) τούτου δὲ συμβαίνοντος δῆλον ὡς ἀνάγκη πάντων ἴσων καὶ ὁμοίων ὄντων, καθ' ὅσον ἂν ἀπορρέῃ τὸ ὑγρόν, κατὰ τοσοῦτον τοὺς φελλοὺς καταβαίνειν καὶ τὰς βακτηρίας κρύπτεσθαι κατὰ τῶν ἀγγείων. (9) ὅταν δὲ τὰ προειρημένα γένηται κατὰ τὸν χειρισμὸν ἰσοταχῆ καὶ σύμφωνα, τότε κομίσαντας ἐπὶ τοὺς τόπους, ἐν οἷς ἑκάτεροι μέλλουσι συντηρεῖν τὰς πυρσείας, ἑκάτερον θεῖναι τῶν ἀγγείων. (10) εἶτ' ἐπὰν ἐμπέσῃ τι τῶν ἐν τῇ βακτηρίᾳ γεγραμμένων, πυρσὸν ἆραι κελεύει, καὶ μένειν, ἕως ἂν ἀνταίρωσιν οἱ συντεταγμένοι: γενομένων δὲ φανερῶν ἀμφοτέρων ἅμα τῶν πυρσῶν καθελεῖν. εἶτ' εὐθέως ἀφεῖναι τοὺς αὐλίσκους ῥεῖν. (11) ὅταν δὲ καταβαίνοντος τοῦ φελλοῦ καὶ τῆς βακτηρίας ἔλθῃ τῶν γεγραμμένων ὃ βούλει δηλοῦν κατὰ τὸ χεῖλος τοῦ τεύχους, ἆραι κελεύει τὸν πυρσόν: (12) τοὺς δ' ἑτέρους ἐπιλαβεῖν εὐθέως τὸν αὐλίσκον, καὶ σκοπεῖν τί κατὰ τὸ χεῖλός ἐστι τῶν ἐν τῇ βακτηρίᾳ γεγραμμένων: (13) ἔσται δὲ τοῦτο τὸ δηλούμενον πάντων ἰσοταχῶς παρ' ἀμφοτέροις κινουμένων."

49 Stymphalos is such a front line polis for an invasion of the Peloponnese, albeit not on the main route in, and the Aineias of Stymphalos who was a general of the Arkadian league may well be our Aineias.

Aineias' key innovation appears to have been the provision of a *menu of user-defined messages*, from which users could select as appropriate, and the rest of the design flows from that idea. Since the containers, corks and rods all had to be made especially for the purpose of telegraphy, in duplicate, triplicate or whatever multiple desired, the officials who commissioned these devices probably added their own preferences to the basic design. Adjustments could and would have been made in practice too; Polybios tells us explicitly that trials were conducted before the device was deployed, and only when the jars were properly synchronised would they be used in earnest. During such trials any modifications, such as increasing the space allocated to each message to reduce the possibility of reading the wrong message (at the expense of a shorter menu), or increasing the outflow aperture to speed up operation (at the expense of a higher risk of getting the wrong message) would have been implemented to the particular users' satisfaction. The wonderful thing about hand-made-to-order goods that we tend to forget in our mass-machine-manufactured world is that the buyer makes decisions about everything, not just whether to take or leave one complete something or other off the shelf.

Aineias' important and valuable innovation is the *concept of a menu of messages from which the user chooses*, rather than a packaged product. And that concept still exists; indeed, it is ubiquitous in our world. For example, menus of anticipated fixed messages have long been used by mariners in the guise of the International flag code and pennant system where, for example, a flag divided vertically into yellow and blue halves means 'I want to communicate with you' (equivalent to Aineias' raised torch to start signalling) and divided diagonally into yellow and red means 'Man overboard'.[50] Turning from salty seadogs to progressive professionals, menus of anticipated options are also common features of modern web technologies, where the user selects from a drop-down menu or a pop-up menu of predetermined choices, the year of card expiry for example. The particular messages given in Aineias' description were offered merely as examples, and to focus on those specifics (as Polybios did; they may even be *his* examples) is to miss the point. The device as instantiated on each occasion would have been as simple or difficult to use, and have as much or as little flexibility, as each individual designer had made it, much as modern websites are. If we find the options inadequate, the problem is the particular options on offer, not the technology that allows or requires us to choose between them.

50 Thanks to J.V. Tucker for this example.

Fire-Signals after Aineias

We need to deal briefly with the post-Aineias history of fire-signals because our source for the Aineian system had a vested interest in it, which impacted on his presentation of the evidence. After describing Aineias' system, Polybios goes on to describe a much more complex system of telegraphy invented by Kleoxenos and Demokleitos, presumably at some time in the late fourth or third century BC, and allegedly perfected by Polybios himself (though he is not specific about his own contribution to this technology), which enabled any message at all to be sent, letter by letter. This came at the cost of much greater complexity, much more equipment, and much more highly skilled operators. The operators needed to be clever as well as dextrous, because Polybios suggests that they first rephrase the message to be sent in the most laconic version imaginable to minimize the amount of signalling required (10.46.4). In Polybios' own words, "in practice it requires care and exact attention".[51] One immediately begins to wonder at the practicality of this system, because the rephrasing alone would have offered rich opportunities for the true message to be lost. Another limitation with Kleoxenos, Demokleitos and Polybios' system is that it could have been reliable only over relatively short distances, because the signallers need to be able to distinguish between multiple torches—up to five—being held in each hand. However sharp-sighted the signallers were, this could only be achieved at significantly shorter distances than that at which a single torch could be seen.[52] Kleoxenos, Demokleitos and Polybios' system looks to me like a good-idea-on-paper that probably found little use in the real ancient world. Since in many respects it is not superior, I cannot imagine it displacing Aineias' system, which remained on the technology shelf for later generations, at least for those individuals who had read his work or excerpted it like Polybios and Aelian.

There are few recorded instances of the use of fire signals in the subsequent period, but that involving Philip V in 207 BC and described by Livy (28.5) could have been implemented by any fire signalling system. We are told that the fire relay stations were organised on high points in Phokis, Euboia and Thessaly so that Philip could receive instant information 'of where his enemies were active'. Aineias' system could have delivered this information as well as and more quickly than Kleoxenos, Demokleitos and Polybios', but old fashioned simple torches waved would have delivered it even more quickly, albeit without the option to relay more specific information that the other systems offered.

51 Polyb. 10.45.6; see also his special pleading in 10.47.

52 See esp. Polyb. 10.45.11.

Signal lamps are still used by the military, primarily to communicate between ships, though now the light is electric not flame, and Morse code enables any message to be sent, though still not especially quickly.

Conclusion

Aineias' *Poliorketika* contains much information on the technology of his times, most of which is the technology of everyday life; there is very little dedicated military technology here, and what we usually find is civilian or household (i.e. 'domestic') equipment pressed into military service. This is true of the fighting force too of course, composed as it was of militiamen, who were also normally occupied with domestic matters. After surveying the range of such domestic equipment pressed into military service, we undertook a detailed analysis of just two examples, drawn from either end of the complexity spectrum: the old and simple ladder, and the then-new and complex telegraph system, the description of which survives in Polybios as a fragment from another of Aineias' works on warcraft.

With respect to the ladder, although this is a very simple technology with no moving parts, it became evident that the object cannot be treated on its own; if we wish to understand its deployment or non-use in a given situation, and in particular, to understand why the Greeks did not often attempt escalade of a walled city, *how it is used* is a key part of the story. In this case, the angle at which a ladder is set up turns out to be not as we would have it. In antiquity ladders were usually set at a significantly lower incline than we would normally employ, and this had consequences. In particular, it suggested a new solution to the functioning of the door-like anti-ladder device that Aineias describes in the *Poliorketika*.

With the torch-and-pot telegraph we are dealing with a technological system that involves moving parts, multiple users, and something that is apparently Aineias' own invention. Unfortunately our knowledge of it comes from someone who claims a hand in an alternative and putatively superior system and who, not surprisingly therefore, finds fault with it (Polybios). This fragment tells the reader how to build the device, and is sufficiently detailed to allow it to be constructed in various forms now, as in antiquity. This system does not redeploy civilian technology in a military context; it needs specially commissioned equipment, and it is the equipment not the commissioning that is special (commissioning is a normal modus operandi in ancient Greek manufacturing). The basic structure would work, and it would enable the transmission of messages much more rapidly than by messenger using any then-existing mode of transport. Aineias' contribution to the simple traditional fire signal system was the concept of a menu of different user-defined messages, and a device by which to select the same one at the sending and receiving relay stations. How easily or otherwise it worked would have depended on each set of users' decisions in their commissioning, adjusting, and employment of it. It has several significant advantages over the more complicated and fiddly system that Polybios later promoted, and there is no intrinsic reason to suppose that Aineias' device was superseded by his. Aineias' concept of a menu of fixed messages, from which users select one, is very familiar to all of us today, being used widely in ubiquitous modern software applications.

Bibliography

Adam, J.P. (1994) *Roman Building: Materials and Techniques*. Batsford, London.

Barends, D. (1955) *Lexicon Aeneium: a Lexicon and Index to Aeneas Tacticus' Military Manual 'On The Defence of Fortified Positions'*. Van Gorcum, Assen.

Bettalli, M. (1990) *Enea Tattico. La difesa di una città assediata (Poliorketika). Introduzione, traduzione e commento*. Ets editrice, Pisa.

Bouras, L. and Parani, M.G. (2008) *Lighting in Early Byzantium*. Dumbarton Oaks, Washington DC.

Casson, L. (1965) 'Harbour and River Boats of Ancient Rome', *Journal of Roman Studies* 55: 31–39.

Casson, L. (1971) *Ships and Seamanship in the Ancient World*. Princeton University Press, Princeton.

Clarke, S. and Engelbach R. (1990) *Ancient Egypt Construction and Architecture*. Dover, New York.

Creasman, P.P. pers. comm. in discussion of what are otherwise thought to be 'pleating boards', *Experiment and Experience: Ancient Egypt in the Present* Conference, Swansea University, 2010

Cuomo, S. (2007) *Technology and Culture in Greek and Roman Antiquity*. Cambridge University Press, Cambridge.

Edgerton, D. (2006) *The Shock of the Old*. Profile, London.

Van de Maele, S. and Fossey, J. (eds) (1992) *Fortifcationes Antiquae*. Gieben, Amsterdam.

Garlan, Y. (1974) *Recherches de poliocétique grecque*. Bibliothèques de l'Ecole française d'Athènes et de Rome—Série Athènes 223, Paris.

Hunter, L.W. and Handford, S.A. (1927), *Αἰνείου Πολιορκητικά. Aineias on Siegecraft*. Clarendon Press, Oxford.

Lienhard, J.H. (2000) *The Engines of our Ingenuity*. Oxford University Press, New York.

Lienhard, J.H. (2006) *How Invention Begins: Echoes of Old Voices in the Rise of New Machines*. Oxford University Press, New York.

Nye, D.E. (2006) *Technology Matters*. The MIT Press, Cambridge Mass.

Pritchett, W.K. (1992) *Studies in Ancient Greek Topography vol. 8*. Gieben, Amsterdam.

Rihll, T.E. (2007) *The Catapult: a History*. Westholme, Yardley.

Waschow, H. (1938) *Geschichte der Belagerungstechnik 1: 4000 Jahre Kampf um die Mauer*. Buch- und Kunst-Druckerei W. Postberg, Bottrop.

Whitehead, D. (1990) *Aineias the Tactician, How to Survive under Siege. Translated with Introduction and Commentary*. Clarendon Ancient History Series, Oxford.

Woolliscroft, D.I. (2001) *Roman Military Signalling*. Tempus, Stroud.

CHAPTER 15

The Reception of Aineias' *Poliorketika* in Byzantine Military Literature*

Philip Rance

Almost all surviving examples of ancient military manuals in Greek are preserved today via six manuscript prototypes (here P, M, A, V, N-E, P-B). These codices were produced between c. 925 and c. 1040 in scriptoria in Constantinople, in five, and possibly all, cases at or in association with the imperial court.[1]

---

* The research for this paper was facilitated by a Humboldt-Forschungsstipendium für erfahrene Wissenschaftler, hosted by the Institut für Byzantinistik, Ludwig-Maximilians-Universität, München, 2009–2011.

1 A brief chronological summary will assist in elucidating the tradition. The multi-period composite *codex Parisinus suppl. gr.* 607 or so-called Mynas codex (= P) contains a corpus of classical poliorcetic texts copied in the second quarter of the tenth century, and is thus the earliest witness to the genre. Although its Constantinopolitan origin cannot be doubted, the circumstances of its production remain uncertain. See Wescher (1867) xv–xxiv, xxxviii; Dain (1933) 19–20, Dain and de Foucault (1967) 380–381 (also 347–349 with further bibliography); Gatto (2010) 121–123; Németh (2011). A later dating accepted by Schindler (1973) 194–195 relies on a palaeographical assessment of a sample text by Rochefort (see *infra* n. 35). *Mediceo-Laurentianus gr.* LV-4 (= M), compiled c. 950, was almost certainly an officially sanctioned project of the palace scriptorium at Constantinople. As the unique witness to the complete text of Aineias' *Poliorketika*, it will be discussed in detail below (pp. 302–310). It has been convincingly argued that *Ambrosianus gr.* 139 (B 119 sup.) (= A) also emanated from court circles c. 959 (see *infra* nn. 18–19). Three other closely related codices were produced in an unidentified monastic scriptorium in Constantinople: two, *Vaticanus gr.* 1164 (= V) and a codex now split and separately bound as *Parisinus gr.* 2442 and *Barberinianus gr.* 276 (II 97) (= P-B), were copied from a common hyparchetype c. 1020. A third codex, similarly now divided into *Neapolitanus gr.* 284 (III C 26) and *Scorialensis gr.* Y-III-11 (281) (= N-E), dates to c. 1040. The relationship of N-E to V and P-B has been disputed, but the prevailing view of scholarship holds that N-E is a direct copy of V; owing to lacunae in V, however, N-E is often a primary witness (see *infra* n. 100). The existence of a fourth, lost manuscript of this family, apparently an eleventh-century sibling of V and P-B, is evidenced by the chance survival of two folios reused as endpapers in *Parisinus Coislinianus* 101, see Wescher (1867) xxviii, xxxviii; Dain (1946) 234–235; Gatto (2010) 107–108. In addition, there is evidence that at least one other manuscript containing classical poliorcetic texts survived into the late Byzantine period: extracts in fourteenth-century *Vindobonensis phil. gr.* 120 (olim 113)

 | DOI: 10.1163/9789004352858_016

They contain varying collections of Greek, Roman and Byzantine texts, which were broadly classified as works relating to tactics and generalship (*taktika*, *strategika*), siegecraft (*poliorketika*) or artillery (*belopoiika*), even if these labels are sometimes loosely applied and rarely exhibit precise equivalence to modern categorisations. In most of the manuscripts individual works are grouped into several corpora, which to some extent reflect contemporary notions of genre and/or periodization, although in certain cases the evidence suggests that two or more texts had been bound together and transmitted collectively from a much earlier date. Overall, the guiding principle of Byzantine editors appears to have been inclusiveness rather than the application of any obvious selective criteria.

The scholarly enterprise invested in assembling, editing and copying the military literature of antiquity followed an era traditionally characterised as the Byzantine 'Dark Age', now more narrowly dated from c. 640 to c. 780, during which there was a sharp decline in all aspects of book production. The causes and nature of this major cultural dislocation cannot be examined here, but a crucial factor was the contraction and evolution of the educated municipal élites of the East Roman Empire, with concomitant changes to their traditional schooling, tastes and perspectives. This transformation was apparent in the lapse of certain categories of literature that were hitherto essential components of Greek literary culture, notably classicizing historiography, philosophy, philology, epigram and panegyric. This did not amount to a general cessation in the writing and copying of books; indeed other genres persisted and even flourished, including doctrinal and liturgical literature, homiletics, hagiography and

---

were copied from an unknown and now lost exemplar, a collateral descendant of the same tradition as P, see Wescher (1867) xxx–xxxii; Dain (1933) 19–21 (misdating *Vindob. phil. gr.* 120 to the sixteenth century); (1946–1947) 35; Gatto (2010) 111–112, 150–153 (repeating Dain's dating error). The only classical military treatise to be transmitted independently of these six codices is Polyainos' *Strategemata* (or *Strategika*), of which the unique prototype is *Mediceo-Laurentianus gr.* LVI-1, a composite codex of diverse content, dated to the late thirteenth/early fourteenth century; purchased in Crete around 1450, its original provenance is unknown. See Bandini (1764–1770 [1768]) II col. 290–294; Dain and de Foucault (1967) 389 (with older bibliography); Schindler (1973) 15–18. The most comprehensive overview of the manuscript evidence remains Dain and de Foucault (1967) 376–390, which assembles the older bibliography, although the analysis is in parts outdated and suffers from errors and inconsistencies arising from posthumous publication. Subsequent studies of individual works or codices offer supplements and correctives, see especially Mazzucchi (1978); Dennis (1981) 19–24, 28–42; Dagron and Mihăescu (1986) 13–25 (though the stemma of Dennis [1981] 33–36; [1985] 5–6 is preferred here); Haldon (1990) 38–39, 47–53; McGeer (1995a) 3–10, 81–86; Rance (2007a) 733–737; (forthcoming a); (forthcoming b).

florilegia.[2] Nevertheless, even if the effects of this 'Dark Age' were less uniform than conventionally assumed, the demise of the high-brow secular media of antiquity appears also to have disrupted or interrupted the tradition of technical military literature. No new works of military science can be dated between the mid-seventh and mid-ninth centuries, and it is unlikely that scriptoria produced many, or possibly any, copies of classical military manuals, or at least one struggles to discern firm evidence of editorial interest in classical treatises, which in this period may have found a readership only among exceptional antiquarians or soldier-savants with the requisite literary education.[3] It must be stressed that even this potential audience remains an optimistic conjecture: little can be said with certainty regarding the number or identity of the consumers of this broad genre in any period of Greek, Roman or Byzantine history, though it seems reasonable to assume that the technical content appealed to a minority taste, and doubtless literacy and expense were at all times constraining factors on distribution and ownership.[4]

In the later ninth century Byzantium witnessed a striking revival of interest in military writing, which persisted until the beginning of the eleventh century. Intellectual and literary activity in this field is manifest in both the composition

---

2 On these developments see Irigoin (1962); Lemerle (1971) 68–108 (with review by I. Ševčenko in *AHR* 79 [1974] 1531–1535); Mango (1975); Treadgold (1979); Cameron (1992); Haldon (1997) 425–435.

3 A doctrinaire tendency in older scholarship calls for a note of caution. The apparent barrenness of the period c. 640–c. 850 with regard to military literature is partly founded on comparisons with developments in other genres, but also harbours a certain circularity of argument, whereby several undated or undatable texts, which could in theory belong to this period, have been assigned an ill-defined late antique date (i.e. pre c. 600) largely on the presumption that they could not possibly be products of the 'Dark Age', and thus reinforcing the overall gloominess of the era (e.g. classically Dain [1946] 108–109). This is the case with the many compositions, both actual and hypothetical, outlined in Dain and de Foucault (1967) 336–340, 346–353, few of which can be dated more precisely than c. 400–c. 900. For similar reasons the possibility of editorial activity in this period cannot be excluded, see *infra* nn. 14, 16. Furthermore, given that Greco-Roman military literature encompasses a diverse range of texts, a distinction might be drawn with regard to non-literary works, such as the *Strategikon* (completed c. 600), expressly written in a plain *koine* comprehensible to senior army officers, and employing the vulgarisms and semi-barbarised jargon of their profession. A more selective conception of the cultural disruption associated with the Byzantine 'Dark Age' allows for the possibility that the vernacular idiom and technical content of the *Strategikon* sustained its relevance and utility during a period of waning interest in classical *taktika*. See *infra* n. 16.

4 Bagnall (2009) 50–64 offers an interesting assessment of the cost of books relative to incomes in late antiquity. For literacy and book-ownership see *infra* n. 7.

of numerous new military handbooks and a resurgence in collecting, editing and adapting the surviving specimens from classical antiquity.[5] This retrospective tendency was one element of a broader enthusiasm for literary and artistic forms of the past, formerly termed the Tenth-Century or Macedonian Renaissance, but now conceived as a longer-term cultural reorientation that began in the late eighth and early nine centuries and thus antedates the patronage of the Macedonian dynasty (867–1056) which marked its apogee.[6] Ancient Greek and Roman treatises exercised a multifaceted influence on Byzantine military culture in this period. As 'classics' of the genre they possessed an intrinsic literary appeal and antique cachet, at least within a high socio-cultural milieu, and could cater to a taste for ethical edification over and above their technical content. They provided the military aristocracy and educated officer class with a conceptual framework and an elevated vocabulary—I cannot improve upon Eric McGeer's expression 'a sort of military katharevousa'—in which to intellectualise and lend respectability to an often crude and brutal profession, and the ability to read and cite classical authorities became an index of a soldier's social and intellectual standing.[7] With the dawning of a golden age of Byzantine military victories and (re)conquest around the middle of the tenth century, such literary posturing also played a rhetorical role in a society with a heightened awareness of antiquity and self-consciously enamoured of the past glories of the *Rhomaioi*, to whose traditions tenth-/eleventh-century Byzantium could lay claim, albeit with some historical conceit.[8] More tangibly, certain classical

5 For a overview of Byzantine military literature, again Dain and de Foucault (1967) 340–374 (with some caution) remains fundamental; see also Hunger (1978) II 323–340; Dagron and Mihăescu (1986) 139–160; McGeer (2008); Sullivan (2010).

6 Irigoin (1962); Lemerle (1971); Treadgold (1979); (1984); Speck (1981); Wilson (1983) 68–147.

7 See McGeer (1995a) 191–194 for the high regard accorded ancient tactical treatises in tenth-/eleventh-century Byzantium. Levels of literacy in the Byzantine army are not easily assessed in any period. The author of the *Strategikon* (c. 600) specifies that, ideally, divisional commanders (or '*merarchs*') should be literate (1.4.16–17: εἰδότας, εἰ δυνατόν, καὶ γράμματα), yet elsewhere he appears to assume functional literacy throughout the entire officer corps, insofar as commands and instructions are routinely addressed 'in writing' (ἐγγράφως) to officers of the rank of *tribunus* and above, cf. 1.6.5; 3.5.122–123, 11.3–4; 7.A.4.6 (cf. 7.A.14; 12.B.10); 7.B.16.40–41; 12.B.pr.7, B.24.30–32. A few of these passages are reprised almost verbatim in Leo's *Taktika* (c. 905): 7.35 (= *Strat.* 3.5.122–123); 8.1 (= 1.6.5), although in other cases the explicit reference to 'written' directives is excised: 12.59 (= 3.11.3–4); 13.4 (= 7.A.4.6) or the passage is omitted altogether. See Browning (1978) 42–44, 47; Rance (forthcoming a) for evidence of book-ownership among the military élite in the later Byzantine period.

8 E.g. Psellos, *Chron.* 7 (Mich. VII) §16 (ed. É. Renauld [Paris 1926–1982] II 181.8–10) praises the military education of the *caesar* John Doukas (died c. 1088): 'As for matters relating to

texts served as models for the form and/or content of contemporary compositions, a theoretical continuum fostered in part by antiquarian or literary objectives pursued by some tradition-conscious authors, but also indicative of actual continuities between ancient and mediaeval warfare, which often permitted a Byzantine author to reprise much older material without compromising practical utility.

The inspiration for and initial phase of this renewed interest in military literature remain obscure, partly owing to the uncertain chronology of several works.[9] Intensification of warfare against the Arabs supplies the historical context and focus of most tenth-century treatises, following a shift in the 920s from a long-term defensive stance to nearly a century of successful Byzantine offensives. In any case, imperial sponsorship is evident from an early date. Leo VI the Wise (r. 886–912) composed or authorised two tactical treatises, the *Problemata* (c. 890s) and *Taktika* (c. 905), in large part bookish and retrospective compilations drawing extensively on classical and late antique models. Similarly derivative opuscula were composed or redacted by his son Constantine VII Porphyrogenitus (r. 913–959), who was probably responsible

---

generalship, he has acquired a knowledge of that science comparable to those ancient and much-sung Caesars, and all the daring deeds and accomplishments of the Hadrians and Trajans and men of the same company …' (Τὰ μὲν οὖν ἐς στρατηγίαν κατὰ τοὺς ἀρχαίους ἐκείνους καὶ πολυϋμνήτους Καίσαρας, καὶ ὁπόσα τετολμήκασί τε καὶ κατωρθώκασιν Ἀδριανοὶ καὶ Τραϊανοὶ καὶ οἱ τῆς αὐτῆς ξυμμορίας …). Similarly, at *Chron.* 3 (Rom. III) § 8 Psellos says that Romanos III (r. 1028–1034) wished to imitate 'such deeds as tradition tells us were accomplished by those Trajans and Hadrians and, still further back, by the likes of Augustus and Caesar' (καὶ ὁπόσα ἀκοῇ ἴσμεν παρὰ τοῖς Τραϊανοῖς ἐκείνοις καὶ Ἀδριανοῖς καὶ ἄνω παρὰ τοῖς Σεβαστοῖς γενόμενα Καίσαρι). Theophanes Continuatus 6.40 (ed. I. Bekker, [CSHB 21] Bonn 1838) 427.12–19 reports that John Kourkouas, on account of his military conquests in the East from the late 920s to early 940s, 'took on the appearance and name of another Trajan or Belisarios' (καὶ ἄλλον Τραϊανὸν ἢ Βελισάριον εἰκάσαι καὶ ὀνομάσαι τοῦτον). At ibid. 5.4 (271.4–5) the same work, sponsored by Constantine VII, states that Basil I, the emperor's grandfather and the founder of the dynasty, staged a triumph in 873 'as the men of old had conducted the triumphs of emperors into most glorious Rome' (ὡς οἱ πάλαι κατὰ τὴν μεγαλόδοξον Ῥώμην τῶν αὐτοκρατόρων τοὺς θριάμβους κατάγοντες). The common source of John Skylitzes, *Synopsis Historiarum* (ed. H. Thurn, [CFHB 5] Berlin 1973) 310.55–57 and Leo Diaconus, *Historiae* 9.12 (ed. C.B. Hase, [CSHB 4] Bonn 1828) 158.6–7 implied a parity between the triumphs of John Tzimiskes in 971 and Camillus in the fourth century BC (Plut. *Cam.* 7.1). On triumphs in the ninth and tenth centuries see McCormick (1986) 131–188.

9 Traditionally, surveys of Middle Byzantine military literature begin with the oeuvre of Leo VI (886–912), but a number of works may date to the preceding half century, including Syrianos Magister's compendium (see below pp. 315–317), the so-called *Apparatus Bellicus* and, conceivably, the *Hypotheseis* (or *Excerpta Polyaeni*).

for the completion and revision of his father's *Taktika* as well as encouraging the composition of other tactical and poliorcetic compendia.[10] The *Excerpta Constantiniana*, the thematic anthologies of historical extracts compiled at the direction of Constantine VII, also included at least four now-lost volumes devoted to warfare.[11] Shortly thereafter, Nikephoros II Phokas (r. 963–969) and his circle complemented this reverence for antique precedent with a series of original tactical monographs, based on contemporary operational practices in the Near East and the Balkans.[12]

The extent and nature of classical influences on the evolution of a tenth-century corpus of military literature was initially shaped by the availability of texts. For example, while the sources and content of Leo VI's projects may in part reflect the emperor's personal tastes, it is difficult to escape the impression that Leo made use of the most accessible source-material, or at least he appears to have favoured works that were already popular in late antiquity. It cannot be mere coincidence that the three chief sources for Leo's *Taktika*—namely Aelian's *Taktike Theoria* (c. 106–113), Onasandros' *Strategikos* (49–57/8) and Maurice's *Strategikon* (c. 600)—are those works best represented in the tenth-/eleventh-century manuscript prototypes. Indeed, the evidence of some codices (V, P-B, N-E) suggests that these three texts were previously transmitted together as a small corpus of military writings, assembled at an unknown date between the early seventh and mid-ninth centuries, and thus may have come

10 Editions: Leo's *Problemata* ed. A. Dain (Paris 1935); Leo's *Taktika* ed. G.T. Dennis (2010), with commentary by Haldon (2014). A dossier of three military texts compiled by Constantine VII is edited with Eng. trans. by Haldon (1990). The so-called *De moribus diversarum gentium* (Στρατηγικὸν περὶ ἐθῶν διαφόρων ἐθνῶν), ascribed to Constantine and found uniquely in codex M (fols. 403, 405 [405, 407]), is published only by Lami (1741–1763 [1745]) VI 1409–1418, though he mechanically reproduced the misfoliation in the manuscript. See further Rance (forthcoming b). For the military works of Leo and Constantine see generally Dain and de Foucault (1967) 353–357, 361–362; Lemerle (1971) 270–274, 292–293; Dagron (1983); Tougher (1997) 167–172; McGeer (2003); Haldon (2014) esp. 13–38.

11 See cross-references in the extant *Excerpta Constantiniana* (ed. C. de Boor [Berlin 1903–1906]) to a lost Περὶ στρατηγημάτων: *De Leg.* I 14.26, 62.31; II 379.26; *De Vit.* I 335.19; II 116.19, 123.26; *De Ins.* 33.8–9, 222.3; *De Sent.* 131.28; Περὶ συμβολῆς: *De Vit.* I 99.9; Περὶ συμβολῆς πολέμων (presumably the same work): *De Ins.* 207.34; Περὶ νίκης: *De Leg.* II 390.3; Περὶ ἀνακλήσεως ἥττης: *De Vit.* I 9.20; Περὶ ἥττης (probably the same work): *De Sent.* 210.15. Lemerle (1971) 280–288 offers the most convenient summary of the shape and composition of the *Excerpta*. See also remarks by Eramo (2007) 133–137.

12 For overviews, with earlier bibliography, see Dagron and Mihăescu (1986) esp. 153–175; McGeer (1995a) esp. 171–195; succinctly Sullivan (2010) 156–159.

into Leo's hands as a pre-assembled collection.[13] Furthermore, of greater significance than the number of surviving manuscript prototypes, none older than the tenth century, is the evidence they contain for the majuscule or uncial tradition of these three texts prior to their transliteration into minuscule (c. 850–900), as witnessed by transcriptional errors characteristic of a misreading of majuscule script. The texts of some Greek and Roman tactical authors (Aineias, Asklepiodotos, Arrian, Polyainos) are known to us via a unique minuscule manuscript, the sole surviving descendant of a single majuscule exemplar of often uncertain date and purity.[14] In contrast, the more numerous witnesses to the texts of Aelian, Onasandros and Maurice descend, directly or indirectly, from at least three majuscule exemplars, that is to say, three different transliterations of each author into minuscule, circumstances indicative of a much more ample majuscule tradition across the 'Dark Age' of c. 640–780. These three works are therefore preserved today in three distinct recensions, which Alphonse Dain distinguished as an 'authentic recension' (M) and an 'interpolated recension' (A, V, P-B, N-E), as well as a third recension, of which no manuscripts now survive but for which the works of Leo VI are the main indirect witness.[15] The existence of 'interpolated recensions' of these three authors

13 For Leo's debt to Aelian, Onasandros and Maurice see the index fontium in Dennis (2010) 669–671 (not exhaustive and requiring minor modifications), with remarks of Vári (1898) 56–63; Dain (1930) 151–154; (1946) 134–147; Rance (2000) 235–244; Haldon (2014) 39–55. For evidence of a corpus comprising these three authors see Dain (1930) 15–18, 167–171; (1946) 122–126, 204–207. Dain was not able to date this corpus more precisely than prior to the transliteration into minuscule in the mid to late ninth century.

14 For Aineias, Asklepiodotos and Arrian's *Ektaxis kata Alanōn* see Dain (1934); (1934–1935); (1935). The transmission of Arrian's *Techne Taktike* is potentially amplified by evidence of an indirect tradition (via Leo's *Taktika* and Nikephoros Ouranos' *Taktika*), not examined by Dain, which may derive from a second majuscule exemplar. I plan to examine this question in a future study. In the case of Polyainos, the uncertainty surrounding the date of the derivative *Hypotheseis* (or *Excerpta Polyaeni*), c. 500–c. 850, leaves open the question of whether a second majuscule exemplar of Polyainos' *Strategemata* survived the Byzantine 'Dark Age'. The tradition is further complicated by two excerpts of the *Strategemata* in *Parisinus suppl. gr.* 607 (90^{v}–91^{r}), dated to c. 925–950, and three others in the late ninth-/early tenth-century *Apparatus Bellicus* (see below pp. 324–331). See Schindler (1973) 187–197, 205–216 (with caution).

15 The manuscripts and textual evolution of Aelian's treatise are examined at length by Dain (1946) esp. 119–127, 153–182, 200–202. For Maurice see Dennis (1981) 19–24, 28–41, with remarks of Rance (2007a) 733 n. 94. The manuscript tradition of Onasandros *Strategikos* requires re-examination; for the present see Dain (1930) 15–17, 151–154, significantly revised by E. Korzenszky in reviews in *PhW* 52.1 (1932) 1–8 and *BZ* 35 (1935) 145–149, with fuller remarks of Korzenszky and Vári (1935) v–xix with stemma codicum at xvii.

in particular, in each case predating their transliteration into minuscule, not only reinforces the codicological evidence for their conjoined transmission, but also opens the possibility that they, unlike other examples of the genre, attracted editorial interest during the course of the 'Dark Age'.[16]

Shortly after the production of Leo's oeuvre, the number and range of available Greco-Roman models seem to have increased. Textual and codicological evidence from the reign of Constantine VII suggests that by the 940s, if not before, the emperor and the staff of the imperial scriptorium were engaged in a programme of researching, compiling and editing collections of ancient military texts. Constantine VII indicates, in a treatise addressed to his son (c. 952), that his efforts to unearth older military documents took him beyond the confines of the palace libraries, whose holdings before his reign appear to have been rather modest, and involved searching the archives of other institutions.[17] The most important outcome of this activity was a number of codices containing corpora of ancient and contemporary military texts. *Mediceo-Laurentianus graecus* LV-4 (= M), the largest collection of Greek, Roman and Byzantine tactical writers, was produced c. 950, almost certainly for use in the palace library (see below). Another, near-contemporary collection of classical and Byzantine

16 Dain (1946) 61–115, 188–202 distinguishes in great detail the numerous accretions and interpolations in the textual history of Aelian's *Taktike Theoria*. For Maurice's *Strategikon*: Dennis (1981) 33–34, 41, with additional remarks in Rance (forthcoming b). In his study of Aelian, Dain attributed these modifications to sixth-century editors, but they cannot be securely dated. All that can be said with confidence is that, for all three treatises, this editorial interference predates the transliteration from majuscule into minuscule (c. 850–c. 900). In the case of Maurice's *Strategikon* (c. 600), at least, this may point to interest in the text during the 'Dark Age', unless we choose to locate this revision in the period immediately after its composition.

17 Const. Porph. Text C.24–39 (Haldon [1990] 94–96) explains to Romanos II the circumstances in which he discovered a treatise by Leo Katakylas in the 930s or 940s: 'Accordingly, having conducted a great deal of research in this field and finding no treatise deposited in the palace, we were at length just able to discover a memorandum on these matters in the monastery called Sigriane' ("Οθεν πολλά περὶ τούτων ἀνερευνήσαντες καὶ μηδεμίαν ὑπόμνησιν ἐναποκειμένην τῷ παλατίῳ εὑρίσκοντες, ὀψὲ καὶ μόλις ποτὲ περὶ τούτων ὑπόνημα ἐν τῇ μονῇ τῇ καλουμένῃ Σιγριανῆς εὑρεῖν ἠδυνήθημεν). Katakylas, a senior military officer under Leo VI, had been a monk at Sigriane, on the southern shore of the Bosphoros, and perhaps passed his final years there; see Haldon (1990) 41–42, 45–46, 50, 180–181. There is no reason to think that ancient texts were more readily available. Previously, Leo, *Taktika* 19.1 alluded to an extensive and unsuccessful search for ancient military authorities, see *infra* n. 97. The evidence for the library resources in the Great Palace is assembled by Lemerle (1971) 269–270 n. 9; Mango (1975) 35 n. 26.

military treatises, *Ambrosianus graecus* 139 (B 119 sup.) (= A), judging by the quality of its parchment, scriptorial elegance, ornamentation and orthographic correctness, emanates from the highest level of production in court circles.[18] In most cases the ancient texts assembled in this codex were paraphrased in tenth-century Greek, presumably with the aim of updating diction and technical vocabulary to facilitate comprehension by contemporary readers. Carlo Mazzucchi has persuasively argued that the codex was prepared by a single expert editor-paraphrast using materials in the library of Constantine VII, around 959, at the direction of Basil Lekapenos (c. 920–c. 986), Constantine's brother-in-law and *parakoimomenos* or chief minister (c. 946–959), in order to promote his candidacy for a military command.[19] Basil's broader participation in Constantine's projects is suggested by his likely involvement in the final redaction of *De Cerimoniis*, uniquely preserved in *Lipsiensis Bibl. Urb. Rep.* I 17 (*gr.* 28), which was produced c. 963–969 in the same scriptorium as codex M.[20] A few decades later, another possible product of an imperial scriptorium was a third manuscript, Dain's so-called 'Mazoneus', now lost but long recognised as the common ancestor of three surviving codices (V, P-B, N-E). It was compiled around 980–1000, apparently by a single redactor, who brought together three pre-existing corpora of military writings, one of ancient and late antique tactical authors, another of Hellenistic and Roman poliorcetic treatises, and a third comprising tenth-century compositions. The three descendants of 'Mazoneus' were copied in a single Constantinopolitan workshop in the first half of the eleventh century and bear close similarities to manuscripts produced in the (as yet unidentified) monastic scriptorium that had earlier employed the famous scribe Ephraim (fl. 940–960).[21] Nevertheless, the content of 'Mazoneus' gives

---

18 Martini and Bassi (1906) I 157–160 (with errors). For a detailed description see Müller (1882) 18–39, and esp. Mazzucchi (1978) 276–284, 310–316, with summaries in Dain (1930) 36–42; (1937) 65–67; Dain and de Foucault (1967) 385; Dennis (1981) 21–22, 39–40; (1985) 6; Cosentino (2000) 245–248; (2001); Leoni (2003) xviii–xxviii; Rance (2007a) 733–736; Bevilacqua (2013). See further bibliography in Pasini (2007) 213.

19 Mazzucchi (1978) esp. 267–282, 292–306; summarised by Leoni (2003) xviii–xx; supplemented by Cosentino (2000) 243–246; Bevilacqua (2013). For Basil Lekapenos generally see Brokkaar (1972), and as a sponsor of scholarship and book production esp. 214–216; also Boura (1989); Eramo (2007) 132–137.

20 See Kresten (2000a) 37–39 with n. 102; (2000b) 474–475; Featherstone (2004) 117–121. Two other manuscripts are inscribed with dedications to Basil, both containing the works of John Chrysostom: *Athos Dionysiou* 70, dated 955, and *Vaticanus Ross. gr.* 5 (olim VIII 108), dated 961, see Diller (1974) 517.

21 For 'Mazoneus' see Dain (1946) 203–222, with additional remarks by Dennis (1981) 20–21, 33–36, 41. The siglum *Mazoneus* (*Maz*) honoured Dain's colleague Paul Mazon. For

reason to suspect a connection with the imperial court, notably the inclusion of two works recently composed within the circle of the late Nikephoros II Phokas (r. 963–969).[22] Overall these developments, by contributing to the preservation and reception of classical works, ultimately permitted their transmission to the early modern West—indeed, in evolutionary terms this may be considered a near-extinction moment for this species: virtually the entire corpus of Greco-Roman military literature surviving today would almost certainly have perished were it not for the interest and endeavours of a small number of patrons, scholars, editors and copyists, for the most part at or associated with the Great Palace in the middle decades of the tenth century.[23]

Within the scope of the tenth-century renaissance in military writing, how well known was the ancient writer Aineias? How should one gauge the currency and reputation of his *Poliorketika*? To take a comparative example from the highest level of expectation: among Byzantine writers and readers by far the most popular military treatise from antiquity was Aelian's *Taktike Theoria*, written c. 106–113 and dedicated to Trajan, an expressly antiquarian study of the organization, internal articulation and manoeuvres of an idealised infantry phalanx, which drew, via an uncertain textual tradition, on a late Hellenistic archetype, presumed to be a lost treatise by the Stoic philosopher Poseidonios of Apameia (c. 135–c. 51 BC).[24] The popularity of the *Taktike Theoria* is

---

the codicological and palaeographical affinities between V, P-B and N-E and the monastic scriptorium associated with Ephraim see Irigoin (1958–1959) II 181–195, esp. 189–190; generally Diller (1974) esp. 516–518; Mazzucchi (1978) 276–281; E. Gamillscheg, *Repertorium der griechischen Kopisten* 3A (Vienna 1997) 82–83.

22 Codices V, P-B and N-E inherited from 'Mazoneus' the texts of *De Velitatione* (Περὶ Παραδρομῆς) and *De Re Militari* (Περὶ καταστάσεως ἀπλήκτου), both composed shortly after 970. The former was explicitly compiled on the instructions of Nikephoros II Phokas and based on his own informal writings, but the final draft was executed by an unnamed redactor and published after the emperor's death. It has been argued on the basis of style and date that the *De Re Militari* was by the same author, see Dagron and Mihăescu (1986) 153–175. Also in favour of a link between 'Mazoneus' and court circles: the text of the *Strategikon* in V, P-B and N-E appears to derive from the same majuscule exemplar or transliteration as the text used by the paraphrast of *Ambrosianus gr.* 139 (B 119 sup.), itself clearly a product of imperial patronage c. 959 (*supra* nn. 18–19), see Dennis (1981) 39–41 with Rance (2007a) 733 n. 94 and in detail in Rance (forthcoming b).

23 Mazzucchi (1978) 280–281.

24 Greek text ed. in Köchly and Rüstow (1853–1855) II.1 241–471; Eng. trans. by Devine (1989). A new edition is forecast as A.M. Devine, 'Aelianus Tacticus. A Critical Edition of Aelian's ΤΑΚΤΙΚΗ ΘΕΩΡΙΑ, with Prolegomena and Translation' in *ANRW* II.37.5 (forthcoming). The text in C. Matthew, *The Tactics of Aelian* (Barnsley 2012), a selective composite of

documented by several criteria: in the frequency of its citation in historical narratives, usually to validate the military education or expertise of emperors and generals;[25] in the greater number of manuscripts available to tenth-century editors and copyists,[26] and in the abundant evidence for late antique and Byzantine editorial interest in the text, witnessed by numerous supplements, interpolations, excerptions and adaptations.[27] It seems that Aelian's work, despite, or perhaps because of, its antiquarian objectives, arcane content and recherché vocabulary, attained an almost fabled renown in the Greek-speaking world by the Middle Byzantine period, and possibly already in late antiquity, broadly akin to the reputation of Vegetius' *Epitoma Rei Militaris* in the late mediaeval Latin West. Similar criteria give reason to believe that Onasandros' *Strategikos* (written 49–57/8) was a popular or fashionable book, while Polyainos' *Strategemata* (c. 161–163) spawned a succession of excerpta and adaptations.[28] In the

---

defective antiquarian editions, is not suitable for academic purposes. On Aelian's sources and inspiration see Dain (1946) 26–40, summarised Dain and de Foucault (1967) 329–331, with additional remarks by Sekunda (2001).

25 E.g. Psellos, *Chron.* 7 (Mich. VII) §16 (ed. É. Renauld [Paris 1926–1982] II 181.12–14) states that his close friend, the *caesar* John Doukas, had learned military science 'from books on tactics and generalship and siegecraft and from what the likes of Aelian and Apollodoros have written' (ἀπὸ τῶν τακτικῶν βιβλίων καὶ στρατηγικῶν καὶ πολιορκητικῶν καὶ περὶ ὅσων οἱ περὶ Αἰλιανὸν καὶ Ἀπολλόδωρον συγγεγράφασιν). Psellos himself compiled an epitome of Aelian's *Taktike Theoria*, see Περὶ πολεμικῆς τάξεως ed. Köchly and Rüstow (1853–1855) II.2 234–238. Similarly, Anna Comnena, *Alexiad* 15.3.6 (edd. D.R. Reinsch and A. Kambylis, [CFHB Series Berolinensis 40.1–2] Berlin-New York 2001) stresses the tactical expertise of her father, Alexios I Komnenos (r. 1081–1118), by remarking that 'he was not unfamiliar with Aelian's *Taktika*' (ἦ γὰρ οὐδὲ τῆς Αἰλιανοῦ Τακτικῆς ἀδαής). See further Dain (1946) 301–302; Rance (forthcoming a).

26 Dain (1946) 119–127, 153–377.

27 Vári (1898) 59–63; Dain (1946) 77–115, 129–182; Zuckerman (1990) 217–219; Loreto (1995) 563–565, 572–574; Kuchma (1998); Eramo (2008) 130–131; Dennis (2010) 669. On Maurice's use of Aelian see Rance (forthcoming b). Aelian is cited as an authority by Leo, *Taktika* 6.25; 7.67–68; *Sylloge Tacticorum* 30.1; 47.20.

28 The most widely cited edition of Onasandros' *Strategikos* is Oldfather (1923) 369–527, though Korzenszky and Vári (1935) is greatly superior. For the manuscript evidence see *supra* n. 15. The Byzantine reception of Onasandros still requires comprehensive study but for specific instances see Vári (1898) 56–59; Oldfather (1923) 350–352; Lowe (1927); Dain (1930) 145–157; (1931) 340–341; Kuchma (1982–1986); Haldon (1990) 54, 179; Krentz and Wheeler (1994) xx–xxi; Rance (2000) 235–244; Taragna (2004) 799–801; Dennis (2010) 671; see also Nicephoros Ouranos, *Takt.* 66, 68–74 (de Foucault [1973] 302–311), with Dain (1937) 40–45, 50; McGeer (1991) 132. The late antique and/or Byzantine adaptations of Polyainos' *Strategemata* are edited in Wölffin and Melber (1887) 427–540 and de Foucault (1949), with

poliorcetic tradition the fullest *Nachleben* belongs to the *Poliorketika* of Apollodoros of Damascus (probably c. 101–102), or more accurately a multilayered compilation transmitted under his name.[29]

With regard to Aineias, in contrast, the corresponding categories of evidence are at best muted if not silent. The manuscript tradition and Byzantine reception of his work will be discussed below, but the meagre testimonia can be dealt with briefly here. Only two late antique or Byzantine works refer to Aineias. First, writing probably in the 550s–560s, the bureaucrat and antiquary John the Lydian included Aineias in a catalogue of ancient Greek and Roman military writers, whom he cited ostensibly as 'attestors' (μάτυρες) to the usage of the Latin word *veterani* (βετερανοί). The demonstrable unreliability of John's citations elsewhere in his work gives cause for scepticism, a suspicion hardened in this instance by the fact that none of the extant Greek authors listed (Aelian, Arrian, Aineias, Onasandros, Apollodoros) uses the term *veteranus* or could ever be suspected of doing so, given their classicizing idiom (and date, in the case of Aineias). It therefore remains doubtful whether John knew the writings of Aineias firsthand or even indirectly; the most that can said with confidence is that John was aware that an author called Aineias had written a military treatise or treatises at an undefined point in antiquity.[30] Second, the compilator(s) of the *Suda* (c. 1000) included a lemma on Aineias, which specifies, on the authority of Polybios, that he wrote on the subject of fire-beacons (περὶ πυρσῶν) and a 'memorandum on generalship' (στρατηγημάτων ὑπόμνημα). As the information in this lexical entry explicitly derives from an extant part of Polybios' *Histories* (10.44.1–2), one of the most frequently excerpted authors in the *Suda*, and adheres closely to the historian's wording (τὰ περὶ τῶν στρατηγικῶν ὑπομνήματα ... διὰ τῶν πυρσῶν), it cannot serve as independent evidence for familiarity with Aineias and his

discussion by Dain (1931); Dain and de Foucault (1967) 337, 340, 364–365; Schindler (1973) 205–225; Krentz and Wheeler (1994) I xvi–xxiii, with additional conjectures by Eramo (2008) 141–145.

29 See Blyth (1992); Sullivan (2000) esp. 4–8, 20–21, 172–177, 190–191, 210–222; Whitehead (2008a) 147–153; (2008b); (2010) 17–18, 28–34.

30 J. Lyd. *De Magistribus* 1.47 (edd. M. Dubuisson—J. Schamp [Paris 2006] I.2 62–63): Ἀδωράτωρας οἱ Ῥωμαῖοι τοὺς ἀπομάχους καλοῦσιν ..., βετερανοὺς δὲ τοὺς ἐγγεγηρακότας τοῖς ὅπλοις (μάτυρες Κέλσος τε καὶ Πάτερνος καὶ Κατιλίνας, οὐχ ὁ συνωμότης ἀλλ' ἕτερος, Κάτων ⟨τε⟩ πρὸ αὐτῶν ὁ πρῶτος καὶ Φροντῖνος, μεθ' οὓς καὶ Ῥενᾶτος, Ῥωμαῖοι πάντες· Ἑλλήνων δὲ Αἰλιανὸς καὶ Ἀρριανός, Αἰνείας, Ὀνήσανδρος, Πάτρων, Ἀπολλόδωρος ἐν τοῖς πολιορκητικοῖς κτλ.). Dismissed by Lammert (1940) 285. For the broader unreliability of John's citations see Bosworth (1983) esp. 267–270.

work in late tenth-/early eleventh-century Byzantium; on the contrary, the absence of additional data arguably points to a lack of knowledge about Aineias in this period.[31]

The Direct Tradition of Aineias: *codex Mediceo-Laurentianus graecus* LV-4

The text of Aineias' *Poliorketika* is uniquely preserved in *codex Mediceo-Laurentianus graecus* LV-4 (159v–181v), hereafter M, from which descend four sixteenth-century copies.[32] This large and high-quality volume of 405 (407) folios, the work of a single scribe, is the most famous and important collection of Greek military authors and a valuable specimen of manuscript production in tenth-century Constantinople.[33] M contains copies of most of the Greek, Roman and Byzantine tactical (distinct from poliorcetic or naval) treatises available at the time of compilation, framed at its beginning and end by the most recent works, which are ascribed to or closely associated with Constantine VII. In total its contents comprise sixteen military texts, for nine of which M is the codex unicus.[34] M was produced c. 950, in part as a continuing expression of the renewed interest in military literature inspired by Leo VI, but also reflective of a broader systematisation of knowledge under Constantine VII (r. 913–959).[35]

31 *Suda* αι 215 (Adler II 172): Αἰνείας· οὗτος ἔγραφε περὶ πυρσῶν, ὥς φασι Πολύβιος, καὶ περὶ στρατηγημάτων ὑπόμνημα. Cf. Polyb. 10.44.1–2, with Oldfather (1923) 204 n. 2; Lammert (1940) 285; Dain and Bon (1967) xiv. See Schettino (1998) 26–27 for the equivalance of στρατηγήματα and στρατηγικά in the *Suda*, with the generic sense of *res militares*. For citations of Polybius in the *Suda* see De Boor (1912) 395–397; (1915–1919) 24–27, 65–81; Becker (1915) 67–71.

32 The four *recentiores* containing the text of Aineias are examined in detail in Dain (1935) esp. 10–32; (1940a); (1941); Dain and Bon (1967) xlviii–li.

33 The codex contains two foliations, one, eighteenth-century (?), in ink in the upper margin, the other, twentieth-century, in pencil in the lower margin. The two foliations differ after 181v owing to the insertion of two blank folios (181a [182] and 188a [190]). As all earlier literature adheres to the first sequence, I have retained this numeration with any divergence noted in parentheses.

34 The contents are catalogued by Bandini (1764–1770 [1768]) II col. 218–238. The initial or prefatory work (1r–2r) is edited as Const. Porph. Text B in Haldon (1990) 82–93 with an analysis of its textual history at 35–53, 66–67; alternatively Speck (1991) esp. 275–288. The concluding text in M (403, 405 [405, 407]) is Const. Porph. *De moribus diversarum gentium*, ed. Lami (1741–1763 [1745]) VI 1409–1418.

35 The dating of M to c. 950, or perhaps a little earlier, is so widely accepted in the secondary

M has attracted the attention of western scholars since the mid-seventeenth century, none more so than the celebrated codicologist Alphonse Dain (1896–1964), the most prolific and influential authority on the Greek tactical tradition during the first half of the twentieth century.[36] Dain demonstrated that, within

literature that I cite here only works containing relevant argumentation: Dain (1946) 184–186; (1954) 43–44; Dain and de Foucault (1967) 361–362; Irigoin (1958–1959) II 178–180; with selected remarks in Mazzucchi (1978) 276–281, 287, and additional arguments in Haldon (1990) esp. 37–39, 45–53, 66–69. A significantly later date is allowed by Schindler (1973) 216–217 on the basis of a palaeographical analysis undertaken at his request by Gabriel Rochefort, who advised that M, while appearing to belong to the mid-tenth century, was in fact produced c. 985. This opinion is not easily compatible with the content and layout of M, the wider textual tradition of Byzantine tactical writing or the relative chronology of several other tenth-century codices. Schindler seems largely unaware of the difficulties. Certainly M cannot be an apograph of a lost Constantinian manuscript, as the texts in M were demonstrably copied from several different majuscule and minuscule exemplars. When Rochefort's palaeographical judgement here gives rise to so many improbabilities, one cannot ignore that his methodology led him to adopt maverick positions on the dating of other Byzantine manuscripts, not least his view that the famous *Palatinus gr.* 23 (the *Palatine Anthology*) was completed c. 1050–1070, a century later than the estimation of any of the many other palaeographers who have examined this codex. See Aubreton (1968) 47 n. 1; (1969) 461; for the *communis opinio* in favour of c. 940–950 see most recently Diller (1974) 520–521; Cameron (1993) 97–99, 108–116, 313–314; Lauxtermann (2003) 83–85. See further criticism of Rochefort's methodology by Lauxtermann (2003) 329, 333 regarding the date of *Parisinus suppl. gr.* 690. See also Rochefort's eccentric dating of *Parisinus suppl. gr.* 607 in Schindler (1973) 194–195. In any case, a dating of M later than the 950s is precluded by studies of the dossier of three military texts compiled by Constantine VII, one of which (Text B) is the preambular work in M (1^r–2^r), while all three (A, B and C) likewise preface *De Cerimoniis* in *Lipsiensis Bibl. Urb. Rep.* I 17 (*gr.* 28) (1^r–21^r), which was produced in the same imperial scriptorium c. 963–969 (see *supra* n. 20). It is generally agreed that Text B, whatever its origin, must have been transcribed into M before Constantine composed Text C, a revised and expanded version of B. Constantine (died Nov. 959) addressed Text C to his son Romanos II, probably on the occasion of his attaining adulthood in c. 952. See esp. Haldon (1990) 36–69, with further remarks by Featherstone (2002) 468–469; (2004) 113. An alternative hypothesis by Speck (1991) strains credulity, but does not controvert the relative chronology of M and the *Lipsiensis*. Correspondingly, it is most unlikely that the copyist of M was instructed to transcribe a preliminary 'pre-edition' of Leo's *Taktika* (as found in M 281^r–400^v [283^r–402^v]) more than a quarter of century after a revised redaction of the same text, as first witnessed by *Ambrosianus gr.* 139 (B 119 sup.), dated c. 959 (see *supra* n. 19), but doubtless produced significantly earlier. See Dain (1946/47) 34–36, 40; Dain and de Foucault (1967) 355–356; Grosdidier de Matons (1973) 229–231; Dennis (2010) x–xi; Haldon (2014) 61–66.

36 Holste (1640 [1715–1719]) 100–103; Bandini (1766) = (1764–1770 [1768]) II col. 218–238;

the manuscript tradition of Greek tactical literature, M is the sole witness to a 'recension authentique', which contains a version of each text closest to its author's original wording and least modified by Byzantine editorial intervention.[37] On the basis of the codex's quality and content, Dain concurred with an earlier assessment by Karl Müller that M was 'due à une enterprise de caractère officiel, à laquelle l'empereur ou son entourage ont pu prendre part'.[38] This view is supported by the more expansive codicological analysis of Jean Irigoin, who drew attention to close affinities, including codical construction, paginal format, script, textual arrangement and decoration, that M shares with four other manuscripts, which likewise contain works by or for Constantine VII and were clearly copied in the same imperial scriptorium. Irigoin envisaged that all five codices were produced during Constantine's lifetime for deposition in the new palace library he established. Even though subsequent research has extended the period of production into the decade following Constantine's death, there is no doubt that this group of manuscripts originated from the emperor's compilatory projects and they were intended for use within the Great Palace.[39] This gilded seclusion would explain why there is no firm

---

Müller (1880) 106–108; Dain (1930) 42–47; (1934–1935) I 344–346; (1935) 6–10; (1946) 183–188, 375–377; Dain and de Foucault (1967) 382–385; Dain and Bon (1967) xxx–xli; Dennis (1981) 19–20, 31–33; Haldon (1990) 38–39, 47–53; Loreto (1995) 565–567. It is necessary to stress again that Dain's final and most influential publications were written largely by others after his death, namely Jules-Albert de Foucault in Dain and de Foucault (1967) and Anne-Marie Bon in Dain and Bon (1967) vii–lvii. Both these posthumous publications rehearse Dain's scholarship over a period of thirty years, sometimes retaining inconsistencies or opinions which Dain had modified or discarded.

37 Dain first used this expression in (1946), though Vári (1917–1922) I xi–xxxv; Korzenszky and Vári (1935) v–vi, xvii had previously drawn similar conclusions.

38 Dain (1946) 183–185 (quoting 185); Dain and Bon (1967) xxxii. See previously Müller (1882) 38, 'die verhältnismässige Reinheit der Textüberlieferung berechtigt uns wohl in ihm ein officielles, vielleicht der kaiserlichen Bibliothek angehöriges Exemplar zu sehen'.

39 For Constantine's new library in the Kamilas cf. Theophanes Continuatus 3.43 (ed. I. Bekker, [CSHB 21] Bonn 1838) 145.6–9, with remarks of Lemerle (1971) 269–270 n. 9; Mango (1975) 35 n. 26. Irigoin (1958–1959) II 177–181 demonstrated the close relationship between M and three codices, similarly produced for use at court: *Turonensis* 980 (*Exc. de Virtutibus et Vitiis*), *Vaticanus gr.* 73 (*Exc. de Sententiis*) and *Lipsiensis Bibl. Urb. Rep.* I 17 (*gr.* 28) (*De Cerimoniis*). It has since been demonstrated that the *Lipsiensis* was compiled c. 963–969, most probably under the auspices of Constantine's brother-in-law and chief minister, Basil Lekapanos, see Kresten (2000a) 37–39 with n. 102; (2000b) 474–475; (2005) 87–98; Featherstone (2004) 117–121. Irigoin also identified affinities with a fourth codex, *Berlin Staatsbib. Phillipps* 1538 (*gr.* 134), thought to have been prepared in the same scriptorium as

evidence of M or the other four manuscripts having been copied or consulted in the Byzantine period.[40] The subsequent history of M up to the beginning of the fifteenth century is unknown, though it had clearly passed out of imperial ownership sometime before c. 1400.[41] Notices of births and deaths inserted into black spaces in M (255^v, 11^v, 12^r) indicate that by 1408 and until at least 1450 the codex was in the possession of the family of Demetrios Laskaris Leontares (died 1431), the general and chief minister of John VII (r. 1390, 1399–1402), Manuel II (r. 1391–1425) and John VIII (r. 1425–1448).[42] In uncertain circumstances in 1491 the distinguished scholar Janus (Giano) Lascaris (c. 1445–1535) purchased M, reportedly at Velestino (ancient Pherai) in Thessaly, during the second of his trips to Greece to acquire manuscripts for Lorenzo de' Medici (1449–1492). Lascaris brought M to Florence, where it became the property of Lorenzo's successors and, following various vicissitudes, was deposited in the Bibliotheca Laurenziana shortly after 1521.[43]

Following a prefatory opusculum closely associated with Constantine VII, codex M embodies three distinct corpora: first, an assemblage of late antique

---

a deluxe presentation copy for Constantine VII, see Cohn (1900) 158–160; Weitzman (1935) 16–18, McCabe (2007) 269–276. For the position of M within Constantine VII's literary and encyclopaedic programme see Dain (1953) esp. 75–77; (1954) 33–47; Lemerle (1971) 292–293; Mazzucchi (1978) 280–281; Wilson (1983) 142–145.

40 Haldon (1990) 36–39, 48–50, specifically with regard to the initial Constantinian work in M (Haldon's 'Text B'), cautiously considers the possibility that the text of the same work in *Lipsiensis Bibl. Urb. Rep.* I 17 (gr. 28), dated c. 963–969, might be 'based upon' M, although he does not press this view nor adopt it in his concluding stemma codicum (at 69). Such a relationship would in any case merely reveal an affinity between two "in house" products of the palace and not demonstrate a wider knowledge of M. More recently Dennis (2010) x–xi infers that the text of Leo's *Taktika* in *Vindobonensis phil. gr.* 275 'may have been copied directly from M', but the present writer, while accepting that these two codices are closely related representatives of the same recension, is not convinced by Dennis' line of reasoning for a direct descent. Previous analyses by Vári (1917–1921) I xxx–xxxv and Dain (1946–1947) 34–35 (neither cited by Dennis) concluded a collateral descent from a common archetype.

41 Irigoin (1958–1959) II 180 plausibly suspects a general dispersal of the imperial library resulting from the capture of Constantinople in 1204. One of M's fellow imperial manuscripts, *Vaticanus gr.* 73, had been re-cut and palimpsested by the fourteenth century.

42 For the marginalia see Schreiner (1975–1979) I 641–649 (98A–B.I.1–3), with remarks at II 445, 449, 477, 484; III 132–133. See further Rance (forthcoming a).

43 Müller (1884) 380 (fol. 48^v 13–15); Knös (1945) 31–51; Dain and de Foucault (1967) 382; Dain and Bon (1967) xl–xli; Fryde (1983) 170–172, 182–187, 201–203; (1996) I 60–63; Kollautz (1985) 98–102; Gentile (1994) 177–194; Jackson (1998), esp. 85–86 (fol. 108^r 22–108^v 6).

and early Byzantine treatises; second, a collection of mainly classical, Hellenistic and Roman authors; while the third corpus comprises compositions by Leo VI and Constantine VII. The second corpus is of interest here: it incorporates six tactical treatises, apparently an attempt to garner all such texts surviving from antiquity, of which Aineias' work is by far the earliest:

Asklepiodotos, *Techne Taktike* (late first century BC)	132^{r}–142^{v}
Aelian, *Taktike Theoria* (c. 106–113)	143^{r}–159^{r}
Aineias, *Poliorketika* (350s BC)	159^{v}–181^{v}
Arrian, *Techne Taktike* (136/7)	182^{r}–195^{v} (183^{r}–197^{v})
Arrian, *Ektaxis kata Alanōn* (c. 135)	196^{r}–197^{v} (198^{r}–199^{v})
Onasandros, *Strategikos* (49–57/8)	198^{r}–215^{v} (200^{r}–217^{v})

To these are appended, somewhat incongruously, the so-called *Rhetorica Militaris* of Syrianos Magister (216^{r}–230^{v} [218^{r}–232^{v}]), traditionally located in the later sixth century but the mid-late ninth is now favoured, and Book VII of the *Kestoi* of Julius Africanus (231^{r}–243^{r} [233^{r}–245^{r}]), compiled c. 227–231. The chronology of the three corpora in M has not been firmly established. Dain tentatively suggested that the centrally placed collection of five Greco-Roman tacticians was already ancient, perhaps first put together in late antiquity and, accordingly, copied en bloc into M from a single exemplar or dossier, although this hypothesis does not accommodate all the evidence or at least requires certain exceptions.[44] It remains possible that this second corpus, like the third and probably the first, was a recent development, perhaps a system of cata-

44 See Dain (1946) 122–126, 186–187; although seemingly contradicted by Dain (1953) 44 and Dain and Bon (1967) xxxii–xxxiii, which state that the assemblage of ancient tacticians is recent. Dain's thesis, which he concedes is largely impressionistic, is undermined by his own analysis of the textual tradition of the six constituent works. Transcriptional errors in M indicate that the copyist transcribed the texts of Aineias, Asklepiodotos and Arrian from a majuscule or uncial exemplar(s); the evidence for Onasandros is less certain. In contrast, the copyist's model for the text of Aelian had already been transliterated into minuscule, circumstances more suggestive of its independent transmission. This does not exclude the possibility that the other four ancient authors already comprised a corpus. For further comments on the textual tradition of Aineias see below (pp. 308–310); for Asklepiodotos: Dain (1934–1935) I 344–347; Arrian: Dain (1934) 158–160; Onasandros: Dain (1930) 42–47; Aelian: Dain (1946) 188–202; with general remarks by Fryde (1983) 193, 202; (1996) I 61–62; Lemerle (1971) 293; Loreto (1995) 566; Rance (2007a) 736–737.

loguing introduced by or for Leo VI to organize material assembled in preparation for writing his own works or, more probably, a slightly later scheme associated with Constantine VII's 'encyclopaedic' programme. In any case, it is significant that Aineias' work, although apparently entitled *Poliorketika* (Πολιορκητικά) in codex M, is grouped here with classical writers on tactics and generalship and not included in a separate corpus of ancient *poliorketika* transmitted in other manuscripts (P, V, P-B, N-E), which, on the whole, comprises texts dealing with the construction and operation of siege machinery, in most cases from the perspective of the besieger.[45] This categorization of Aineias appears to reflect editorial criteria rather than an accident of transmission, and corresponds to distinctions that tenth- and eleventh-century authors drew between the tactical/strategic and technological branches of this broad genre.[46]

The received text of Aineias' *Poliorketika* shows no obvious signs of modification or interpolation to its ancient form, except for the insertion of chapter-headings, sometimes rather crudely, which were possibly imported into the text from earlier marginal notation. This paratextual apparatus was not an original feature but reflects late antique and Byzantine editorial practices, as more

45 The title of Aineias' surviving work is uncertain. In codex M the convoluted formula τακτικὸν ὑπόμνημα περὶ τοῦ πῶς χρὴ πολιορκουμένους ἀντέχειν in the superscription (159ᵛ) is clearly descriptive rather than titular. The subscription reads Αἰνείου πολιορκητικά (181ᵛ). Even if, as seems likely, both these designations are late interpolations, they are potentially instructive with regard to tenth-century perceptions of the work. For discussion of the authentic title see Schöne (1911) xiii–xiv; Dain and Bon (1967) xii–xiii; xxii–xxiii; Whitehead (2002) 16–17, favouring Πολιορκητικὴ βίβλος.

46 For modern attempts to locate Aineias' *Poliorketika* within the genre see Lammert (1940) 284; Whitehead (2008a) 143–144. In Byzantine usage τακτικά and στρατηγικά and their cognates are widely attested as both titles and generic labels for diverse works connected, even tangentially, with military affairs. The terms τακτικά and στρατηγικά were often used loosely and interchangeably, and without exact equivalence to modern conceptions of 'tactics' and 'strategy', see Rance (forthcoming b). Nevertheless, tenth- and eleventh-century sources differentiate texts of primarily technological content as a distinct subgenre. E.g. in a reading-list of military instruction compiled for his son, Constantine VII distinguishes 'books on generalship' (βιβλία στρατηγικά), 'books on engines' (βιβλία μηχανικά) and 'artillery manuals' (βελοποιϊκά), see Text C.106.196–199 in Haldon (1990) with remarks at 210–211. Cf. similarly Psellos, *Chron.* 7 (Mich. VII) §16 (ed. É. Renauld [Paris 1926–1982] II 181.12–13): ἀπὸ τῶν τακτικῶν βιβλίων καὶ στρατηγικῶν καὶ πολιορκητικῶν. The tenth-century anonymous *De Re Militari* 27 (Dennis [1985] 316.8–13) explicitly eschews discussion of siege engines and refers the reader to a separate branch of scientific literature. For similar sentiments in classical treatises: e.g. Onas. 42.3–4; Front. *Strat.* 3.pr. with remarks by Whitehead (2008a) esp. 144–146.

clearly identified in the textual history of other ancient military treatises.[47] The broader reception of Aineias' *Poliorketika* assists in clarifying the chronology of these accretions. Identical or similarly-worded headings are affixed to excerpts of Aineias' work incorporated into the so-called *Apparatus Bellicus*, a late ninth-/early tenth-century Byzantine compendium transmitted in other, unrelated codices (V, N-E, P-B).[48] As the text of these excerpts and the text in M demonstrably descend from two different majuscule archetypes, the introduction of the chapter-headings necessarily antedates the divergence of the two majuscule traditions. According to Dain's calculation, this occurred before c. 400 (see below).[49]

Although written in a clear minuscule script, the text of Aineias' work in M is perhaps the most badly corrupted and lacunose of any Greek military treatise (possibly accepting Arrian's *Ektaxis*), and a glance at the apparatus criticus in the Budé edition reveals the many *loci desperati* and the layers of scholarly restorations and conjectures, dating back to the *editio princeps* of Isaac Casaubon (1609), with which successive editors have sought to repair the damage. Dain believed that the most serious complications originated in the exemplar copied by the scribe rather than his own transcriptional error. It seems that the copyist mechanically reproduced the text precisely as it appeared before him, even where it made little or no sense, and there is no evidence of his editorial intervention or attempt at correction. Where he was able, he preserved lex-

---

47 For example, this is similarly the case with the received text of Onasandros' *Strategikos*, divided into 42 chapters, whose headings contain lexical and syntactical usages not found in the text, including terms otherwise unattested before late antiquity, e.g. τὰ ἄπληκτα (ch. 9), σφενδονιστής (17), ἐγκρύμματα (22). In the 'authentic' tradition (M), although Onasandros' treatise is divided into chapters, there is no trace of chapter-headings. These are present in the 'interpolated' tradition (V, P-B, N-E), but more often as marginalia than within the text. See further Petrocelli (2008) 137–138.

48 See remarks of Schöne (1911) x–xi. Identical chapter-headings: e.g. *App. Bell.* 49 Πυλωρικόν = Aineias 28 Πυλωρικά; *App. Bell.* 50 = Aineias 29.1–10 Ὅπλων λάθρᾳ εἰσκομιδή; *App. Bell.* 54 = Aineias 37 Ὑπορυσσόντων γνῶσις καὶ κώλυσις; similar wording at e.g. *App. Bell.* 45 Πῶς ἡμεῖς ἐμπρήσωμεν = Aineias 33 Ἔμπρησις; *App. Bell.* 51 Περὶ κρυφίας ἐπιστολῶν εἰσπομπῆς = Aineias 31 Περὶ ἐπιστολῶν κρυφαίων; *App. Bell.* 56 Δολίευμα = Aineias 39 Δολεύματα.

49 See Dain and Bon (1967) xxxii–xxxiii. Previously Oldfather (1923) 19; Hunter and Handford (1927) xxxix n. 4, and even Dain (1935) 6 n. 2, who certainly knew better, assumed that the chapter-headings must predate the early third century, but this line of reasoning was based on the misconception that the *Kestoi* of Julius Africanus (c. 227–231) incorporates excerpts from Aineias' *Poliorketika*, whereas in fact excerpts of both Aineias and Africanus occur together in the ninth-/tenth-century *Apparatus Bellicus* (see below pp. 327–331). See the correction by Bettalli (1990) 7.

ical, orthographic and syntactic peculiarities of the original, but he employed a supralinear symbol (resembling a capital sigma) to mark passages he could not understand or decipher with confidence, and left blank spaces corresponding to lacunae in his model. Given this degree of scribal diligence, Dain explained the exceptionally poor state of the text of the *Poliorketika* in M by supposing that the single ancient exemplar available to the copyist was already gravely defective owing to previous careless transcription and physical damage.[50] This is not to say that the scribe did not introduce new errors; indeed more recent studies of other texts transmitted in M, where additional manuscript witnesses are available for comparison, have drawn a somewhat less favourable opinion of his capabilities.[51] Most conspicuous among the errors attributable to the scribe is the mistaken ascription of Aineias' work to Aelian, which apparently occurred when the scribe (or a rubricator) conflated its superscription with the subscription of the preceding text, Aelian's *Taktike Theoria*. It is perhaps significant that additional confusion occurs in the rubrication elsewhere in M.[52]

---

50 Dain and Bon (1967) xxxiii, xxxvi–xl.

51 In his analysis of the manuscript tradition of the *Strategikon*, Dennis (1981) 31–33 concluded that the majority of errors unique to M originate with the copyist rather than his model. See similarly Dennis (1985) 6.

52 The superscription (159^{v}) reads Αἰλιανοῦ τακτικὸν ὑπόμνημα περὶ τοῦ πῶς χρὴ πολιορκουμένους ἀντέχειν. The subscription to Aineias' work (181^{v}) exhibits corresponding confusion: Αἰνείου πολιορκητικά· ἢ Αἰλιανοῦ καθὼς ἡ ἀρχή, followed by an erasure of 13 letters. The latter appears to combine an authentic ascription and a misguided editorial attempt to the reconcile conflicting attributions. See Dain (1935) 7–8; Dain and Bon (1967) xxx–xxxiv; Whitehead (2002) 4–7. The error in the superscription in M could not have been present in an earlier manuscript used by the copyist, as his model for the text of Aineias was demonstrably a majuscule exemplar, while he copied the text of Aelian from a miniscule manuscript (see *supra* n. 44); it is highly unlikely that the two manuscripts of different periods were conjoined. Furthermore, comparable confusion occurs earlier in M in the superscription to the *Strategikon* (3^{r}), which uniquely ascribes the work to Urbicius (ΟΥΡΒΙΚΙΟΥ), a historical impossibility, rather than Mauricius (ΜΑΥΡΙΚΙΟΥ), as found in all other manuscript witnesses and the indirect tradition. The stemma codicum of the *Strategikon* strongly suggests that the copyist's exemplar (ε) correctly read ΜΑΥΡΙΚΙΟΥ, see Förster (1877) 455–459; Dain (1968) esp. 136; Dennis (1981) 17–18, 41; Rance (forthcoming b). All previous literature has held the copyist responsible for these blunders, but the book titles and chapter-headings in M are rubricated and it is uncertain whether the scribe himself was the rubricator or, as was often the case, a separate specialist performed this task on the scribe's instructions, an additional link in the chain of production that could only have increased the likelihood of error. Additional problems with the rubrication in M (71^{r}) occur in the so-called *De Militari Scientia* or Müller/Müllerische Fragment, in which the headings of chapters 7–9 are muddled or duplicated, see ed. Müller (1880) 119–120.

Nevertheless, there can be no doubt that in the case of Aineias' *Poliorketika*, and some of the other ancient treatises assembled in the second corpus, the scribe encountered particular difficulties arising from the age, form and condition of his exemplar. Of relevance here is the wider scholastic development, beginning in the later eighth century, whereby minuscule script superseded majuscule or uncial, with the result that all extant Greek texts written before c. 800 have undergone a process of transliteration from the old to the new script, for secular works typically c. 850–c. 950, after which the majuscule copies were ignored or discarded as redundant.[53] The fact that the text of Aineias' work in M does not contain a single transcriptional error that would typically or exclusively arise from a misreading of minuscule indicates that the exemplar available to the copyist was written in majuscule and M is itself the transliteration.[54] Furthermore, in a magisterial display of codicological expertise, Dain showed that the incidence and spacing of certain lacunae left by the copyist of M were indicative of physical damage in his exemplar. On the basis of a well-established chronology of book formats in antiquity and the patterns of damage typically suffered by manuscripts, Dain was able to calculate the dimensions and layout of the text in the exemplar and even in its ancestor, and thus reconstructed two prior stages of the text's transcriptional prehistory. Dain demonstrated that the exemplar used by the copyist of M must have been a codex, almost certainly of papyrus rather than parchment, and, judging by its format, most probably dating to the first half of the fifth century. This exemplar in turn was a descendant of a papyrus volumen with a layout suggestive of the later second century. In consequence, Dain argued persuasively that the copyist of M had been instructed to transliterate the text of Aineias' *Poliorketika* from a late antique majuscule papyrus codex, presumably discovered or identified by systematic research in the palace library or another Constantinopolitan repository.[55] Even if some details of Dain's analysis cannot be substantiated, it seems that the sole copy of Aineias' *Poliorketika* available to the staff of the imperial scriptorium was an especially decayed and physically damaged text, which had for around five centuries attracted none of the editorial attention or fresh transcriptions that would typically signal a work's popularity and contemporary or

53 See generally Lemerle (1971) 113–127; Wilson (1983) 63–68; in Greek tactical writing: Dain (1946) 119–122; Dennis (1981) 31–38; Rance (2007a) 736–737.

54 Dain (1954) 43–44; Dain and Bon (1967) xxxiii, xxxvi–xl. Within the Greek tactical tradition an even earlier specimen of a transliteration is *Parisinus suppl. gr.* 607, dated c. 925–950 (*supra* n. 1). In the absence of a more recent analysis of this aspect of the codex see Wescher (1867) xx–xxi.

55 Dain and Bon (1967) xxxiii–xxxvi.

recent readership. The codicological evidence for the direct tradition, therefore, like the scarcity of testimonia, suggests that Aineias was a little-known and little-read author who was lucky to survive extinction.

The 'Poliorcetic Tradition', Philon of Byzantion and Syrianos Magister

Dain noted with surprise that, despite their shared subject matter, no trace of Aineias' *Poliorketika* can be discerned in Byzantine manuals on siege warfare. Dain had in mind specifically two anonymous compilations, the so-called *De Obsidione Toleranda*, probably written in the first quarter of the tenth century, and a document he called *Mémorandum sur la défense des places*, both of which drew on a common source, now lost and difficult to date (Dain's '*Antipoliorceticum*').[56] Nor can direct knowledge of Aineias' treatise be demonstrated in the case of two other works relating to siegecraft, the tenth-century anonymous *Parangelmata Poliorketika* (formerly 'Heron of Byzantion') and a surviving section of the military compendium of Syrianos Magister (formerly 'Anonymus Byzantinus'), traditionally dated to the later sixth century but in recent studies reassigned to the ninth. Nevertheless, given the broader textual tradition from which these last two works descend, it is worth considering the possibility that Aineias-derived material might have flowed into Byzantine poliorcetic writing via an indirect route.[57]

The main conduit for this indirect tradition was Philon of Byzantion, a writer on mechanics and civil and military engineering active c. 200 BC.[58] Little is known of his life and career: Philon mentions sojourns in Rhodes and Alexandria, where he consulted artificers in military workshops, and he implies that

56 Dain and Bon (1967) xlii–xliii. Editions: *De Obsidione Toleranda* ed. van den Berg (1947); repr. with Eng. trans. in Sullivan (2003a); *Mémorandum*: ed. with Fr. trans. in Dain (1940b) 124–130; with Eng. trans. in Sullivan (2003a) 145–149. Dating: Sullivan (2003a) 139–141. For the textual interrelationships see Dain (1940b) 131–136; summarised in Dain and de Foucault (1967) 349–350, 359, 366–367, with additional observations by van den Berg (1947) 19–30.

57 For overviews of Byzantine siegecraft and poliorcetic manuals see McGeer (1995b); Sullivan (1997); (2003b).

58 See Dain and de Foucault (1967) 323–324 with older bibliography; Garlan (1973); (1974) 281–285; Whitehead (2016). Subsequent references to Philon's *Paraskeuastika* and *Poliorketika* are cited after Garlan (1974) 291–404 followed in parentheses by the conventional numeration of Thévenot (1693) 79–104.

he was somewhat younger than the celebrated Alexandrian inventor Ktesibios (c. 270–230 BC), whose discoveries in pneumatics and ballistics he refined, though their personal acquaintance is uncertain. Philon's major work was a large technological compendium entitled *Mechanike Syntaxis*, apparently the first of its kind, in eight or nine books, now mostly lost, though its content and arrangement can be surmised from internal references. Surviving sections are addressed to Ariston, an unidentified correspondent to whom each volume was sent independently upon completion in accordance with a previously outlined programme of enquiry.[59] The extant Greek text seems to preserve most of the military content of the compendium, probably a reflection of late antique and/or Byzantine editorial tastes and perspectives, namely book IV (*Belopoiika*) in its entirety, together with substantial excerpts or abridgements of VII (*Paraskeuastika*) and VIII (*Poliorketika*), which older scholarship considered a single book (formerly misnumbered V). Philon's *Belopoiika* contains technical specifications for the construction of artillery, principally torsion-powered catapults, in which he critiques recent developments and proposes improvements to both standard designs and famous inventions. The *Paraskeuastika* and *Poliorketika* constitute a comprehensive guide to defending and attacking a Hellenistic city. The *Paraskeuastika*, the only treatise on fortification to have survived from antiquity, concerns the design, layout and construction of walls, towers, battlements and outworks, as well as the organization of manpower and provisions, while the *Poliorketika* considers engineering, equipment and tactics used by both besieger and besieged.[60] In all three volumes Philon variously combines written sources and empirical knowledge obtained from technical informants and/or personal expertise. Previous scholarship has drawn attention to close correspondences, in wording, content and/or outlook, between books VII and VIII and Aineias' *Poliorketika*, sufficient to confirm Philon's extensive but selective use of Aineias' treatise.[61]

Philon's dependence on Aineias appears also to have extended to a lost volume of the *Mechanike Syntaxis*. Towards the end of his *Poliorketika*, Philon states that he intends to write a further book 'on the sending of secret let-

59 For convenient summaries of Philon's life and works: Garlan (1973) 16–19; (1974) 281–285; Whitehead (2016) 20–25.

60 Text: ed. with Fr. trans. and comm. in Garlan (1974) 291–404; partial Eng. trans. in Lawrence (1979) 67–107, and now a complete Eng. trans. and comm. in Whitehead (2016), which appeared too late to be fully integrated into this chapter.

61 Lammert (1940) 286–287; Garlan (1974) 366, 374–376, 378–379, 383–386, 392–393, 401–403; Whitehead (2016) esp. 17, 35–59, with index locorum at 464–466.

ters' (περὶ ἐπιστολῶν τῶν κρυφαίως ἀποστελλομένων).[62] This book, conventionally regarded as the next or ninth volume, has not survived; in fact there is no certainty that it was ever written, though some confidence can be gleaned from the observation that all other such projections in Philon's extant writings foreshadow an actual publication. Given Philon's familiarity with Aineias' *Poliorketika*, an obvious point of reference for this topic was chapter 31, a lengthy self-contained excursus on cryptography, or more precisely steganography, the earliest known formal treatment of the subject, which constitutes almost an eighth of Aineias' treatise, perhaps indicative of his particular interest or expertise in this field.[63] In a preview to his forthcoming study (D.76–82 [102.29–50]), Philon outlines four methods of sending a secret message: the first appears to be of his own devising, but the second is a variation of a technique described by Aineias, while the remaining two so closely correspond to Aineias' text in substance and vocabulary as to leave no doubt that he was Philon's source.[64] If this preliminary sample accurately reflects the eventual shape and character of Philon's book on secret messages, it seems likely that it was heavily indebted to Aineias' chapter 31 as both a conceptual model and a source of technical information. From this modest conclusion no line of enquiry can extend beyond guesswork. In a study printed in 1932, Jean-René Vieillefond inferred that modified fragments of Philon's lost book, identifiable by evidently Aineias-derived content, may be traced in certain tenth-century Byzantine military compilations, though he based his analysis on an incomplete knowledge of largely unpublished manuscripts. Vieillefond's view was endorsed by Dain, seemingly without further enquiry, and is sporadically cited in subsequent studies. It will

62 Philon D 82 (102.47–50), 'There are many other ways of sending messages secretly, as I shall show in the exposition on the sending of secret letters' (Πολλοὶ δὲ καὶ ἄλλοι τρόποι εἰσὶ τῶν κρυφαίως ἀποστελλομένων γραμμάτων, ὡς δηλώσομεν ἐν τῷ εἴδει τῷ περὶ ἐπιστολῶν τῶν κρυφαίως ἀποστελλομένων). It is uncertain whether the latter is a title or merely descriptive.

63 For commentaries on Aineias 31 see Bettalli (1990) 302–312; Whitehead (2002) 183–192, and studies by Jenkins (2006) 51–59; Debidour (2006), citing older bibliography on ancient cryptography at 234.

64 Philon D.77 (102.31–36) offers his own proposal whereby a message is written on a sunhat (καυσία) in a type of 'invisible ink', whose lettering is revealed to the recipient once he has subjected the hat to chemical treatment. Philon D.78 (102.37–39), a message written on a thin membrane stitched inside the hat, is a variation of Aineias 31.23, where the means of concealment is clothing. Philon D.79 (102.39–40), a message hidden inside the sole of a shoe, and D.80 (102.40–47), a message written on a bladder inflated inside an oil flask, derive from respectively Aineias 31.4–5 and 10–13. Philon's dependence on Aineias here was recognised as early as Casaubon (1609) 80, 84, 86–87. See also Vieillefond (1932a) 30–33 (with caution); Garlan (1974) 402–403; Whitehead (2016) 386–390.

be shown in a following section that this hypothesis, however tantalising, is demonstrably false; indeed Vieillefond himself noted inconsistencies in his reconstruction which pointed to alternative explanations.[65]

While the fate of Philon's book on secret messages remains wholly obscure, there can be no doubt that his *Paraskeuastika* and *Poliorketika* became sources for two of the above-mentioned Byzantine writers, the author of the *Parangelmata Poliorketika* and Syrianos Magister. The first is in many ways the more straightforward. *Parangelmata Poliorketika* is the conventional title for a largely derivative poliorcetic compendium by an unknown compiler and commentator. The unique manuscript prototype originally lacked an indication of author and title, but a later Byzantine editor interpolated a false ascription to a 'Heron', hence the now largely obsolete modern pseudonyms 'Heron of Byzantion' and 'Heron the Younger'. Internal references suggest that the author was writing in Constantinople in the late 930s–940s.[66] His expressed objective was to rework, supplement and update selected classical authorities on siegecraft with a view to making ancient texts more accessible to a contemporary non-specialist readership.[67] For this purpose he had access to a manuscript containing a corpus of Hellenistic and Roman poliorcetic writers, similar in format to the three corpora of military texts we previously encountered in codex M (see above). This poliorcetic corpus, preserved today in four manuscript prototypes, included the surviving military sections of Philon's *Mechanike Syntaxis*.[68] Accordingly, insofar as the *Parangelmata Poliorketika* selectively reprises Philon's *Paraskeuastika* and *Poliorketika*, the tenth-century compiler became a potential heir to mate-

65 See below pp. 347–356. Vieillefond (1932a), endorsed by Dain (1931) 341–342; (1937) 54, 64–65; (1939) 45, 53–54, 61; summarised Dain and Bon (1967) xlii, xliv, xlvii, 72 n. 1; Dain and de Foucault (1967) 339, 352–353. Dain further asserted that material from Philon's lost book on cryptography had passed into Polyainos' *Strategemata*, though he offered no grounds or argumentation, and it remains ambiguous whether he envisaged a direct or indirect transmission, see Dain (1939) 53 n. 4; Dain and de Foucault (1967) 339, 352; Dain and Bon (1967) xlii.

66 For authorship, date and historical context: Sullivan (2000) 2–4, 15–21 citing older bibliography.

67 For the author's aims and methodology: Sullivan (2000) 4–15.

68 The corpus of Hellenistic and Roman *poliorketika* is transmitted in two collateral branches of the manuscript tradition. The first is represented chiefly by P (c. 925–950), and more tenuously as excerpts in fourteenth-century *Vindobonensis phil. gr.* 120. Philon's works are now lost from the corpus in P, doubtless owing to their terminal position and the chaotic history of this composite codex, see *supra* n. 1. The second branch is witnessed by early eleventh-century V, P-B and N-E. The *Parangelmata Poliorketika* is affiliated to the first branch of this tradition, see Dain (1930) 19–20; Sullivan (2000) 4–5.

rial that Philon had in turn drawn from Aineias' writings.[69] This second-hand inheritance remained both meagre and illusive, however, largely owing to the compiler's use of Philon's work primarily as a source for offensive devices and techniques employed by besieging forces, a perspective that necessarily minimised the degree of content originating in Aineias' defence-orientated *Poliorketika*.[70]

The case of Syrianos Magister is more complicated, both in its textual transmission and modern bibliography. Syrianos was the author of three extant military texts, unconnected in the manuscripts and at present published separately, which were until recently deemed anonymous and/or unrelated, largely owing to complications in twentieth-century scholarship, even though numerous preceding studies since the seventeenth century had consistently pointed to their textual unity.[71] The three items—*De Re Strategica*, *Rhetorica Militaris* and *Naumachiae*—are now recognised as surviving elements of a comprehensive and well-structured compendium of military science, which was apparently sliced into thematic sections by tenth-century editors. The largest and most familiar part, known by the modern title Περὶ Στρατηγικῆς or *De Re Strategica*, and hitherto ascribed to 'Anonymus Byzantinus' or 'the Sixth-Century Anonymous' owing to the loss of its beginning in the manuscript prototypes, covers diverse aspects of land warfare, including the construction and defence of fortified sites, deployment, tactics, weaponry and training.[72] *Rhetorica Mil-*

---

69 For Philon-derived content in the *Parangelmata Poliorketika* see e.g. Garlan (1974) 365, 377, 386, 393, 395–396, 398–399, 401; Sullivan (2000) 156–157, 167–171, 175, 177–182, 185–186, 216, 223, 225–226, 228–231, 247; Whitehead (2016) 19–20, 412–416, 476.

70 One possible instance, though if so a distant echo, occurs at *Parang. Polior*. 39.8, where the author refers to fire-bearing caltrops (πυροφόρων τριβόλων) thrown by the defenders in order to fasten incendiaries to the planking of a siege tower. This passage is based on Philon C.41 (94.8–10): καὶ πυροφόρους ὡς πλείστους καὶ τριβόλους καιομένους στιπτύῳ περιειλι<γμένους, cf. also C.55 (95.8); D 49 (100.20–21); D 104 (104.19–21). Garlan (1974) 386 draws a comparison with Aineias' description (33.2) of large pestle-shaped pieces of wood studded with iron spikes used to deliver incendiary material on to the enemy's mantlets. See additional remarks and bibliography in Sullivan (2000) 216–217; Whitehead (2016) 207–208, 304. Another example, with differences of detail, may be Aineias 32.8 > Philon A.76 (85.23–29) > *Parang. Pol.* 11.9–15, relating to concealed trenches designed to collapse under the weight of engines; see Garlan (1974) 365–366; Sullivan (2000) 179; Whitehead (2016) 205–206.

71 See most recently Zuckerman (1990) esp. 210–217 with summary of older scholarship; augmented with additional proofs by Cosentino (2000) 248–262; see also Rance (2007a); Eramo (2008); (2012).

72 The text has been edited twice: Köchly and Rüstow (1853–1855) II.2 1–209, with Germ.

itaris, long attributed to the same 'Anonymous', combines an exposition of rhetorical principles with historically orientated exempla of military oratory, and thus uniquely spans the two genres of instructional manuals on rhetoric and warfare.[73] A third section, usually entitled *Naumachiae*, is the oldest technical treatment of naval warfare in Greek. It alone bears an ascription to Syrianos Magister, but recognition of the common parentage of all three parts necessarily assigns the other two to this author.[74] All three sections of the compendium draw selectively on classical sources, several of which have been identified.[75] Syrianos' detailed consideration of topics such as fortifications (9–13), fire-beacons (8) and river-crossings (19) has prompted some scholars to infer engineering interests or expertise, though these chapters form a small fraction of the whole oeuvre and were certainly inspired by earlier literature.[76] Syrianos was essentially retrospective and antiquarian in method, if not in purpose, and the structure, content and idiom of his compendium appear to have been shaped more by literary antecedents and cultural milieu than firsthand experience or technical knowledge.[77] When Constantine VII advised his son on suitable reading matter for a commander-in-chief, he recommended 'historical books, above all Polyainos and Syrianos'. If this compendium is meant,

---

trans. and notes (311–355), as 'Byzantini anonymi Peri Strategikes'; Dennis (1985) 10–135, with Eng. trans., as 'The Anonymous Byzantine Treatise on Strategy'.

73 See now the new edition with It. trans. and comm. by Eramo (2010 [2011]), which supersedes Köchly (1855–1856).

74 For text of the *Naumachiae* and Eng. trans. see Pryor and Jeffreys (2006) 455–481, which supersedes Dain (1943) 41–55. In the unique manuscript witness, *Ambrosianus gr.* 139 (B 119 sup.), the first folio of the *Naumachiae* is lost, but on the verso of the preceding folio (332v) Dain was able to decipher a reversed trace of the words Ναυμαχίαι Συριανοῦ Μαγίστρου, see Dain (1937) 66–67.

75 Syrianos' sources are summarised in Rance (2007a) 703–706, with additional remarks in Eramo (2008) esp. 125–134.

76 For presumed engineering expertise and/or interests: e.g. Köchly and Rüstow (1853–1855) II.2 38; Kuchma (1980) 73, 89; but doubted by Zuckerman (1990) 215; Cosentino (2000) 278. Syrianos' treatment of fortifications and fire-beacons will be discussed in more detail below pp. 318–319. In his chapter on river-crossings (19.22–55) Syrianos critiques a design for a floating bridge attributed to Apollodoros of Damascus, cf. Apoll. *Polior.* 189.1–193.5, with remarks by Blyth (1992) 142, 157; Whitehead (2010) 132–136; Eramo (2008) 133–134.

77 Dain and de Foucault (1967) 343 labels Syrianus 'un stratège en chambre', with similar assessments by Vári (1906) 70; Zuckerman (1990) 215; Cosentino (2000) 278; Rance (2007a) 703–706, 716–719. For a slightly higher estimate of the author's utility: Müller (1882) 48; Kuchma (1980) 73–74, 89–90. For suggestions concerning Syrianos' cultural milieu see esp. Cosentino (2000) 277–280.

the emperor presumably alludes to the predilection of both authors for adapting ancient sources and citing historical exempla.[78] Far more problematic is the date of Syrianos' work. The largest section, *De Re Strategica*, which has attracted the greatest interest, was conventionally dated to the reign of Justinian (527–565), following the surmise of its first editors, Herman Köchly and Wilhelm Rüstow (1855).[79] Upon closer scrutiny, the evidence they adduced largely evaporates or at best does not preclude a later date. The only certain termini merely locate the work between the mid-sixth century and the beginning of the tenth century, when it was known to Leo VI in the composition of his *Taktika* (c. 905).[80] In recent years a growing chorus of scholars has raised objections to the traditional dating, drawing attention to features incompatible with the sixth century and more congruent with a Middle Byzantine context, and a date of composition in the mid to late ninth century has gained favour in recent works of reference. If this view is correct, and it so far stops short of a new orthodoxy, the compendium of Syrianos Magister becomes the earliest known work in the ninth-/tenth-century revival of Byzantine military literature.[81]

---

78 Const. Porph. Text C.198–199 in Haldon (1990) 106 (= *De Cerimoniis*, app. II, ed. J. Reiske [Bonn 1829–1830] 467.8–9): βιβλία ἱστορικά, ἐξαιρέτως δὲ Πολύαινον καὶ Συριανόν. For Constantine's familiarity with Syrianos' *Rhetorica Militaris* see McGeer (2003) 112–114; Eramo (2007) 138–141, 160.

79 Köchly and Rüstow (1853–1855) II.2 37–38; followed by e.g. Dain (1937) 55–56; Dain and de Foucault (1967) 343; Hunger (1978) II 327; Kuchma (1980) esp. 73–75; Dennis (1985) 2–3; Zuckerman (1990); (1994) 361, 364.

80 The earlier terminus is affirmed by an incidental allusion to the generalship of Belisarios (*De re strat.* 33.35), active c. 530–559, though there is no certainty whether the reference is contemporary or historical, see Cosentino (2009) 263; Rance (2007a) 709–711. Leo VI's use of both *Rhetorica Militaris* and *Naumachiae* in the composition of his *Taktika* (c. 905) excludes a later date, see Dagron (1983) 226–232; Cosentino (2000) 260–261 with n. 79; Pryor and Jeffreys (2006) 180–181, 388, 395–396, 401–402; Haldon (2014) 18–21, 44–45, *contra* Trombley (1997) 272. Manuscripts of the 'Laurentian' recension of Leo's *Taktika* contain a scholiast's marginal source-notice at proem. 6 (Dennis [2010] 6 app. crit.) which includes Συριανοῦ, see Vári (1898) 66–68; (1906) 68; (1917–1922) I xxxiii n. 1; Dain (1946) 150–151; (1946–1947) 35; Dain and de Foucault (1967) 373.

81 All recent studies have contested the traditional sixth-century dating. This was first challenged by Baldwin (1988), whose objections deny a date before the 630s. For subsequent proponents of a ninth-century context see Lee and Shepard (1991) 25–30; and at length Cosentino (2000) esp. 262–280; (2001) 84–87. In Rance (2007a) I adduced additional positive arguments in favour of composition in the ninth century, which, in my view, make a sixth-century date hard to defend. The sixth-century dating is doubted by McGeer (1995a) ix, 348 n. 31; (2008) 910; while a ninth-century date is now endorsed by Pryor and Jef-

Syrianos provides a lengthy treatment of defensive arrangements in a frontier zone, including intelligence gathering, fire-signals, and the construction and fortification of a sample fortress or city, together with its defence against a formal siege by a well-resourced opponent equipped with machinery and engineering capabilities (*De Re Strat.* 6–13). It seems that a section on offensive siegecraft has been lost, presumably when the compendium was dismembered.[82] In the chapters concerning the defence of a city (12–13), specific correspondences of language and/or substance provide firm evidence of Syrianos' familiarity with Philon's *Paraskeuastika* and *Poliorketika*.[83] There are, however, no conclusive instances of Aineias-derived content transmitted into Syrianos' text via Philon's writings. This is in large part due to Syrianos' highly selective use of Philon's two books, chiefly in relation to the construction of fortifications, a focus that reduces the potential for second-hand borrowings from Aineias' *Poliorketika*. So much can be said with confidence. Older scholarship, beginning with Köchly and Rüstow, entertained the possibility that Syrianos might have enjoyed direct access to Aineias' treatise, though they did not develop this thesis beyond a tentative identification of superficial similarities in Syrianos' chapters on sentries and fire-signals (7–8).[84] The only scholar to pursue this proposal was Friedrich Lammert, who in 1940 drew attention to

freys (2006) 180; Haldon (2008) 474; Mecella (2009) 96–98; Sullivan (2010) 151–152. To my knowledge, no counter-arguments have been presented.

82 At 13.61–62 Syrianos observes 'we have explained in the section on siegecraft how mining devices called tortoises are constructed' (τὰς δὲ διωρυκτῆρας καλουμένας χελώνας ὅπως γίνονται ἐν τοῖς πολιορκητικοῖς ἡμῖν ἐκπεπόνηται), a cross-reference that does not correspond to any part of the extant text. Again, with regard to the danger posed by ships to seawalls, cf. 11.12–13, 'this is made clear in the section on siegecraft' (ὥστε φανερὸν τοῦτο ἐν τοῖς πολιορκητικοῖς). For this lost section see Lammert (1940) 283; Rance (2007a) 704 n. 10.

83 The most comprehensive analysis is now Eramo (2008) 131–133 (whose examples can be supplemented with Philon C.4 [91.9–10] > Syr. 13.115–120); see also Garlan (1974) 342, 361, 377, 380; Ravegnani (1983) 29–30; Whitehead (2016) 19, with index locorum at 481.

84 Köchly and Rüstow (1853–1855) II.2 13, 'Nicht unwahrscheinlich ist es übrigens, daß unser Anonymus [Syrianos] selbst bei der Bearbeitung einiger seiner Capitel noch den Aineias benutzt hat. Irren wir nicht, so erinnern an diesen namentlich die Abschnitte "von den Wachen" und "von den Fanalen" (VII und VIII)'; similarly II.2 38, 'zu den ἀρχαιότεροι, die er VIII, 6 bei Gelegenheit der Fanale anführt, möchte vorzugsweise Aineias selbst gehören'; see also II.2 315. Again at II.2 38, '… auch die πρὸ ἡμῶν, deren Anstalten gegen die Breschschildkröten er [Syrianos] XIII, 16.17 erwähnt, möchten bis auf diesen [Aineias] zurückgeführt werden können'. Müller (1882) 48 gives qualified approval; similarly Lammert (1940) 283. Upon closer inspection, the respective treatments of sentries by Aineias (6) and Syrianos (7), which both contain a large number of recommendations, diverge almost entirely and coincide at only two points: that sentries should be posted through-

certain parallels between Aineias' *Poliorketika* and Syrianos' long chapter on defensive siege warfare (*De Re Strat.* 13), which were, in his view, suggestive of a source-relationship. It must be stressed that the affinities which Lammert all too easily identified in several Greek and Latin texts on siegecraft relate to commonplace defensive techniques widely documented in ancient and mediaeval historical sources and need not depend on or imply a textual connection. In fact, the correspondences he alleged between Aineias and Syrianos do not survive closer scrutiny: none entails an exact correlation and in most cases the two texts barely approximate; there are often significantly different contexts or conflicting details, and verbal parallels are negligible. More to the point, and fatal to Lammert's argumentation, in many of these passages Syrianos' source is demonstrably Philon and not Aineias.[85]

out the night (Aineias 22.5; Syr. 7.8–9) and that men of property have a vested interest in a city's security (Aineias 22.15; Syr. 7.6–7); even here there are differences of detail.

85 See Lammert (1940) 283–286, partly inspired by Köchly and Rüstow (1853–1855) II.2 38. See also Kuchma (1980) 78–81, who reiterates some of Lammert's observations. The closest parallel is Syr. 13.62–69, where Syrianos describes methods used by 'our predecessors' (οἱ πρὸ ἡμῶν) to destroy enemy tortoises (cf. also 13.105–114); this passage exhibits some broad similarities to Aineias 33.1–2, but there is significant divergence and no verbal correspondence. For other alleged points of resemblance cf. Syr. 12.37–46, 13.24–42 describes ditches dug parallel to the wall to intercept mines; Lammert cites Aineias 37 but verbal parallels indicate that Philon is Syrianos' source: e.g. Philon C.7 (91.20–22): … ὀρυκτέον ἐστὶν ἀνὰ μέσον τοῦ τείχους καὶ προτειχίσματος ἱκανὴν τάφρον ἰσηλίκην κατὰ βάθος τῶν θεμελίων τῷ κατὰ γῆς; Syr. 12.37–40: Καλὸν δὲ καὶ πρὸ τούτου τάφρον ἀνορύττειν ὥστε δυσὶ τούτοις, προτειχίσματι καὶ τάφρῳ, τὸ τεῖχος φυλάττεσθαι … τὸ δὲ βάθος πλέον ἢ ὁπόσον ἂν ἔχοι τὸ βάθος τῶν θεμελίων (see Whitehead [2016] 278–279). Syr. 13.49–60 advises building an internal counter-wall to screen a breach, as does, very briefly, Aineias 32.12, but again Philon is Syrianos' source: cf. Philon C.18 (92.22–27): Ἀντοικοδομητεον τέ ἐστι καὶ ταύτῃ τρίγωνον τεῖχος ⟨ἔσω⟩ τὸ ἔμβολον ἔχον > Syr. 13.51–60: οἰκοδομήν τινα ἔνδοθεν … τὸ δὲ σχῆμα τοῦ τείχους ἔστω τρίγωνον, μιᾶς λειπούσης πλευρᾶς (see Garlan [1974] 380; Dennis [1986] 136, fig. 3; Whitehead [2016] 288). Syr. 13.69–91 recommends matting *hung over* the battlements to protect the defenders against missiles fired by besieging troops accompanying the mantlets; Lammert attempts to draw parallels with Aineias, but his prescriptions are significantly different, cf. 32.1 (taut sails *erected on top of* the battlements as a screen against missiles fired from towers or masts); 32.9–10 (similar screens *projecting inside* the walls to prevent missiles from harming the inhabitants); 33.3–4 (flame-resistant covering for wooden walls). Syr. 13.115–120 recommends netting suspended from the battlements to absorb stones; Lammert cites Aineias 32.3, but the latter prescribes sacks of absorbent material hung down to cushion the blows of a ram; in any case Philon is undoubtedly Syrianos' source: Philon C.4 (91.9–10): ἐκ τῶν σχοινίων πλέξαντας δίκτυα > Syr. 13.115–120: σχοινία εἰς σχήματα δικτύων πλέξαντες. Syr. 13.121–135 explains the use of a heavy stone, suspended on ropes from a projecting crane, to crush

Beyond Syrianos' unproven debt to Aineias' *Poliorketika*, Lammert formulated an altogether more elaborate hypothesis. Resurrecting an idea originally floated by Müller in 1882, Lammert proposed that Syrianos not only made use of the surviving *Poliorketika* but also had access to lost books by Aineias, now known chiefly through cross-references in his extant text. In short, in Lammert's opinion, it was 'at the very least highly probable' that Syrianos' wide-ranging military compendium was modelled on and drew extensive content from Aineias' complete oeuvre.[86] Lammert's hypothesis was largely overlooked (or misunderstood) in subsequent scholarship on Aineias, but its recent endorsement in a study of Syrianos warrants here a brief examination of the merits of his case.[87] Lammert sought to match sections of Syrianos' compendium with corresponding elements of Aineias' lost writings. Accordingly, he identified Syrianos' *Rhetorica Militaris* with Aineias' vague allusion to ἀκούσματα (38.5), which Lammert construed as the title of a formal treatise on military oratory, an interpretation that is far from certain.[88] By a similar line of reasoning, Syrianos' chapters on defensive arrangements, including fire-signals, fortifications and provisioning (chs. 6, 8–13), might be indebted to Aineias' lost 'Book on Preparations' (Παρασκευαστικὴ βίβλος), while his chapters on

rams or tortoises; Lammert cites Aineias 32.5–7, a much briefer notice lacking the detailed technical description of Syrianus; Philon C.10 (91.33–38) is equally brief but offers closer verbal parallels.

86 Müller (1882) 48–49 offered little more than tentative musings. Lammert (1940), esp. 280–288, argues that it is 'zum mindesten sehr wahrscheinlich' (280) that the various parts of Syrianus' compendium have a close textual relationship with Aineias' writings, both lost and extant, which Lammert appears to conceive as the underlying source for a nexus of military texts, including Philon's *Mechanike Syntaxis* VII (*Paraskeuastika*) and VIII (*Poliorketika*) and Vegetius' *Epitoma rei militaris* IV (on siegecraft and naval warfare). Along similar lines, H. Schöne, 'Aus der antiken Kriegschirurgie', *BJ* 118 (1909) 1–11 at 7 n. 3, had previously speculated that a passage of Aineias' lost Παρασκευαστικὴ βίβλος was the ultimate source of corresponding material in *App. Bell.* 37 (now considered an authentic excerpt from the *Kestoi*) and Philon B.53 (90.14–24). On these passages see Vieillefond (1970) 334; Garlan (1974) 374.

87 See Cosentino (2000) 278–289, 'It has been correctly stated that Syrianus wrote his … [compendium] to imitate the work of Aineias the Tactician' (citing Lammert [1940] 81–82). Cosentino envisages Syrianos' motivation as literary mimesis. Eramo (2008) 127–129 reviews the arguments with a more caution.

88 Lammert (1940) 281–282. Cf. Aineias 38.5: ἐν τοῖς ἀκούσμασι γέγραπται. This obscure expression has been differently interpreted as an internal cross-reference (to 26.8–10) or an allusion to a separate work of uncertain character; see discussion in Dain and Bon (1967) xvi–xvii; Bettalli (1990) 12 n. 36, 331; Whitehead (2002) 14, 201–202; Eramo (2008) 129.

sentries and encampment procedures (chs. 7, 26–29) could derive from a projected 'Book on Encampments' (Στρατοπεδευτική βίβλος). Likewise Syrianos' *Naumachiae* becomes the main heir to a lost treatment of naval warfare which, Lammert supposed, originally followed Aineias' *Poliorketika*.[89] As Lammert's hypothesis mostly concerns lost or putative books by Aineias, his argument amounts to little more than conjectural analogies with the scope and content of Syrianus' extant compendium. Upon closer inspection the alleged associations between the two opera soon begin to unravel. It is now understood that the theoretical component of Syrianos' *Rhetorica Militaris* is a selective simplification of rhetorical handbooks by or ascribed to Hermogenes of Tarsus (c. 160–c. 230), which were influential in the theory and instruction of oratory in late antiquity and the Middle Byzantine period. There is, in contrast, no trace of a fourth-century BC rhetorical antecedent.[90] Similarly, it has since been demonstrated that Syrianos' prescription for a fortified encampment, which clearly reflects Roman practice, is partly based a lost late antique monograph on castrametation, written in Greek before the late sixth century, which can be reconstructed from other Byzantine sources.[91] Somewhat more promising, at least initially, is Syrianos' extended treatment of fire-beacons used to warn of enemy incursions across the frontier (ch. 8), in which he alludes to a more sophisticated signalling system 'as has also been discussed by some of the more ancient authors (ὥσπερ καὶ παρά τισι τῶν ἀρχαιοτέρων εἴρηται)'. Lammert inferred that Syrianos had read Aineias' Παρασκευαστική βίβλος, which, as Aineias specifies (7.4), included a discussion of fire-signals. The argument is doubtful: this section of Aineias' lost book is cited *in extenso* by Polybios (10.44–47), who indicates that its content did not correspond to the mechanism or purpose Syrianos attributes to unnamed ancient authorities.[92]

89 Lammert (1940) 280–283. Aineias refers to his Παρασκευαστική βίβλος at 7.4 (fire-signalling), 8.5, 21.1, 40.8, and to a planned Στρατοπεδευτική βίβλος at 21.2. The existence of a lost naval treatise is surmised from the final line of the extant work at 40.8, where Aineias appears to begin a discussion of this subject, see *infra* n. 94. The evidence for these and other lost books by Aineias is variously assessed by Oldfather (1923) 8–10; Hunter and Handford (1927) xii–xiv; Dain and Bon (1967) xii–xviii; Bettalli (1990) 10–12; Whitehead (2002) 13–16.

90 Hermogenes is cited at *Rhet. Mil.* 3.2, 25.2. Syrianos' use of the *corpus Hermogenianum* is examined by Zuckerman (1990) 219–223; Eramo (2008) 125–126; (2010) 36–38; (2010 [2011]) 16–18, 121–123 *et passim* 115–194.

91 Zuckerman (1994) 373–385 esp. 374.

92 See Lammert (1940) 280–283, and previously Köchly and Rüstow (1853–1855) II.2 13. Cf. Syr. *De Re Strat.* 8.16–17. For Aineias (7.4) on fire signals and the Παρασκευαστική

Most tantalising is the prospect that Syrianos' *Naumachiae* might preserve material from a lost naval treatise dating to the fourth century BC.[93] Unfortunately, nothing can be said with confidence about which written sources, if any, Syrianos consulted during the composition of this section. Even Aineias' authorship of such a work remains in doubt: the received text of his *Poliorketika* breaks off just where he begins a discussion of naval deployment (40.8), but it is impossible to determine whether a large independent treatise or a single chapter has been lost.[94] It is worth noting that Philon, whose knowledge of Aineias' *Poliorketika* has been demonstrated, concludes his *Poliorketika* C and D with a consideration of naval forces, but these appendices merely comprise brief remarks about the possibility of seabourne attacks against a coastal city rather than a comprehensive treatise on naval warfare *per se*.[95] Whatever the extent of the naval section lost from Aineias' *Poliorketika*, we can be certain that it had already come adrift from the majuscule exemplar used by the copyist of M, apparently before c. 400.[96] If Syrianus did have the good fortune to possess

βίβλος cf. Polyb. 10.44–47 with discussion by Hunter and Handford (1927) 120–124; Bettalli (1990) 228–231; Whitehead (2002) 13–14, 111–113. Syrianos attributes to 'some more ancient authors' a system for signalling the size of an enemy force by lighting one beacon per thousand men. Polybios' quotation of Aineias' lost book describes a much more complex mechanism for transmitting detailed information about the composition of hostile forces by timing fire-signals using a water clock. It is of course possible that Polybios cites only part of a wider treatment. The Byzantine textual tradition relating to fire-signals is examined by Pattenden (1983); Zuckerman (1994) 361–366.

93 First conjectured by Müller (1882) 48; elaborated in Lammert (1940) 280 n. 1, 281–282, 288. This view was recently endorsed by Cosentino (2000) 260, 262, 279–280, who discerns (262) in Syrianos' *Naumachiae* 'tactics of the Athenian fleet dating back to the 5th century BC, together with manoeuvres such as the *diekplous* and the *periplous*', which were long obsolete by the time of writing (cf. Syr. *Naum.* 9.24–27, 35–40). For discussion of Byzantine naval tactics see Pryor and Jeffreys (2006) 399–402, who question the practicability of some aspects of contemporary naval treatises.

94 Aineias 40.8, '... I shall move on to naval deployment. Of a naval force there are two types of expedition ...' (... περὶ ναυτικῆς τάξεως δίειμι. Ναυτικοῦ δὲ στρατεύματος δύο εἰσὶ στόλοι ...). The various conjectures concerning Aineias' treatment of naval matters are summarised by Whitehead (2002) 14–15, 207.

95 Philon C.51–62 (94.36–95.32); D.101–111 (103.49–104.46).

96 Dain and Bon (1967) xvii. Excerpts of Aineias' *Poliorketika* incorporated into the late ninth-/early tenth-century *Apparatus Bellicus* descend from a separate majuscule tradition, which must have diverged as early as c. 400 from the exemplar used by the copyist of M (see below p. 336). These excerpts derive solely from the *Poliorketika*, presumably transmitted in isolation, and there is no evidence to suggest that the excerptor had access to more of Aineias' writings than are preserved in the direct tradition.

a naval treatise by Aineias, he appears to have been the only Greek, Roman or Byzantine writer ever to have seen a copy. Certainly Leo VI was less fortunate: writing c. 905 and familiar with Syrianos' (in my view, recent) *Naumachiae*, the emperor complained that he had failed to track down a single 'ancient' text on naval warfare.[97]

In the absence of external evidence, any relationship between Syrianos' compendium and lost or conjectural books by Aineias remains unprovable. Indeed, when there is no compelling evidence of Syrianos' direct knowledge of Aineias' extant *Poliorketika*, his familiarity with lost books must remain in the realm of speculation or wishful thinking. Syrianos' demonstrable dependence on other, more recent sources in most of the allegedly Aineias-derived chapters or sections not only rules out or casts doubt on Lammert's hypothesis in each particular instance, but also undermines the overall rationale of Lammert's effort to portray Syrianos' compendium as a mirror of Aineias' oeuvre. Furthermore, it does not seem plausible to suppose that lost books of Aineias survived into the ninth (or even sixth) century and were uniquely available

97 Leo, *Taktika* 19.1, '... we have found no regulations on this subject among the ancient tacticians' (οὐδὲν μὲν ἐν τοῖς παλαιοῖς τακτικοῖς περὶ αὐτῆς κεκανονισμένον εὑρόντες), but only 'scattered references' (σποράδην ἀνέγνωμεν). Leo seems to contradict this statement later in his exposition of naval warfare (19.72), by making an obscure allusion to 'the assembled volume of ancient tactics and stratagems' (τῷ συνειλημμένῳ τῶν τε ἀρχαίων τακτικῶν καὶ στρατηγημάτων βιβλίῳ), possibly a composite codex of military texts or a version of Polyainos' *Strategemata* (see below in this note). For Leo's written sources on naval warfare see Pryor and Jeffreys (2006) 176–179; Haldon (2014) 389–392 with subsequent commentary. Leo's use of Syrianos' *Naumachiae*: *supra* n. 80. In references to his sources Leo routinely distinguishes 'the Ancients' (οἱ ἀρχαῖοι or παλαιοί), i.e. classical writers, from 'the more recent' (οἱ νεωτέροι) late antique and Byzantine authors. Syrianus, whether sixth- or ninth-century, belongs to the second category, Aineias, if known to Leo at all, very much to the first. There are no other witnesses to the existence of a naval treatise in Greek from the classical period. In classical *taktika* discussion of naval warfare is conventionally postponed or passed over, cf. Asklep. 1.1; Aelian, 2.1; Arr. *Takt.* 2.1. The stratagem collections of Frontinus and Polyainos contain historical exempla relevant to Greek and Roman naval operations, but there is no thematic section devoted to this topic. In the *Hypotheseis* (c. 500–c. 850), a topically arranged adaptation of Polyainos' *Strategemata*, exempla relating to naval warfare are collected together in the last two chapters (57: Ναυμαχικά; 58: Κατάληψις παραλίων τόπων καὶ πόλεων). The chapters concerning the navy in Veg. *Epit.* 4.31–46 belong to a separate Latin literary tradition that in part drew on different, non-military genres, including the *libri navales* of M. Terrentius Varro (first century BC). Of Vegetius' sixteen 'naval' chapters only three (44–46) concern warfare; the others treat more general matters such as shipbuilding, navigation, tides, winds and astronomy.

to Syrianos without leaving the slightest trace in the rest of late antique and Byzantine military literature or the manuscript tradition. Nor is such a survival consistent with the preceding examination of both the codicological evidence and the testimonia, which point to the tenuous transmission of a single ancient and decayed copy of the *Poliorketika*, already transmitted independently by c. 400, and apparently the sole surviving work of a little-known author. Ultimately, there is no evidence that any late antique or Byzantine author, editor, excerptor or paraphrast had at his disposal more of Aineias' writings than survives today; indeed for a credible witness to Aineias' lost books one must go back as far as Polybios in the second century BC.[98]

The Excerpted Tradition: The *Apparatus Bellicus*

The earliest, largest and most important collection of excerpts from Aineias' *Poliorketika* is found in the so-called *Apparatus Bellicus*, an anonymous and untitled compendium dating to the late ninth or early tenth century.[99] Before examining these excerpts it is necessary to clarify the textual history and character of the *Apparatus Bellicus*, especially as this text has been the subject of great confusion in modern scholarship on Aineias. For an overview of the sometimes complex textual interrelationships discussed in this and the following section readers are referred to the diagram at the end of the chapter. The text of the *Apparatus Bellicus* is preserved in two closely related manuscript prototypes, *Vaticanus gr.* 1164 (= V) and *Barberinianus gr.* 276 (olim II 97) (= P-B). They were both copied c. 1020 from a common hyparchetype (Dain's '*Mazoneus*'), a collection of military texts compiled by an unknown redactor c. 980–1000. The relationship of V and P-B to a third eleventh-century manuscript, *Scorialensis gr.* Y-III-11 (281) (= N-E), has been disputed but it is now generally acknowl-

98 Polyb. 10.44–47 quotes from the Παρασκευαστικὴ βίβλος in a passage critical of Aineias' recommendations concerning fire-signals (cf. Aineias 7.4), see Dain and Bon (1967) lxi–lxiii. Much later, c. 106–113, Aelian, *Taktike Theoria* 3.4, contrasts Aineias' and Polybios' definitions of ἡ τακτική, but it seems most likely that Aelian merely rehearses another critique by Polybios and did not know Aineias' lost work directly, see Lammert (1940) 281; Dain and Bon (1967) lxi.

99 On the *Apparatus Bellicus*: Dain (1937) 59–63; (1939) 46–53; Dain and Bon (1967) xliv–xlvi; Dain and de Foucault (1967) 359–361 with older bibliography; Vieillefond (1970) 79–83, 189–198; Zuckerman (1994); Mecella (2009) 87–98, 111–113. For internal indications of date: Zuckerman (1994) 364, 388; with codicological arguments in Vieillefond (1970) 82, 193–194 n. 6.

edged that N-E is a copy of V, executed c. 1040. The text in N-E is of critical value insofar as it fills substantial lacunae in V. All three manuscripts were produced in the same monastic scriptorium in Constantinople.[100] N-E became the ancestor of numerous *recentiores*, which provided the basis of the earliest scholarship.[101] Although the *Apparatus Bellicus* was known to western scholars

100 For *Mazoneus* see *supra* nn. 21–22. With regard to the interrelationship of V, P-B and N-E uncertainty has been fuelled by uncritical repetition of discredited theses. There is nothing to be said in favour of the older contention that N-E antedates V and P-B, as first alleged by de Rochas and Graux (1879) 100–102, and reiterated in Schneider (1908–1912) I 6–7; Garlan (1974) 285. That N-E is a copy of V was first stated by Schöne (1893) v–vii, repeated (1911) xiii–xiv. This view is affirmed by the editors of numerous other texts transmitted in V, P-B and N-E, see Vári (1901) xii–xv, xx; (1917–1922) I xxxi–xxxii; Korzenszky and Vári (1935) v–vi with stemma at xvii; van den Berg (1947) 5–9; Dennis (1981) 21, 35–36; (1985) 4–6, 140–141; (2010) xii; Zuckerman (1994) 360; Mecella (2009) 87–88. Dissent from this stemma originates in Dain (1930) 34–36 (with stemma post 171), repeated (1933) 19–20, who concluded that N-E and V descend independently from a common exemplar, while P-B has a more remote connection. Vieillefond (1932b) xxxii–xxxiv concurred with Dain's stemma, and was followed uncritically by Lammert (1951) 362. Mazzucchi (1978) 285–287 n. 60 also cites Dain (1930) but does not exclude the possibility of Schöne's stemma. The arguments of Dain (1930) and Vieillefond (1932b) were irrefutably demolished by E. Korzenszky in her successive reviews of their works in *PhW* 52.1 (1932) 1–8, and *BZ* 35 (1935) 145–149. Accordingly, Vieillefond (1970) 80–83, 198 retracted his earlier stemma and conceded that N-E is indeed a copy of V. Dain also revised his earlier position, but in (1946) 237–240, reiterated (1946–1947) 41, he argued for a even less plausible stemma, in which P-B and V are correctly identified as siblings, but N-E has a more remote connection. Dain's primary interest in this instance was Aelian's *Taktike Theoria*, however, for which N-E supplies a mere one-folio fragment, an insufficient sample from which to draw definitive conclusions, and Dain himself acknowledged the need for further research. Dain's shifting opinions were posthumously summarised by de Foucault in the influential Dain and de Foucault (1967) 386–387, with additional errors, and seemingly unaware of the eccentricity and intrinsic difficulties of both of Dain's stemmata. More recently, Dagron and Mihăescu (1986) 15, 21–25, following Dain's 1946 stemma, not only grant N-E authority independent of V but choose to make N-E the basis of their edition. They do not acknowledge the considerable body of literature arguing to the contrary. Most recently, Gatto (2010) 153–157, 159 has devised a far more complex stemma, whereby N-E and P-B descend independently from *Maz* (Gatto's γ) via two hypothetical intermediaries, respectively ε and ζ, while V descends via hypothetical η, a product of contamination between ε and ζ. As Gatto's analysis is based on a small textual sample, examined in isolation from the rest of the codices and secondary literature, it remains to be seen whether his model can be applied to other texts in V, P-B and N-E.

101 The *recentiores* derived from N-E are listed by Vieillefond (1932) xliii–xlv; (1970) 84 n. 11; supplemented by Mecella (2009) 87 n. 10.

from the late fifteenth century, its conventional modern title was inspired by a brief notice by Gabriel Naudé in 1637, who reported an 'opus ... de bellico apparatu' in several manuscripts in Rome, Munich and Paris.[102] Three centuries later Vieillefond and Dain revived this designation for the compendium in various studies relating to its structure and sources.[103] There is no modern critical edition of the complete text. The *editio princeps* was prepared by Philippe de la Hire from a highly defective secondary manuscript and published in Melchisédech Thévenot's *Veterum mathematicorum opera* in 1693, together with insightful annotations by Jean Boivin.[104] A second edition, originally drafted by Jan van Meurs (Meursius) and published over a century later by Giovanni Lami in 1746, is the most recent publication of the entire work, though most of its constituent parts have since been re-edited separately.[105]

---

102 Naudé (1637) II (IV§8) 520–521, 'Iulii Africani opus magnum de Cestis nullibi gentium comparet, sed qui eius partem faciebant libri quintus & sextus de bellico apparatu, cum Romae in Vaticana, Palatina, Barberina, Farnesiana, tum Monacis, Parisiis, atque alibi servantur ... Sed praeter opus illud de bellico apparatu, quod Casaubonus videtur quodammodo innuere, transcriptum suisse ex libris Aeneae Tactici scriptoris antiquissimi sane, & probatissimi; quodque in Bavarica non duobus libris, sed capitibus LXXIX distinguitur'.

103 Vieillefond (1932a); (1932b); Dain (1931) 336–337; (1937); (1939). Dain (1931) 336–337 specifies that he and Vieillefond had adopted the title *Apparatus Bellicus* after Naudé's 'de bellico apparatu', but this expression was in fact coined earlier by Isaac Casaubon, whose writings Naudé cites. To my knowledge its earliest usage is in I. Casaubon (ed.), *C. Suetoni Tranquilli de XII. Caesaribus libri* VIII (Paris 1606), Animadversiones p. 26.34.

104 Thévenot (1693) 275–316 (owing to an error in pagination 279 is followed by 290, with further mispagination 291–298). The text is not the work of Thévenot, who died in 1692, but of his collaborator Philippe de la Hire, who simply reproduced the text he found in *Parisinus gr.* 2439 (formerly 2706), a poor sixteenth-century apograph of N-E (Bon in Dain and Bon [1967] xliv n. 1 erroneously states *Parisinus gr.* 2435; see correctly Martin [1854] 337; Vieillefond [1970] 85). At Thévenot (1693) 339–360 Jean Boivin de Villeneuve appended valuable annotations with *variae lectiones* from *Parisinus gr.* 2441 (formerly 1996 Colbert) and 2445 (2173).

105 Jan van Meurs (1579–1639) collated two *recentiores*, descendants of N-E, available to him in the Netherlands, one belonging to André Schott (1552–1629), the other in the Universiteitsbibliotheek, Leiden. Giovanni Lami (1697–1770) discovered a copy of van Meurs' collation in the University Library at Uppsala and published it in Lami (1741–1763 [1746]) VII 897–984, with annotations which included *variae lectiones* and Boivin's conjectures from the *editio princeps* of 1693, together with van Meurs' and Lami's own improvements and some readings from codex M (for chs. 1–30). The editions of La Hire–Thévenot and van Meurs–Lami are assessed by Martin (1854) 337–339; Vieillefond (1932b) xxv–xxvii; (1970) 84–86. The modern editions of selected sections will be cited below.

The *Apparatus Bellicus* is a composite work, comprising, according to the La Hire–Thévenot numeration, 79 chapters of unequal length (1–77 with 56b and 72b).[106] Clear internal seam lines mark where the text has been stitched together, sometimes rather crudely, from several distinct components, each of which derives from different source-materials and exhibits differing authorial or editorial methodologies. These distinctions must be examined in further detail if the textual history of the *Apparatus Bellicus* and its Aineias-derived content are to be correctly understood. Four sections can be distinguished. The first section, chapters 1–30, consists of lengthy extracts from Book VII of the *Kestoi* of Julius Africanus, a wide-ranging miscellany of technical and scientific knowledge, including medicine, veterinary science, pharmacology, weaponry, tactics and hunting. Written c. 227–231, the *Kestoi* ultimately comprised 24 books, now mostly lost or known only though excerpts in Byzantine compendia.[107] A superscription (τάδε ἔνεστιν ἐν τῷ ⟨ζ⟩ τῶν Ἀφρικανοῦ Κεστῶν) at the beginning of the *Apparatus Bellicus* and a corresponding subscription or *explicit* after chapter 30 ('Ιουλίου Ἀφρικανοῦ Κεστὸς ζ') signal the unity and autonomy of these chapters. The prior independence of this section is confirmed by the occurrence of the same collection of excerpts from *Kestos* VII also in codex M (231^{r}–243^{r} [233^{r}–245^{r}]), in which it is transmitted as a self-contained treatise alongside more strictly military works from classical antiquity (see above). The text of the *Kestoi* in chapters 1–30 of the *Apparatus Bellicus* differs from the text in M only by the interpolation of certain glosses; otherwise the editorial principle in this section seems to have been the preservation and transmission of Africanus' original wording.[108] The second section of the *Apparatus Bellicus*, chapters 31–59, reflects a different objective and methodology. It comprises a mixture of heterogeneous and unattributed excerpts on military science, drawn from three sources: Aineias' *Poliorketika* (chs. 38, 45–46, 48–58), Polyainos' *Strategemata* (35–36, 59) and an otherwise unknown source, which, by its style, content and character, has long been recognised as another part(s) of the *Kestoi* of Julius Africanus, to which these chapters are a unique witness

106 The numeration of the La Hire–Thévenot edition is conventionally cited in most of the secondary literature and is retained here for the sake of convenience. The complex textual history of the chapter divisions, numeration and headings, together with inconsistencies between the text and the pinaces in the codices, are discussed by Vieillefond (1970) 82–83, 99–100; Mecella (2009) 88–89.

107 See now Wallraff and Mecella (2009) with earlier bibliography.

108 The text of *App. Bell.* 1–30 is edited by Vieillefond (1970) 102–187 (as I 1–20); see also a selective re-edition in Mecella (2009) 117–131.

(31–34, 37, 39–44, 47).[109] In contrast to the first section, chapters 31–59 are short, lightly paraphrased extracts; where comparison with the original work is possible, the text of each extract generally coincides with its model, but with simplifications of vocabulary and syntax, some deletion of specific historical or geographical references, and occasional insertion of personal opinion.[110] The third section of the *Apparatus Bellicus*, chapters 60–72, differs in source and method again, comprising unattributed extracts from the *De Re Strategica* of Syrianos Magister. On the whole, in contrast to the second section, these are rather long extracts from a single text, sometimes arranged in a different order than in the original (as witnessed in M and A), but, with the exception of certain omissions, transcribed without textual modifications, including the retention of historical information.[111] In a fourth section, chapters 72b–77, the methodology changes yet again: some chapters (72b–74) appear to be recent and original contributions, most probably by the compilator of the *Apparatus Bellicus*, in part reflecting tactical deployments documented in tenth-century military manuals, but influenced by the content and language of earlier treatises (notably the 'interpolated recension' of Aelian's *Taktike Theoria* and possibly Maurice's *Strategikon*). Another chapter (75) partly reprises a lost late antique monograph on castrametation, while the concluding chapters (76–77) return to Syrianos' *De Re Strategica*, but here as freely adapted and/or paraphrased extracts of varying length, supplemented by the compilator's own lengthy digression.[112] The *Appa-*

109 The extracts from Aineias' *Poliorketika* are most recently edited in Dain and Bon (1967) 93–101. The extracts from Africanus' *Kestoi* are found in Vieillefond (1970) 201–213 (as II 1–12) and now Mecella (2009) 131–143. Of the three chapters derived from Polyainos' *Strategemata*, chs. 35 and 36 are edited (but not identified) in Vieillefond (1932b) 58; whence ch. 35 is reprinted in Mecella (2009) 91 n. 34, while ch. 36 is reprinted in Dain (1939) 27 and again, in a muddled format, in Schindler (1973) 224–225. For ch. 59 one must still consult Thévenot (1693) 306 or Lami (1741–1763 [1746]) VII 961 (= ch. 60 in Lami's numeration).

110 See remarks of Vieillefond (1970) 189–193; Mecella (2009) 90–93, 95.

111 See remarks of Köchly and Rüstow (1853–1855) II.2 8–13; Dain (1937) 55–58; Zuckerman (1994) 361. The text of *App. Bell.* 60–72 is available only in Thévenot (1693) 306–312 or Lami (1741–1763 [1746]) VII 961–973 (= chs. 61–73). Readings from *App. Bell.* 60–72 are included in Dennis' edition of the *De Re Strategica* ([1986] 5–6 with app. crit.), but he does not inform the reader that the version of *De Re Strategica* transmitted in codices V, P-B and N-E is in fact the *Apparatus Bellicus* and not a direct tradition.

112 Chs. 72b–76 ed. and Fr. trans. in Zuckerman (1994) 361–373, with discussion at 361, 373–379; summarised by Mecella (2009) 94–95. For the text of *App. Bell.* 77 see Thévenot (1693) 315–316 or Lami (1741–1763 [1746]) VII 979–980 (= ch. 78).

ratus Bellicus is clearly a product of the resurgence of interest in military literature in the late ninth and tenth centuries, and the text variously reflects the prevailing scholarly traits of 'encyclopédisme', excerption, paraphrase and redaction. The identity and milieu of the author, together with the purpose and intended readership of his compilation, remain matters of conjecture, though his access to diverse source-materials, both ancient and recent, make a location in Constantinople almost certain and a connection with court circles likely.[113]

The authorship and character of the *Apparatus Bellicus* have long been the subject of confusion and uncertainty, which predate its publication and owe more to academic inertia than scholarly argument. The superscription τάδε ἔνεστιν ἐν τῷ ⟨ζ⟩ τῶν Ἀφρικανοῦ Κεστῶν at the beginning of the work, together with the subscription Ἰουλίου Ἀφρικανοῦ Κεστὸς ζʹ at the end of the first section (chs. 1–30), variously misled early editors and commentators into believing that the entire conglomeration belonged to the *Kestoi* of Julius Africanus, and thus it is ascribed in the editions of La Hire–Thévenot and van Meurs–Lami.[114]

113 Köchly and Rüstow (1853–1855) II.2 13 inferred that the *Apparatus Bellicus* had been assembled by Leo VI or upon his instructions as a preliminary dossier of source-material for his own military writings, though this conclusion was largely based on the now discredited belief that Leo was the author of the (then still unpublished) *Sylloge Tacticorum* (c. 950), on which see below pp. 341–342. Zuckerman (1994) 364 opines, 'On pourrait imaginer qu'il fit partie de l'entourage du savant empereur Léon VI, ou, plus prudemment, évoquer le gout pour la science militaire qui se réveille à cette époque'. See also Mecella (2009) 97–98, whose attractive hypothesis lays emphasis on the recent publication of Syrianos' compendium (accepting its new ninth-century dating) as a stimulus to the compiler to offer his own selective appraisal of and supplement to contemporary military science.

114 This erroneous assumption can be traced back at least as far as Casaubon (1609) *passim*. Naudé (1637) II (IV§8) 520 refers to 'Iulii Africani opus magnum de Cestis nullibi gentium comparet, sed qui eius partem faciebant libri quintus & sextus de bellico apparatu ... servantur'; presumably Naudé (or an earlier copyist) misread ζ as ϛ in the subscription after ch. 30, which he further misconstrued as a heading to the subsequent material. Accordingly, he inferred that the following chapters (31–77) belonged to *Kestos* VI, and the preceding section (1–30) to *Kestos* V. La Hire in Thévenot (1693) 303, who did not print the subscription or signal a division in the text, assigned the entire compilation to Julius Africanus on the basis of the superscription. Van Meurs–Lami (1741–1763 [1746]) VII 949–950 emended ζ to β in the subscription, believing it to be a heading to the following chapters. This numeration was followed by Köchly and Rüstow (1853–1855) II.2 8–13. Martin (1854) 354–355, 363, although aware that the entire collection did not derive from the *Kestoi*, also misinterpreted the subscription to *Kestos* VII as a heading to the *Kestoi*-derived excerpts in the second section (in his view, chs. 31–37) and accordingly identified the preceding chapters (1–30) as *Kestos* VI; see similarly Kroll (1917) 118. Finally, Vieillefond

This misconception cast a long and seemingly inescapable shadow over subsequent research, whereby Aineias' treatise on siegecraft, quickly identified as the ultimate origin of certain chapters, came to be regarded as one of the sources of Africanus' miscellany. This view was reinforced by successive editions of Aineias' *Poliorketika* throughout the later nineteenth and early twentieth centuries, which routinely contained an appendix of what purport to be Aineias-derived excerpts found in the *Kestoi*, even if some editors expressed reservations about the nature and identity of this excerpt-collection.[115] Simultaneously another, equally venerable but more enquiring strand of scholarship correctly recognised that material from the *Kestoi* was merely one ingredient of the *Apparatus Bellicus*. As early as the *editio princeps* in 1693, Jean Boivin observed that considerations of content and style made Africanus' authorship of the whole compilation an impossibility.[116] This view was reiterated and expanded in some nineteenth-century studies, and in particular by Köchly and Rüstow (1855), whose provisional analysis identified the latter half of the collection (chs. 31–77) as a 'Sammelsurium' of excerpts taken partly from Africanus' *Kestoi*, partly from the works of Aineias, Polyainos, Anonymus Byzantinus (now Syrianos) and unknown others.[117] It was not until 1932, however, that Vieillefond's comprehensive assessment of the manuscript evidence isolated the authentic excerpts of the *Kestoi* in the *Apparatus Bellicus* and demonstrated that the only textual connection between Africanus and Aineias is that excerpts

(1932b) xxxvi–xxxviii; (1970) 39–41, 82, 99 correctly identified the formula Ἰουλίου Ἀφρικανοῦ Κεστὸς ζʹ placed after ch. 30 as a subscription or *explicit* to the preceding material. An identical subscription to *Kestos* XVIII is preserved in *P. Oxy*. 412.55: Ἰουλίου Ἀφρικανοῦ Κεστὸς ιηʹ; ed. Vieillefond (1970) 291; see now Mecella (2009) 89–90; Hammerstaedt (2009).

115 The convention begins with Hercher (1870) viii, 121–127, who edited the excerpts on the basis of two *recentiores*: *Parisinus gr*. 2441 and *Monacensis gr*. 195. See subsequently Schöne (1911) 115–123; Oldfather (1923) 19, 24–25, 206–225; Hunter and Handford (1927) xli–xlii, 240–251. The only exception is Hug (1874). For reservations see Schöne (1911) x, 'sive is Iulius Africanus ipse fuit sive recentioris aetatis homo'; ix, '... adieci capita inde excerpta, quae in Iulio Africani Cestis leguntur. Quae quamquam dubitari vel potest vel debet, num ab ipso Iulio Africano huic operi inserta sint'.

116 See Boivin's annotations in Thévenot (1693) 339, 357.

117 Köchly and Rüstow (1853–1855) II.2 8–13, who first identified *App. Bell*. 60–72 as extracts from the *De Re Strategica* of the Anonymus Byzantinus (Syrianos), whose (in their view) sixth-century date precluded the possibly that these chapters belonged to the *Kestoi*. Separately Martin (1854) 344, 355–363, although labouring under certain misconceptions, recognised that the *Apparatus Bellicus* was a composite work, of which only part (in his view chs. 1–37) derived from Africanus' *Kestoi*.

of both are found together in this Byzantine compilation.[118] Nevertheless, old misconceptions continue to linger and have proved especially tenacious in Anglophone scholarship, one suspects owing in some measure to the enduring convenience and popularity of William Oldfather's parallel edition and English translation of Aineias for the Loeb Classical Library (with appended 'Excerpts from Aeneas in the Κεστοί of Julius Africanus'), often reprinted but unrevised since its first publication in 1923, and thus predating Vieillefond's monograph.[119]

Although the compilatory character of the *Apparatus Bellicus* is beyond dispute, its genesis and early history remain obscure. In his final analysis (1970), Vieillefond inclined to the view that the compendium initially comprised only its first two components, which he styled 'Recueil A' (chs. 1–30) and 'Recueil B' (31–59), while the subsequent chapters (60–77) were a slightly later accretion. Potentially in favour of this thesis is the fact that only excerpts from the first two sections found their way into the so-called *Corpus Perditum* (see below), the lost common source of two later Byzantine compilations, the *Sylloge Tacticorum* (c. 950) and the *Taktika* of Nikephoros Ouranos (c. 1000). In contrast, material from chapters 60–77, which includes the most recent and original contributions, is entirely absent.[120] According to this view, the compiler of the

118 Vieillefond (1932b) esp. xxxix–xl; also (1932a) esp. 28–29, 31 n. 1; (1970) 189–193; see also Zuckerman (1994) 359–361; Mecella (2009) 88–96. To my knowledge, the only attempt to contest Vieillefond's analysis is Lammert (1951), as previously (1940) 285, chiefly in relation to *App. Bell.* 72b–74, which Lammert sought to assign to the *Kestoi*. His argumentation is countered by Zuckerman (1994) 364 n. 16, 385–388. This controversy is of no immediate concern.

119 See Oldfather (1923 [repr. 1928, 1948, 1962, 1977, 1986]) 19, 24–25, 'clearly derived from [Aineias]', with the text of the excerpts at 206–225. For the continued error see e.g. Dennis (1986) 4 n. 9, 'the *Apparatus Bellicus* of Julius Africanus'.

120 To judge from both the fragments and descendants of the *Corpus Perditum*, its compiler drew material from twenty-five chapters of the *Apparatus Bellicus* spread throughout 'Collection A' (1–30) and 'Collection B' (31–59), namely *App. Bell.* 2–4, 9, 14–16, 31–44, 51, 53–54, 56, but took nothing from *App. Bell.* 60–77. More specifically, Dain demonstrated that when the compiler of the *Corpus Perditum* wished to incorporate material from Syrianos' *De Re Strategica* he did not make use of the extracts in the *Apparatus Bellicus* (chs. 60–72, 76–77) but was dependent on another, somewhat different collection of extracts of Syrianos' work. In comparison to the *Apparatus Bellicus*, this other Syrianos-derived collection appears to have contained fewer extracts, or at least shows no trace of those extracts which the author of the *Apparatus Bellicus* found most important or stimulating. See Dain (1937) 55–58 with stemma after 91; (1939) 45–46, 58–59 with stemma at 56. From the content of the *Corpus Perditum*, therefore, one could infer that its compiler had at his

Corpus Perditum drew on a shorter 'first edition' or *editio minor* of the *Apparatus Bellicus*, which differed from the version later transmitted in the direct tradition, as now witnessed by codices V, P-B and N-E. Mindful that the evidence does not permit conclusive demonstrations, I have preferred this reconstruction to alternative propositions.[121] As a separate but closely related issue, both Vieillefond and Dain, with different and sometimes shifting arguments, maintained that 'Collection A' (1–30) and 'Collection B' (31–59) were independent entities before their incorporation into the *Apparatus Bellicus*. The prior autonomy of 'Collection A' seems beyond doubt: as observed above, codex M contains the same collection of extracts from *Kestos* VII, which evidently circulated as a self-contained treatise during the mid-tenth century. Furthermore, in the manuscript prototypes of the *Apparatus Bellicus* a clear break, framed with decorative bands, separates 'Collection A' and 'Collection B'.[122] The derivation

---

disposal a version of the *Apparatus Bellicus* that had not yet acquired chs. 60–77, hence his resort to another source. Mecella (2009) 112 reverses this line of reasoning, arguing that the compiler's apparent neglect of chs. 60–77 does not signal their absence but merely his desire to avoid duplication of Syrianos-derived material that had already come into his hands by a different route. If that were the case, we must assume that the compiler, for some reason, not only preferred this other excerpt-collection but also found nothing of additional interest when he later encountered the more numerous and extensive extracts of Syrianos in the *Apparatus Bellicus* (chs. 60–72, 76–77) nor anything of value in the other, most recent chapters (72b–76) towards the end of this compendium.

121 See Vieillefond (1970) 193–198, esp. 193–194 with n. 6, modifying his earlier analysis in (1932b) xxxviii–xliii, li–liv, and revising aspects of a previous reconstruction by Dain (1939) 46–53, who was also of the opinion that the compiler of the *Corpus Perditum* knew the *Apparatus Bellicus* in 'un état antérieur'. Note that the stemma at Vieillefond (1970) 198 does not accurately depict the conclusion favoured in the text, i.e. 'Collection A' and 'Collection B' were 'plus vraisemblement' already conjoined when they became a source for the *Corpus Perditum*. Alternative models of textual evolution have been proposed. Zuckerman (1994) 360–361 regards *App. Bell.* 31–79 'comme un tout, comme un recueil achevé au moment où, vers la fin du xe siècle au plus tard, il fut intégré à l'*Apparatus Bellicus*'. Zuckerman claims to discern similar editorial principles in *App. Bell.* 77 to those observed in 'Collection B' (chs. 31–59), but there are in fact significant differences: the excerpts of Syrianos in *App. Bell.* 77 are on the whole shorter and open with the formula Ὅτι …, a typical practice of Byzantine excerption, but not found in 'Collection B'. Vieillefond's view is also contested by Mecella (2009) 96 n. 53, 112, who contends that the *Apparatus Bellicus* is a organic and unitary composition by a single author, but her reconstruction of the genesis of the compilation (see esp. 96–98) nowhere disproves Vieillefond's model nor convincingly overcomes the difficulties of her own.

122 Vieillefond (1932) xxxii, xxxvi–xxxvii; (1970) 82, 189–190; summarised with additional remarks by Mecella (2009) 89–90 and illustration at 144.

or former independence of 'Collection B' is less certain. Dain argued that its origin lay in an earlier, now lost excerpt-collection, probably compiled in late antiquity, which he styled *'Exercitationes'*. Although later research has shown Dain's original line of reasoning to be partly flawed, other considerations can be adduced in favour of his basic hypothesis.[123] First, as previously described, markedly different compositional and editorial strategies distinguish 'Collection B' (31–59) from both 'Collection A' (1–30) and the subsequent chapters (60–77), divergences more plausibly traced to a change in source-material than to the shifting taste and methodology of the compiler of the *Apparatus Bellicus*. Second, 'Collection B' duplicates material found in 'Collection A', in such a way as to suggest that two pre-existing collections have been fused together with minimal or no revision.[124] Third, although not conclusive, lexical changes introduced by the excerptor of 'Collection B' appear to reflect late antique rather

123 See Dain (1939) 47, 50–53, repeated Dain and Bon (1967) xliii; Dain and de Foucault (1967) 339 with stemma at 352. Dain first hypothesised the existence of the *Exercitationes* in order to account for the occurrence of identical material in 'Collection B' of the *Apparatus Bellicus* and in the *Corpus Perditum*, as witnessed by the latter's two surviving descendants, the *Sylloge Tacticorum* and Nikephoros' *Taktika*. Owing to certain misconceptions, Dain inferred that the *Apparatus Bellicus* could not be the source of the *Corpus Perditum*, but surmised that both works had independently drawn on divergent versions of an earlier, now lost except-collection, which Dain called *'Recettes practiques* ou ... *Exercitationes'*. Schindler (1973) 188, 224 rehearses this view uncritically. Dain's rather convoluted arguments do not require detailed rebuttal. It suffices to observed that their deficiencies were exposed by Vieillefond (1970) 195–198, who conclusively demonstrated that the *Corpus Perditum* was dependent on both 'Collection A' and 'Collection B' of the *Apparatus Bellicus*, which were most probably already conjoined. Vieillefond, nevertheless, retained the concept and terminology of Dain's *Exercitationes*, only now deprived of its original raison d'être and assigned a reduced and simpler role as the source of 'Collection B'.

124 The duplication occurs in only one instance but is nonetheless highly instructive: the concluding section of ch. 21 (Thévenot = 22 Lami) (Ἦχου κλοπή) and ch. 47 (Ἀκοῆς θήρα) both derive from the same chapter of *Kestos* VII, but while 21 furnishes the full and unmodified text, 47 is an abridged excerpt with simplified vocabulary. The differing approach is underlined by the fact that the excerptor of 'Collection B' interpolated an introductory observation in 47, in which he expresses his scepticism concerning the practicability of the ruse described and justifies its inclusion purely on the basis of paradoxographical interest. This doublet provides a unique specimen of the *Kestoi* where we have both the original text and an excerpt and can thus assess, albeit on the basis of a small sample, the methodology of the excerptor of 'Collection B'. This in turn offers a useful comparandum with regard to the excerpts from Aineias' *Poliorketika*. See Kroll (1917) 118; Vieillefond (1970) 161, 211–213 (= I 16 and II 12), with remarks at 192; Mecella (2009) 89, 97 n. 61.

than tenth-century usage.[125] Collectively these internal features offer modest support to the view of Dain, albeit formulated on different grounds, that the constituent excerpts of 'Collection B' (31–59) were not created and/or assembled by the ninth-/tenth-century compiler of the *Apparatus Bellicus* but were reproduced by him from an existing excerpt-collection. For convenience I have retained Dain's coinage *Exercitationes*.[126] There is, of course, no way to determine whether 'Collection B' was identical in scope to *Exercitationes* or merely represents a selection of its content. In the circumstances, the chronology of *Exercitationes* also remains elusive; Dain's intuition of a late antique date seems plausible but cannot be proved. If, alternatively, 'Collection B' was first put together in the ninth or tenth century, then its pedigree becomes all the more intriguing, since this scenario requires that a second majuscule manuscript of Aineias' *Poliorketika*, in some measure superior to the exemplar of codex M, survived the 'Dark Age' and was known to the excerptor, but otherwise left no trace on scholarship in this field during the same period (see below).

In total fifteen chapters of 'Collection B' of the *Apparatus Bellicus* are paraphrased excerpts from Aineias' *Poliorketika*, for the most part grouped in a contiguous sequence (38, 45–46, 48–58 including 56b). Recognition of the origin of these excerpts predates even the publication of the *Apparatus Bellicus*. Casaubon first noted certain parallels of content and language in the commentary to his *editio princeps* of Aineias' *Poliorketika* in 1609.[127] Reiterating

125 E.g. *App. Bell.* 45 substitutes μάγγανον for χελώνη at Aineias 33.1, presumably with the equivalent meaning of a mobile shed or mantlet. From the first century μάγγανον signified a hoist or pulley, but by the fourth century had become also a generic designation for any 'machine' or 'engine' with a military application. This generic usage of μάγγανον, later μαγγανικόν, which pertained throughout late antiquity, allows for an equivalence of μάγγανον and χελώνη; e.g. *Chronicon Paschale* (ed. L. Dindorf, [CSHB 5] Bonn 1832) 725.1–5, written c. 630, where μαγγανικά is used as a collective term for siege machinery, encompassing specifically 'the siege towers and the mantlets' (τοὺς πυργοκαστέλλους καὶ τὰς χελώνας); cf. similary 719.15–22: ὀλίγα μαγγανικὰ καὶ χελώνας. In tenth-century military documents, however, μάγγανον and μαγγανικόν, together with various compound terms, typically refer to artillery, either as a general designation or denoting specific categories, see Huuri (1941) 82–89, 127–141, 212–215; Haldon (1999) 137–138; (2000) 269, 272–277.

126 See likewise Vieillefond (1970) 190–196, 198; endorsed by Mecella (2009) 96–97, 111 with stemma at 114.

127 Casaubon (1609) 66, 75–78, 84, 88, 90–91, 93, 102, 104, 108, 110, 112, inferring that 'Julius Africanus' (see *supra* n. 114) had excerpted Aineias' work. Casaubon does not specify in which manuscripts he consulted the *Apparatus Bellicus*. Naudé (1637) II (IV§8) 521 tentatively reports Casaubon's conclusion: 'Sed praeter opus illud de bellico apparatu,

and expanding Casaubon's remarks, Boivin identified all fifteen excerpts in his annotations to the La Hire–Thévenot edition of the *Apparatus Bellicus* in 1693.[128] Subsequent commentators confirmed this list of excerpts.[129] The fifteen chapters were most recently edited by Dain and published posthumously in 1967, the first and only edition of Aineias' work in which these excerpts are correctly attributed to the *Apparatus Bellicus* and not to the *Kestoi* of Julius Africanus.[130] The correspondence between the excerpts and the original text of Aineias (in parentheses) may be tabulated as follows:

38	(= 34)	Πῶς πῦρ δυνάμεθα σβέσαι
45	(= 33.1)	Πῶς ἡμεῖς ἐμπρήσωμεν;
46	(= 33.4)	Πρὸς καιομένας πύλας
48	(= 22.24–25)	Περὶ κλεψύδρας
49	(= 28.1–4; 29.12)	Πυλωρικόν
50	(= 29.1–10)	Ὅπλων λάθρᾳ εἰσκομιδή
51	(= 31.4–5)	Περὶ κρυφίας ἐπιστολῶν εἰσπομπῆς
52	(= 31.17–19)	Ἕτερον ἄλλο πανουργότερον
53	(= 31.31–32, 23, 14, 33, 24)	Ἕτερα περὶ τούτου παρὰ τῶν παλαιῶν πραχθέντα
54	(= 37.1–5)	Ὑπορυσσόντων γνῶσις καὶ κώλυσις
55	(= 37.8–9)	Περὶ τοῦ τοὺς ὑπορύσσαντας μὴ βλάπτεσθαι
56	(= 39.1–2)	Δολίευμα
56b	(= 39.3–4)	Ἄλλο δι' οὗπερ ὅσους ἂν θέλωμεν τῶν πολεμίων κατάσχωμεν
57	(= 40.1, 4–5)	Πῶς δι' ὀλίγων ἀνθρώπων μεγάλης πόλεως φυλακὴ γενήσεται
58	(= 40.6–7)	Περὶ τοῦ στρατιώτας ὀλίγους ὄντας πολλούς φαίνεσθαι

quod Casaubonus videtur quodammodo innuere, transcriptum suisse ex libris Aeneae Tactici scriptoris antiquissimi sane, & probatissimi ...'.

128 Boivin in Thévenot (1693) 339, 352–355, concluding at 357, 'si quidem caput XLV. & sequentia (uno excepto) ad LIX. ex Aeneae Poliorcetico excerpta esse manifestum est'. Boivin (352) also correctly identified the source of *App. Bell.* 38.

129 Martin (1854) 356 with n. 1, 363, 403; Köchly and Rüstow (1853–1855) II.2 9, 13; Kroll (1917) 118; Vieillefond (1932) xxxix; (1970) 191–193. For editions of the excerpts see *supra* n. 115.

130 Dain and Bon (1967) 93–101, as 'Apparatus Bellici qui vulgo dicitur Κεστοί Iulii Africani Capita ex Aenea Excerpta'.

Nineteenth-century editors, especially Rudolf Hercher, recognised and exploited the immense editorial value of the text preserved in these excerpts, even when a paraphrase of the original, as a potential corrective to corruption in the direct tradition witnessed by codex M. The transmission of seemingly correct readings in the excerpts where the text in M contains an evident misreading of majuscule script, and vice versa, indicates that the excerpted and direct traditions descend from two different majuscule archetypes. Their common ancestor necessarily antedates the exemplar used by the copyist of M, which was, according to Dain's calculation, a papyrus codex of c. 400.[131] The availability of two majuscule traditions, at least for those sections of Aineias' work excerpted in the *Apparatus Bellicus*, affords modern editors a more hopeful prospect for pursuing the textual tradition closer to the author's original wording. It also indicates that at least one other manuscript of Aineias' work survived into late antiquity. In any case, it is revealing that the excerptor, like the copyist of M, appears to have known only Aineias' *Poliorketika*, again transmitted in isolation, and none of his other books.

Generally the text of the excerpts adheres closely to the original, with some light rewording in the form of transposition or paraphrasis. Editorial interventions are, on the whole, restricted to introducing greater clarity through minor linguistic simplifications and occasional abridgement. The most extensive modifications typically occur at the beginning and end of excerpts, a general characteristic of excerption practices in Byzantine compendia.[132] The excerptor seldom interpolates additional details and in only two instances appears to impose his own voice.[133] He routinely deletes references to specific

131 For the dating see Dain and Bon (1967) xliii n. 4. Schöne (1911) ix had previously surmised that the excerpts 'ex exemplo aliquo Aeneae transcripta sunt, quod M codice multo fuit vetustius'. My perusal of the apparatus criticus in Dain and Bon (1967) reveals: correct or better majuscule readings in *Apparatus Bellicus*: e.g. πλήθει *App. Bell.* 49 : πάθη M 28.3; ἐπάνω, τὰ *App. Bell.* 55 (emended ἐπάνω, ταῦτα Tur.) : σπανιώτατα M 37.9. Correct majuscule reading in M: e.g. ἐλλιμενισταὶ M 29.5 : ἐνλιμενισταί *App. Bell.* 50; κάδους M 40.4 : κλάδους *App. Bell.* 57.

132 See the opening passages of e.g. *App. Bell.* 50 (= Aineias 29.1–3) and 51 (= 31.1–4), and the endings of e.g. *App. Bell.* 50 (= 29.10–11) and 57 (= 40.4–5). The chief exception is *App. Bell.* 53 (= Aineias 31.31–32, 23, 14, 33, 24), which exhibits a more radical cut-and-paste methodology with severe abridgement towards the end. For the general principles and criteria followed by Byzantine excerptors see Roberto (2009).

133 Additions: e.g. *App. Bell.* 40 (= Aineias 33.1): additional reference to ships and siege-towers; *App. Bell.* 48 (= 22.24–25): appends a clarification concerning the orifice of a water clock; *App. Bell.* 50 (= 29.10): additional reference to the use of ladders; *App. Bell.* 57 (= 40.1, 4–5): supplementary defenders include 'old men and boys' in addition to women. Excerptor's

people, places or events, perhaps judging the historical exempla adduced by Aineias to be obscure or redundant, and with a view to presenting generally applicable precepts.[134] Aineias' vocabulary in the excerpted passages is mostly non-technical, but the excerptor occasionally introduces contemporary terminology by means of substitution (e.g. μάγγανον for χελώνη), gloss (e.g. τείχει ἢ χάρακι) or addition (e.g. σκάλα), or substitutes late nominal forms (e.g. μικρὰ ἀσπιδίσκια for μικρὰ ἀσπίδια).[135] Overall such lexical modifications are rare and not diagnostic of the excerptor's methodology; on the contrary, elsewhere he retains what must have been recherché terms (e.g. ἐλλιμενισταί) and archaic or obsolete concepts (e.g. θώρακας λινέους, 'linen cuirasses').[136] Corresponding editorial practices characterise the remaining fifteen excerpts in 'Collection B' that derived from other sources, namely the three chapters from Polyainos' *Strategemata* (35, 36, 59) and twelve from Africanus' *Kestoi* (31–34, 37, 39–44, 47).[137] The excerpts of Aineias' *Poliorketika* share common themes, both with

---

opinion: see the introduction of first person ἀποτρέπω in *App. Bell.* 49 (= Aineias 28.3) and παραινῶ in 53 (= 31.33), though without altering the meaning of the text.

134 E.g. *App. Bell.* 54 and 55 respectively excerpt Aineias 37.1–5 and 37.8–9, omitting the intervening exemplum of Amasis at Barca (37.6–7). *App. Bell.* 57 excerpts Aineias 40.1, 4–5, omitting the exemplum of Dionysios (40.2–3) and the reference to Datamas and Sinope (40.4). At the end of *App. Bell.* 53 the excerptor reduces Aineias 31.33 to a terse sentence, deprived of the three exempla that follow (Astyanax, tyrant of Lampsacus, and events at Thebes and Mytilene), and likewise 31.24 is severed from a report concerning the Lokrian women at Ilion. With a similar purpose, *App. Bell.* 50 recasts an account of unspecified historical events at Aineias 29.1–10 into a less detailed general prescription. An exception is *App. Bell.* 53, which retains a reference to the use of dogs to carry secret messages in Epirus and Thessaly (31.31), perhaps because an indication of the specific breed(s) was deemed significant.

135 *App. Bell.* 45 = Aineias 33.1: μάγγανον for χελώνη; *App. Bell.* 58 = 40.6–7: τείχει glossed ἢ χάρακι; *App. Bell.* 50 = 29.10: adds διὰ σκάλης; *App. Bell.* 50 = 29.6: ἀσπιδίσκια for ἀσπίδια.

136 *App. Bell.* 50 retains ἐλλιμενισταί at Aineias 29.5, while at *App. Bell.* 49 the excerptor redeployed this term as a substitute for λιμενοφύλακάς τε καὶ ἀποστολέας at 29.12, even though usage of ἐλλιμενιστής, a contractor who collected harbour dues, is otherwise attested only in Dem. *Contra Phorm.* 34 and Pollux, *Onom.* 6.128. Similarly, *App. Bell.* 50 retains θώρακας λινέους at Aineias 29.4, though it is unclear whether 'linen cuirasses' would have been comprehensible to a tenth-century reader. Other omissions may reflect the unfamiliarity or obsolescence of ancient practices: e.g. *App. Bell.* 50 (= Aineias 29.1–10) omits a reference to customs officials affixing seals (κατεσημήναντο) to containers of imported merchandise (Aineias 29.5) as well as a story about improvising defensive equipment using osiers (29.11; note that 29.12 is appended to *App. Bell.* 49).

137 Polyainos: *App. Bell.* 35 = *Strat.* 3.9.60; *App. Bell.* 36 = *Strat.* 4.4.3; *App. Bell.* 59 = *Strat.* 3.9.19. These three chapters have likewise undergone simplification of syntax and language,

one another and with the other material assembled in 'Collection B', which presumably reflect, however dimly, the excerptor's original purpose in compiling the *'Exercitationes'*, notably an interest in stratagems and counter-stratagems, covert operations and secret messages, and the military applications of scientific knowledge, especially pyrology. If these excerpts were all that survived today of Aineias' *Poliorketika*, the original context of a city under siege would still be easily discerned and continues to provide a unifying framework, but the excerptor's selection suggests that the stratagemical dimensions of Aineias' work most appealed to a contemporary reader.[138]

Later Developments in the Excerpted Tradition: The '*Corpus Perditum*', *Sylloge Tacticorum* and *Taktika* of Nikephoros Ouranos

The '*Corpus Perditum*' is a designation coined by Dain in 1937 to describe a military compendium, compiled around the late ninth or, more probably, early tenth century, which is now lost almost in its entirety, but whose scope and sources can be traced in its two surviving progeny, the *Sylloge Tacticorum* (c. 950) and the *Taktika* of Nikephoros Ouranos (c. 1000). In this regard, Dain developed a hypothesis first delineated a few years earlier by Vieillefond, who recognised that corresponding structure and content in the two extant compilations could only be explained by inferring their independent use of a common source, Vieillefond's '*Corpus* X', which had in turn drawn material from, among other sources, the *Apparatus Bellicus*.[139] The degree of correspondence

some abridgement and deletion of references to people, places or events; thus three historical stratagems, one ascribed to Antipater, two to Iphicrates, become anonymous, generally applicable prescriptions. See Köchly and Rüstow (1853–1855) II.2 11; Vieillefond (1970) 191; Schindler (1973) 224–225; Mecella (2009) 91. This is similarly the case with the *Kestoi*-derived excerpts, though these exhibit a greater tendency to preserve specific references and sometimes source citations found in the original text, see now Mecella (2009) 91–93, 95.

138 Wheeler (1981) 77–78; (1988) 13–14 also identifies traits in Aineias' work characteristic of the stratagemical branch of ancient Greek tactical writing.

139 The earliest published exposition of the 'Corpus X' is Dain (1931) 337–344, but Dain acknowledges his debt to forthcoming Vieillefond (1932a) esp. 28–29 and (1932b) lii–liv. The hypothesis was substantially revised and expanded in Dain (1937) 51–81; (1939); summarised in Dain and Bon (1967) xlvi–xlvii; Dain and de Foucault (1967) 352–353; and revisited and nuanced by Vieillefond (1970) 193–198. At a much earlier date Jean Boivin, in Thévenot (1693) 360, had already recognised that extracts from the *Apparatus Bellicus* are found adapted in Nikephoros' *Taktika* (which he ascribed to Constantine VII,

between the two witnesses lent persuasive weight to this line of reasoning, but the hypothesis was substantiated by the survival of five short chapters of this lost work inserted into the margins of the unique manuscript prototype of the *Sylloge Tacticorum*, the fourteenth-century *Mediceo-Laurentianus gr.* LXXV-6. In a marginal note (103^r) the scribe drew the reader's attention to the fact that six chapters and part of a seventh (68–73 with 74) were missing from his exemplar of the *Sylloge Tacticorum*, presumably owing to physical damage. The same scribe subsequently sought to restore the resulting lacuna in the *Sylloge Tacticorum* by transcribing into the margin of this and neighbouring folios (102^v–104^r) the text of five corresponding chapters (numbered 68–72) from a similar work, which he had located in another manuscript, apparently recognising its close kinship with the text either side of the lacuna and its conformity to the sequence of chapter-headings listed in the pinax at the beginning of the *Sylloge Tacticorum*. In doing so, the scribe fortuitously furnished a sufficient sample of text to allow Dain to demonstrate that these marginal substitute chapters belonged to a lost military compendium, which had drawn material from the *Apparatus Bellicus*, and was in turn the common source of the *Sylloge Tacticorum* and Nikephoros' *Taktika*, in short the '*Corpus* X' previously hypothesised by Vieillefond. It follows that this compendium, which Dain rebranded *Corpus Perditum*, still existed, at least in part, in the fourteenth century.[140] Fortunately, for the present purposes, one of the five surviving chapters derives ultimately from Aineias' *Poliorketika* (see below).

With the exception of these five chapters, the *Corpus Perditum* must be reconstructed from its two descendants. On the basis of the parallel content and arrangement of these two independent witnesses, Dain established that the *Corpus Perditum* comprised five components, which occur in the same sequence in the *Sylloge Tacticorum* (chiefly chs. 56–102) and the *Taktika* of Nikephoros Ouranos (chs. 75–173), namely: 1) extracts of the *De Re Strategica* of Syrianos Magister; 2) the so-called '*Cynegeticus* of Urbicius', in fact the final chapter of Maurice's *Strategikon* (12.D);[141] 3) the *Apparatus Bellicus* or a pre-

---

in accordance with scholarly misconceptions of his day). Boivin supplies *variae lectiones* from the unpublished text of the *Taktika*, citing *Parisinus* 3220 (now gr. 2530). Boivin's observations inspired further remarks by Martin (1854) 338, 362–363, 385–387.

140 See at length Dain (1939) 11–42, 70–72; rehearsed with additional proofs by Mecella (2009) 107–111.

141 In its earliest extant form, this tract on large-scale hunting as a method of training cavalry is appended to Maurice's *Strategikon* (12.D), though eccentricities of style and vocabulary betray its prior independence. Detailed argument will follow in Rance (forthcoming b); it suffices here to observe that the designation '*Cynegeticus* d'Urbicius' first surfaced in

edition thereof, principally the extracts of the *Kestoi*; 4) the so-called '*De Secretis Epistolis*', previously mentioned as a putative adaptation of Philon of Byzantion's lost book on military cryptography with Aineias-derived content; and 5) a version of the so-called *Hypotheseis* (or *Excerpta Polyaeni*), a collection of abridged and partly paraphrased excerpts from Polyainos' *Strategemata*, reconfigured in a thematic rather than prosopographical/ethnological format (c. 500–850).[142] The arrangement of these five elements is tabulated as follows:[143]

	Nikephoros' *Taktika*	*Sylloge Tacticorum*
1. Syrianos Magister	75–87	24–27, 37.
2. '*Cynegeticus*' = *Strategikon* 12.D	88	56
3. *Apparatus Bellicus*	89–111 (+ 172–173)	57–75 (+ 53.4)
4. *De Secretis Epistolis*	112–118	76
5. *Hypotheseis*	124–171	77–102

Dain reconstituted the *Corpus Perditum* as a compendium of 87 chapters, whose numeration will be followed here for the sake of convenience, though without complete confidence. The general lines of Dain's analysis cannot be disputed, though some details of his argumentation and conclusions require modification, particularly with regard to the so-called *De Secretis Epistolis* (see below).

Our concern here is to identify and assess Aineias-derived content transmitted from the *Apparatus Bellicus*, via the *Corpus Perditum*, into the *Sylloge Tacti-*

Dain (1937) 58 and was popularised by posthumous Dain and de Foucault (1967) 341–342, 352–353, 372, but both the title and ascription are entirely without manuscript authority and arise from Dain's misconstrual of earlier scholarship on the *Strategikon*. See Rance (2007b) 196–197.

142 For Philo's lost treatment of cryptography see above pp. 312–314 and below pp. 347–356. *Hypotheseis*: ed. Wölffin–Melber (1887) 427–504 (as '*Excerpta Polyaeni*'), with Eng. trans. in Krentz and Wheeler (1994) II 850–1003; see Dain (1931); (1937) 73–81, 84–86; (1939) 54–58, 61–65; Dain and de Foucault (1967) 337, 340, 365; Schindler (1973) 205–218.

143 Dain (1937) 54–65, 73–86; (1939) 43–65, summarised with supplementary observations by Mecella (2009) 100–101, 105–106. Chs. 119–123 of Nikephoros' *Taktika* constitute a self-contained corpus of earlier texts relating to naval warfare, for which the *Sylloge Tacticorum* offers no parallel, and which presumably was not found in the *Corpus Perditum*, see Dain (1937) 65–73; (1943) 89–104; omitted by Pryor and Jeffreys (2006) 183.

corum and Nikephoros' *Taktika*. The amount and condition of this material differ in the two compilations, in large part owing to the different objectives and editorial methodologies of their authors. The *Sylloge Tacticorum* is an anonymous compendium of 102 chapters, preserved in a single fourteenth-century manuscript prototype, *Mediceo-Laurentianus gr.* LXXV-6 (72^r–73^v, 74^v–116^r).[144] The authorship and precise date of composition have not been demonstrated but the work undoubtedly belongs to the first half of the tenth century.[145] Still awaiting comprehensive textual analysis, it is a curious 'cut-and-paste' mélange of modified excerpts from classical and Byzantine military treatises, interwoven with the compiler's critiques of ancient practices and tactical prescriptions for contemporary application. The primary focus is deployments, tactics and weaponry to be used by combined specialized forces in field operations against an unspecified enemy, presumably the Arabs. Material drawn from the *Corpus Perditum* predominately, though not exclusively, comprises the second half of the compilation (chs. 56–102).[146] The author's methodology varies somewhat throughout the work, but on the whole he freely adapted his sources, typically transposing and reconfiguring chapters, inserting additional information and personal opinion, and often abridging the text, which he substantially remodelled in a classicizing idiom.[147] His interests and approach restrict the amount of material originating in Aineias' *Poliorketika* to just two chapters. First, chapter 70 was a discussion of flame-retardant measures based on Aineias' chapter 34. Lost owing to the aforementioned lacuna in *Mediceo-Laurentianus gr.* LXXV-6, its title is preserved in the pinax, while its substance, if not wording, can be surmised from the substitute chapter of the *Corpus Perditum* (31) transcribed into the margin (102^v) and by comparison with the corresponding chapter at Nikephoros' *Taktika* 104. The surviving title aptly illustrates the author's more elaborate style and diction: Πῶς ἂν τὸ ὑγρὸν καλούμενον σβεσθείη πῦρ, καὶ

144 Ed. Dain (1938). For *Laurentianus gr.* LXXV-6 see Bandini (1764–1770 [1770]) III 147–151. An annotated Eng. trans. by Chatzelis and Harris (2017) appeared just as this volume was going to press.

145 Chronological termini: the *Sylloge* (e.g. chs. 35–36, 38–39, 44) drew material from Leo's *Taktika* (c. 905) but was itself a source for the *Praecepta Militaria* ascribed to Nikephoros II Phocas (963–969), see McGeer (1995a) 184–188. See further discussion of historical context in Chatzelis and Harris (2017).

146 For a provisional identification of its sources see the marginal notations to Dain (1938), and Dain and de Foucault (1967) 357–358 citing older bibliography, with additional observations by Mecella (2009) 98–101.

147 See sample texts and discussion in Dain (1937) 42–45, 55, 57–59, 84–86, 90; (1939) 48, 51, 66–70; Dain and Foucault (1967) 357. Mecella (2009) 101 analyses content derived from Africanus' *Kestoi*.

πῶς ἐμβληθὲν ξύλοις ἢ τοίχοις τούτων οὐχ ἅψαιτο, where both the *Corpus Perditum* (31) and *Taktika* 104 read Πῶς δυνάμεθα σβέσθαι πῦρ.[148] Second, chapter 76, concerning methods of sending secret messages, includes abridged adaptations of four techniques described in Aineias' chapter 31 (see below for the so-called *De Secretis Epistolis*). Other connections alleged between the *Sylloge Tacticorum* and Aineias' writings turn out, on closer scrutiny, to be mistaken.[149]

The other descendant of the *Corpus Perditum* is the *Taktika* of Nikephoros Ouranos, a vast, largely derivative encyclopaedia of military science, compiled c. 1000 by a distinguished general, courtier and diplomat during the reign of Basil II (976–1025).[150] Comprising 178 chapters, the *Taktika* is the last and by far the longest extant treatise to be written in the genre of Greek military manuals. It incorporates elements of almost every earlier work of Greek, Roman and Byzantine tactical literature, and is in every sense the culmination of a self-conscious literary tradition stretching back to the fourth century BC. The *Taktika* has never been edited in full: today sections are preserved in three manuscripts, *Monacensis gr.* 452 (c. 1350–1360) (= N), *Oxoniensis Baroccianus* 131 (c. 1250–1280) (= O) and *Constantinopolitanus gr.* 36 (c. 1430s) (= K), none containing the whole text, but from which collectively the work can be reconstituted almost in its entirety.[151] The three manuscripts bear witness to two stages of its textual history, an 'Oxford recension' (O and K), apparently Nikephoros' original composition, and a slightly later 'Munich recension' (N), characterised by a selective metaphrasis whereby an editor sought to remedy perceived vulgarisms in diction and syntax.[152] The authentic title and author,

148 See Dain (1939) 60; Dain and Bon (1967) xlvii, 102 (note that at xlvii Bon mistakenly refers to ch. 68 instead of 70 and appears [n. 4] to misapply the title of ch. 70 to 74).

149 At Dain and Bon (1967) xlvi–xlvii, Bon states that *Corp. Perd.* 38 was the source for *Syll. Tact.* 74, concerning spontaneous combustion (partially lost), and claims that '[e]n dernier ressort, c'est un souvenir du' Aen. 33.1, and thus, of necessity, transmitted via *App. Bell.* 45. The identification is incorrect; see correctly Dain (1939) 21–24, 35, 61, where *Corp. Perd.* 38 (> *Syll. Tact.* 74 and Nik. *Takt.* 111) derives from *App. Bell.* 44, an excerpt from the *Kestoi*, and has no connection to Aineias' work. See now these texts juxtaposed at Mecella (2009) 142 (T19) with discussion at 108–109.

150 The foundational study is Dain (1937); see also McGeer (1991); (1995a) 79–86 *et passim*; Trombley (1997) 261–274 (with several errors); Mecella (2009) 101–107.

151 For the three manuscripts, descendent *recentiores* and early editions see Dain (1937) 11–13, 93–102; summarised by de Foucault (1973) 282–284, with additions and corrections to the dating of *Oxoniensis Baroccianus* 131 in Wilson (1978) 157–179 and McGeer (1995a) 81–86. Summary in Mecella (2009) 102–104. Further research on these manuscripts will be presented in Rance (forthcoming a), including my recent recognition of excerpts in a fourth manuscript, *Vindobonensis phil. gr.* 120 + 112 (c.1350s), overlooked in previous studies.

152 Dain (1937) 102–107, 128–130; McGeer (1995a) 83–86.

together with the pinax, are preserved only in codex K, unknown to scholarship before 1888. In the absence of this vital witness, western scholars from the early seventeenth to early twentieth centuries mistakenly attributed the *Taktika* to an emperor Constantine (usually assumed to be VII or VIII) on account of a spurious ascription that had permeated the manuscript tradition from the 1560s onwards.[153] It was not until 1937, on the basis of codex K, that Dain correctly determined the author, period, scope and sources of the *Taktika*.[154] Dain's life-long ambition to produce a full edition was not realised by his death in 1964, and since then only short unconnected segments (totalling 49 chapters) have been published.[155] In terms of content and source-material, Dain identified four subdivisions within the text; of interest here is the third (chs. 75–173), which appears to derive almost entirely from the *Corpus Perditum*.[156] Where the constituent sources of the *Corpus Perditum* are extant, comparison with the text of the *Taktika* indicates that, on the whole, Nikephoros made only occasional and minor changes to the order of material and sought to introduce

153 The complex tangle of fabrication and editorial guesswork is unravelled by Dain (1937) 98–100, 107–128, 136–140, with stemma at 131; summarised Mecella (2009) 103–104.

154 Dain did not examine codex K in person but knew the text only at some remove, through a photographic reproduction of an apograph made in 1888 by Felix Beheim-Schwarzbach, then a student at Kiel University, and deposited by him in 1919 in the Universitätsbibliothek, Freiburg im Breisgau (now *codex Friburgensis Brisgoviae* 706). For K see Blass (1888) 225, 662; Deissmann (1933) 72–73; Dain (1937) 13, 89–90, 93–95; Dain and de Foucault (1967) 389; McGeer (1995a) 81–82; Rance (forthcoming a).

155 A critical edition of the complete text of the *Taktika* is in preparation by the present author. Modern critical editions are available for the following chapters: chs. 56–65 ed. and Eng. trans. McGeer (1995a) 88–163; chs. 63–74 ed. and Fr. trans. de Foucault (1973); ch. 54 ed. and Eng. trans. Pryor and Jeffreys (2006) 571–605; chs. 89–100, 102–103, 105–111 ed. Mecella (2009) 117–143; chs. 104, 172–173 ed. Dain and Bon (1967) 102–104; chs. 119–123 ed. Dain (1943) 89–104; ch. 176 ed. and Fr. trans. in Zuckerman (1994) 381–382. A provisional selective edition of chs. 2, 4–8, 11–42.38, as '*Recensio Constantiniana*', can be found in the lower register of Vári (1917–1921), based on codex N and its descendant *Heidelbergensis* 52, which Vári erroneously identified as an independent witness. Dain (1937) 42–43, 57, 59, 65, 79, 84–86, 88 also contains provisional editions of chs. 1.1–7, 83, 88.1, 115, 127.1, 131.6[7], 174, 175.2, 177.10–12; similarly Dain (1939) 66, 68–69 for chs. 78.1–2, 163.1–3. Of purely antiquarian value is I. Meursius (van Meurs), *Constantini Porphyrogennetae imperatoris Opera in quibus Tactica primum prodeunt* ... (Leiden 1617) (chs. 1–14), also based on *Heidelbergensis* 52, revised and expanded by Lami (1741–1763 [1745]) VI 1211–1409 (chs. 1–53), who consulted in addition *Veronensis* 127, another descendant of N.

156 For the quadripartite structure of the text see Dain (1937) 51–87, followed by de Foucault (1973) 281–283; with important modifications by McGeer (1991) 131–134; (1995a) 80–81; summarised by Mecella (2009) 104–106.

a lower linguistic register, characterised by the substitution of contemporary diction and vocabulary and a tendency towards greater prolixity. In comparison to the author of the *Sylloge Tacticorum*, therefore, Nikephoros' *Taktika* more faithfully preserves the wording, arrangement and substance of the *Corpus Perditum*.[157]

Dain identified three ultimately Aineias-derived chapters in the *Taktika*.[158] The first of these is of particular interest. We have seen that *Taktika* 104 originally found a counterpart at *Sylloge Tacticorum* 70, now lost owing to the defective exemplar available to the copyists of *Mediceo-Laurentianus gr.* LXXV-6. As previously observed, however, in an effort to fill this lacuna a scribe resorted to transcribing five relevant chapters of the *Corpus Perditum* into the margins of this codex and in so doing supplied the text of the common ancestor of *Taktika* 104 and *Sylloge Tacticorum* 70 (= Dain's *Corp. Perd.* 31 [folio 102^v marg. n° 69]).[159] These exceptional circumstances offer a unique opportunity to chart the transformation of Aineias' text over three successive redactions:

Aineias 34	*App. Bell.* 38	*Corp. Perd.* 31	Nik. Our. *Takt.* 104
	Πῶς πῦρ δυνάμεθα σβέσαι	Πῶς δυνάμεθα σβέσθαι πῦρ	Πῶς δυνάμεθα σβέσθαι πῦρ
Ἐὰν δέ τι οἱ πολέμιοι πειρῶνται ἐμπιμπράναι ἰσχυρᾷ	Ἐὰν ἡμῶν οἱ πολέμιοι πυρὶ κατασκευαστῷ τεῖχος ἤ τι ἕτερον	Ἐὰν οἱ πολέμιοι διὰ πυρὸς σκευαστοῦ τεῖχος ἢ ἕτερόν	Ἂν ἐμπυρίσωσιν οἱ πολέμιοι διὰ πυρὸς σκευαστοῦ (ἤγουν διὰ

157 See generally Vieillefond (1932a) 34–35; Dain (1937) 83, 90; (1939) 48–51. Late vocabulary: e.g. *Takt.* 114: μολχάμιον ἢ χάσδιον instead of βιβλίον, and εἰς τὸ ἐπώμιον τοῦ ὑποκαμίσου instead of ἐπὶ ὤμου τοῦ χιτωνίσκου; 172: σοῦδα for τάφρος; 173: διαβασίδια, apparently a hapax, in place of παρόδους; 172 (tit.) and 173 (tit.): κάστρον. The source-chapter for *Takt.* 113 is missing from the received text of the *Apparatus Bellicus* (see below pp. 345–346), but we can be certain that Nikephoros substituted φούσκα for κύστις in the *Corpus Perditum*, as κύστις occurs in both Aineias 31.10–13 and *Syll. Tact.* 76.2. See similarly *Takt.* 113: φουσκῶ or ἀποφουσκῶ, where Aineias 31.10–12 and *Syll. Tact.* 76.2 read φυσῶ. See also *Takt.* 113: ῥόγιον, cf. Aineias 31.10–12: λήκυθος (> Philon D.80 [102.40–47]); *Syll. Tact.* 76.2: ἀγγεῖον. Prolixity: e.g. see below pp. 353–354 for the juxtaposition of *Takt.* 115 and *App. Bell.* 53.3. See also Mecella (2009) 106–107 for remarks on the style and language of Nikephoros' *Taktika* based on the larger sample of excerpts from the *Kestoi* of Julius Africanus.

158 Dain (1937) 23, 37, 62–63; (1939) 18, 39, summarised in Dain and Bon (1967) xlviii, with editions of texts at 102–104 (with correction: *Taktika* 172 = Aineias 37.1–5 not 37.1–4).

159 Dain (1937) 23, 62–63; (1939) 18–20, 35, 39–40, 60.

Aineias 34	*App. Bell.* 38	*Corp. Perd.* 31	Nik. Our. *Takt.* 104
σκευασίᾳ πυρός, σβεννύειν χρὴ αὐτὸ ὄξει· οὐ γὰρ ἔτι ῥᾳδίως ἐξάπτεται. Μᾶλλον δὲ τὸ προαλείφειν· τούτου γὰρ πῦρ οὐχ ἅπτεται. Τοὺς δὲ κατασβεννύντας ἀπὸ τῶν ὑψηλοτέρων ἔχειν περὶ τὸ πρόσωπον ἔρυμα, ἵνα ἧσσον ὀχλῶνται προσαϊσσούσης αὐτοῖς τῆς φλογός.	ἐμπρήσουσι, πῶς σβέσαι δυνάμεθα; σβέσομεν αὐτὸ συντόμως καταχέοντες ὄξος. Τοὺς δὲ σβεννύντας ἀπὸ τῶν ὑψηλοτέρων δεῖ περὶ τὸ πρόσωπον ἔχειν ἔρυμά τι, ἵνα ἧσσον ὀχλῶνται προσαϊσσούσης αὐτοῖς τῆς φλογός.	τι ἐμπρήσουσι, σβέσομεν αὐτὸ ὄξος καταχέαντες.	τοῦ λαμπροῦ) τεῖχος ἢ ἄλλο τί ποτε, σβέσομεν αὐτὸ καταχύνοντες ὄξος.
	Εἰ δὲ σὺ προγνῷς τὰ μέλλοντα καίεσθαι, χρῖσον ἔξωθεν ὄξος, καὶ τούτοις οὐ πρόεισι πῦρ.	Εἰ δὲ προγνῷς τὰ μέλλοντα καίεσθαι, χρῖσον ὄξος ἔξωθεν καὶ τούτοις ὁ πῦρ οὐ προάψεται.	Ἂν δὲ προνοήσῃς τὸ ποῦ μέλλει γενέσθαι ὁ ἐμπυρισμός, χρῖσον ἔξωθεν ὄξος· καὶ οὐ μὴ ἅψηται ἐκεῖ τὸ πῦρ.

In addition, Dain recognised that *Taktika* 172 and 173, which concern defensive ditches and countermeasures to a besieger's mining operations, are also excerpts of Aineias' work transmitted via the *Apparatus Bellicus* (*Takt.* 172 = Aineias 37.1–5 > *App. Bell.* 54; *Takt.* 173 = Aineias 39.1–2 > *App. Bell.* 56). It is unclear why 172 and 173 are detached from the other chapters that derive (via the *Corpus Perditum*) from the *Apparatus Bellicus*, which are otherwise all assembled in a single block at *Taktika* 89–111. Their terminal position, together with the absence of corresponding material in the *Sylloge Tacticorum*, prompted Dain to suspect that 172 and 173 had become detached from the version of the *Apparatus Bellicus* known to the compiler of the *Corpus Perditum* but somehow still managed to pass into Nikephoros' *Taktika* by an alternative, if ill-defined route. This argument, though not impossible, is unpersuasive.[160] In general the final chapters of the *Taktika* (172–178) are a hotchpotch of extracts from diverse

160 Dain (1937) 82–83. See also doubts expressed by Mecella (2009) 105 n. 107, 111. Given the greater selectivity exercised by the author of the *Sylloge Tacticorum* with regard to his source-materials, the absence of direct parallels to *Taktika* 172 and 173 is not a compelling argument for the omission of these two chapters from their common source, the *Corpus*

sources with no unifying theme or obvious connection, as if Nikephoros (or a posterior editor) had amassed any left-over or previously omitted material into a miscellaneous appendix.[161]

Finally, we should also note that Dain's misconceptions regarding the sources of the *Corpus Perditum*, and in particular the putative *De Secretis Epistolis* (= *Taktika* 112–118), caused him to overlook or misinterpret other remnants of Aineias' work in Nikephoros' *Taktika*. It will be demonstrated below that the *De Secretis Epistolis*, allegedly derived from Philon's lost book, is a fantasy. The mundane reality is that *Taktika* 112–115, which follow directly after the block of chapters (89–111) derived from the *Apparatus Bellicus*, likewise derive from extracts of Aineias' *Poliorketika* found in the *Apparatus Bellicus* (indeed, in his original study, Vieillefond explicitly acknowledged this derivation in the case of 112).[162] This brings the total of Aineias-derived chapters in Nikephoros' *Taktika* to seven (104, 112–115, 172–173). It seems unlikely, however, that Nikephoros would have known the ultimate origin of this material, which came into his hands as dismembered, decontextualised and anonymous excerpts via at least three successive adaptations, if the foregoing reconstruction is correct (Aineias' *Poliorketika* > '*Exercitationes*' > *Apparatus Bellicus* > '*Corpus Perditum*' > *Taktika*). A source-notice which Nikephoros (or a redactor) affixed to the beginning of his *Taktika*, and now preserved only in *Constantinopolitanus gr.* 36, assembles a catalogue of those τακτικὰ ἤγουν στρατιγικά (sic) allegedly consulted by Nikephoros during the compilation of his treatise. The inclusion of the names of ancient historians (e.g. Diodorus, Cassius Dio, Plutarch) and famous historical figures (e.g. Hannibal, Alkibiades, Artaxerxes) implies that '*taktika* or *strategika*' is used here in the widest sense of 'military authorities' and not with the restricted meaning of technical manuals. In any case, the absence of Aineias from this partly fanciful, name-dropping list, clearly compiled to magnify Nikephoros' erudition and reading, suggests that by this date the earliest of extant Greek military writers had ceased to be even a name.[163]

Perditum. As Dain (1939) 48, 51 observed, this situation pertains elsewhere, e.g. *Taktika* 89 (= *App. Bell.* 3 + 4) and 110 (= *App. Bell.* 15 + 16), neither of which has a counterpart in the *Sylloge Tacticorum*.

161 Dain (1937) 37, 82–89; Zuckerman (1994) 380–383.

162 Nik. *Takt.* 112 = *App. Bell.* 51 (< Aineias 31.4); *Takt.* 113 (< Aineias 31.10–13); *Takt.* 114 = *App. Bell.* 53.2 (< Aineias 31.23); *Takt.* 115 = *App. Bell.* 53.3 (< Aineias 31.14).

163 *Codex Constantinopolitanus gr.* 36 (= Topkapı Sarayı Müzesi Kütüphanesi G.İ. 36) 1^r. The source-notice reads: Τακτικὰ ἤγουν στρατιγικὰ Ἀριανοῦ, Αἰλιανοῦ, Πέλοπος, Πολυαίνου, Ὀνοσάνδρου, Ἀλκιβιάδου, Ἀρταξέρξου, Συριανοῦ, Ἀνίβα, Πλουτάρχου, Ἀλεξάνδρου, Διοδώρου, Δίω-

De Secretis Epistolis

We return to the hypothesis that the *Corpus Perditum* also incorporated, at some remove, elements of Philon's lost book (IX) 'On the sending of secret letters' (Περὶ ἐπιστολῶν τῶν κρυφαίως ἀποστελλομένων), which allegedly served as a conduit for material extracted from Aineias' well-known treatment of this subject (ch. 31). As previously discussed, towards the end of his *Poliorketika* Philon foreshadows the content of a book he intends to write on cryptography. He briefly describes four methods of sending a secret message, three of which are taken from or inspired by Aineias' work. To judge from this preview, Philon's book, if in fact written, was heavily indebted to Aineias' chapter 31.[164] In 1932 Vieillefond proposed that modified and paraphrased remnants of Philon's lost book survive in the *Sylloge Tacticorum* (ch. 76) and Nikephoros' *Taktika* (chs. 112–118), from which an outline of its substance, if not its wording, could be partially reconstructed. The corresponding content and broadly parallel sequence of these extracts in the *Sylloge Tacticorum* and *Taktika* indicated that this material had been transmitted via their common source, the *Corpus Perditum* (Vieillefond's 'le *Corpus*', later '*Corpus* X').[165] At the time of Vieillefond's study, however, both the *Sylloge Tacticorum* and the *Taktika* were largely unpublished and improperly understood, and his analysis depended on an incomplete knowledge of the manuscripts.[166] Vieillefond's thesis was endorsed by Dain, though seemingly without closer scrutiny and clearly harbouring some misconceptions. Dain conjectured that the text of

---

νος, Πολυβίου, Ἡρακλείτου, Μαυρικίου, Νικηφόρου καὶ ἄλλων τινῶν, συλλεγὲν παρὰ Νικηφόρου μαγίστρου τοῦ Οὐρανοῦ ἀπὸ πολλῶν ὡς εἴρηται ἱστορικῶν ἐν ἐπιμελείᾳ πολλῇ (based on my autopsy of the manuscript, Oct.–Dec. 2013). See also Dain (1937) 13, reprinted Dain and de Foucault (1967) 371–373; an apparently corrected text is found in Dain (1946) 150–151 (Ἀρριανοῦ, Ὀνησάνδρου, Ἀννίβα). For Dain's indirect knowledge of this codex see *supra* n. 154.

164 See Philon D.76–82 (102.29–50) with preceding comments at pp. 312–314.

165 Vieillefond (1932a) esp. 28–35.

166 Vieillefond (1932a) 28–29 knew the *Sylloge Tacticorum* from its unique prototype, *Mediceo-Laurentianus gr.* LXXV-6, and the edition of *Sylloge Tacticorum* 76 in Wölffin and Melber (1887) 509–511. Vieillefond himself edited *Taktika* 112 and 113 from *Monacensis gr.* 452 (140r), but his unfamiliarity with the other witness, *Oxoniensis Baroccianus* 131, had serious implications for his conclusions, especially as it alone preserves the following two chapters (114 and 115). Furthermore, Vieillefond's study predates Dain's identification of the authorship, title and scope of the *Taktika*, which Vieillefond termed 'l'*Ecloge*' and ascribed, in accordance with sixteenth-century humanist speculation, to Constantine VIII (r. 1025–1028), see *supra* n. 153.

Philon's book on secret messages had been revised at an unspecified date, apparently in late antiquity, and this adaptation, which Dain styled *De Epistolis Secreto Mittendis* or *De Secretis Epistolis*, now also lost, later became the fourth component of the *Corpus Perditum*. Accordingly, in both the *Sylloge Tacticorum* (76) and the *Taktika* (112–118) this material is located in the same position, following directly after chapters derived from the *Apparatus Bellicus* and preceding chapters derived from the so-called *Hypotheseis* (see above).[167] The studies of Vieillefond and Dain are periodically cited in subsequent literature, but the veracity of their hypothesis has never been tested, largely owing to the continued lack of a full critical edition of *Taktika* 112–118.[168] These chapters are now edited from all manuscript witnesses in the appendix below.

Upon detailed inspection, however, the hypothesis rapidly unravels and it becomes difficult credit that Dain ever subjected this material to a thorough examination. To outline the facts of the case: *Taktika* 112–118 describe ten methods of sending a secret message; nine of these are also found at *Sylloge Tacticorum* 76, together with an additional five. Where the original sources can be identified (see below), it is clear that the *Taktika*, as elsewhere, closely reproduces the arrangement and language of their shared ancestor, the *Corpus Perditum*, while the author of the *Sylloge Tacticorum* substantially modified this source, rearranging content, adding material and often severely abridging the text, which he recast in his own classicizing idiom. A further peculiarity of the textual revision in *Sylloge Tacticorum* 76 is the author's decision to transform most of the fourteen cryptographic techniques into pseudo-historical exempla by fancifully inserting the names of famous generals or even invented persons. In consequence, methods of secret communication that demonstrably derive from Aineias' chapter 31 become stratagems attributed to Sulla, Harpagos and an otherwise unattested 'Pseudagoras'.[169] This curious desire to historicize (or 'stratagemize') anonymous general prescriptions by the addition of

---

167 Dain (1931) 341–342; (1937) 54, 64–65; (1939) 45, 53–54, 61; summarised in Dain and Bon (1967) xlii, xliv, xlvii, 72 n. 1; Dain and de Foucault (1967) 339, 352–353.

168 E.g. cited with approval by de Foucault (1967) 182–183; Bettalli (1990) 302–303, 305; accepted by Krentz and Wheeler (1994) xxii; Mecella (2009) 100 n. 75, 105, 111 n. 139.

169 Aineias 31.4 > *Syll. Tact.* 5 (Harpagos); Aineias 31.10–13 > *Syll. Tact.* 76.2 (Sulla); Aineias 31.23 > *Syll. Tact.* 76.7 (Pseudagoras). Similarly, other methods of secret communication derived from another source(s), and which are anonymous in the corresponding passages of Nikephoros' *Taktika*, are assigned to an 'Onias' (*Syll. Tact.* 76.11 = *Takt.* 118.1) and a 'Choirillos' (*Syll. Tact.* 76.13 = *Takt.* 118.3), both names otherwise unknown to ancient military history.

nomina ficta, a propensity already noted by Vieillefond and Dain, is found elsewhere in the *Sylloge Tacticorum*.[170] In the case of chapter 76, this inventiveness was clearly inspired by the form and content of the following chapters (77–102), which originate in the so-called *Hypotheseis*, an adaptation of Polyainos' stratagem collection.[171] In any case, whatever the origin and textual history of the shared material on cryptography in *Sylloge Tacticorum* 76 and *Taktika* 112–118, the latter is by far the more reliable and more complete witness to the shape and wording of their common source. The corresponding content, together with identified sources, is tabulated below:

	Takt.	***Syll. Tact.***	**Source**
A message hidden in the sole of a shoe	112	76.5	Aineias 31.4
written on a bladder inflated inside an oil flask	113	76.2	Aineias 31.10–13
stitched into clothing	114	76.7	Aineias 31.23
written under the wax of a writing-tablet	115	76.8?[172]	Aineias 31.14

170 See remarks of Vieillefond (1932a) 34. Similarly, Dain (1937) 84–86 demonstrates the interpolation of invented historical information at *Syll. Tact.* 80.5. In another instance of pseudo-historicizing, the author appears to have assigned a number of unattributed stratagems to 'Merops' (Μέροψ), a name otherwise entirely unattested outside mythology, cf. *Syll. Tact.* 77.1–2, 95.4, 9, 99.3, with remarks of Dain (1931) 332–333; (1938) 118, 134–135, 140 app. crit.; Lammert (1952) 1435; Eramo (2008) 143–144. More generally, Krentz and Wheeler (1994) xxii–xxii identify other contrived or fraudulent features of the *Sylloge Tacticorum*.

171 In the *Sylloge Tacticorum* a subheading is inserted before ch. 76: Στρατηγικαὶ παραινέσεις ἐκ πράξεων καὶ στρατηγημάτων παλαιῶν ἀνδρῶν Ῥωμαίων τε καὶ Ἑλλήνων καὶ λοιπῶν ἐν κεφαλαίοις κη' ('Recommendations to generals based on the deeds and stratagems of the ancient Romans and Greeks and others in 28 chapters'). Even though only twenty-seven chapters in fact follow, the position of the caption clarifies that the author grouped the cryptographic exempla in ch. 76 with the following 'stratagemical' chapters (77–102) derived from an adaptation of Polyainos' work. In consequence, ch. 76 became ch. 1 of Melber's so-called 'Leonis imperatoris *Strategemata*' (i.e. *Syll. Tact.* 76–102, which Melber published separately as if an independent work), see Wölffin and Melber (1887) 509–511. Vieillefond (1932a) 28 n. 3; Dain (1931) 342; (1937) 65 n. 1 stressed that ch. 76 has a different origin to the subsequent Polyainos-derived material. Accordingly, Dain (1938) 116 app. crit. assumed that the subheading has been misplaced and should in fact occur between chs. 76 and 77, but the author's modifications to the content of ch. 76 suggest its intentional amalgamation with the following chapters, and Krentz and Wheeler (1994) xxii rightly stress the affinity.

172 The MS of *Syll. Tact.* 76.8 reads: Ἄλλος ἐν ψυκτῆρι κηρῷ τὸν χάρτην ἑλίξας ἐξέπεμψεν, which

(cont.)

	Takt.	*Syll. Tact.*	Source
The story of Histiaios and the tattooed slave	116		Herodotos 5.35.3
A message hidden inside a hollowed-out stone	117	76.9	Source unknown
hidden in a woman's hair	118(1)	76.11	Source unknown
hidden in a horse's tail	118(2)	76.12	Source unknown
placed inside a small reed concealed in the mouth	118(3)	76.13	Source unknown
The story of Harpagos and Cyrus the Great	118(4)	76.4	Herodotos 1.123

The five additional methods briefly mentioned in the *Sylloge Tacticorum* are: a message wrapped around a candle (76.1); inserted in a hole bored into a wooden rod (76.6); placed under a horse-blanket (76.3); inside the head of a spear (76.10); inside the aorta of a sheep's heart (76.14). It is uncertain whether these five were also present in the *Corpus Perditum*, and thus for some reason omitted by Nikephoros, or represent inventive additions by the author of the *Sylloge Tacticorum*. The general character and predilections of both writers favour the latter view.

As the table shows, of the ten methods of cryptography found in Nikephoros' *Taktika* four originate in Aineias' *Poliorketika* 31, two are extracts from Herodotos, and four are of unknown origin. The hypothesis of Vieillefond and Dain immediately encounters a major linguistic hurdle: if this material, at some point in its previous history, had been transmitted in a book by Philon, even accepting that it might have undergone successive revisions by late antique and Byzantine redactors, one would expect some uniformity of style and language, if not exact correspondence to the terse syntax and distinctive lexis of Philon's

appears to mean 'Another man sent out [a message] in a cooler (a wine-cooler?) having rolled the paper in wax'. The meaning remains somewhat unclear, at least from a tenth-century Byzantine perspective. The author of the *Sylloge Tacticorum* presents the various cryptographic methods in ch. 76 in roughly the same order as they occur in Nikephoros' *Taktika*. If that is the case here (note that preceding *Syll. Tact.* 76.7 = *Takt.* 114), then the corresponding passage to *Syll. Tact.* 76.8 is Nik. *Takt.* 115 (= Aineias 31.14 > *App. Bell.* 53.3), which describes how a message was written on the wooden panel of a writing-tablet (πινακίδιον in *Takt.*; δέλτος in Aineias and *App. Bell.*) and covered over with fresh wax. Both the sequential position and the central importance of wax suggest that *Syll. Tact.* 76.8 evolved from this method, which is otherwise missing from the *Sylloge Tacticorum*, even though it was demonstrably present in the author's source, the *Corpus Perditum* (whence *App. Bell.* 53.3 = Nik. *Takt.* 115). The author may have misunderstood the ancient context of the original prescription or, given his inventiveness, chose to devise an alternative version.

extant works. Certainly in Philon's preview of his book on cryptography, wherever he drew information from Aineias' chapter 31 he entirely reconfigured this source-material to conform to his own stylistic criteria, to the extent that the original wording of Aineias cannot be traced.[173] In stark contrast, the individual chapters at *Taktika* 112–118 still very clearly bear the linguistic stamp of the diverse sources from which they originate, a characteristic which casts doubt on the assumption of their prior unity or association within an earlier treatise. Detailed examination of individual chapters raises additional obstacles. For example, the fallacy of Dain's hypothesis is exposed by the Herodotean extract at *Taktika* 116, concerning Histiaios and the tattooed slave. Noting that Herodotos' famous tale (5.35.3) had also inspired a passage in Aineias' chapter on cryptography (31.28–29), Dain asserted that when writing *Taktika* 116 Nikephoros 'empruntera à Énée par l'intermédiaire de Philon'. Comparison of the texts of Aineias, Nikephoros and Herodotos, however, irrefutably rules out the involvement of Aineias and his alleged intermediary, and points to a more direct connection between *Taktika* 116 and the historian's work:[174]

Nikephoros, *Taktika* 116	**Herodotos 5.35.3**	**Aineias 31.28**
ὁ Ἱστιαῖος θέλων γράψαι πρὸς τὸν Ἀρισταγόραν ἵνα *ἀποστατήσῃ* ... καὶ *ἀνέμεινεν* ἵνα ἐκβῶσιν αἱ τρίχες. ὡς δὲ ἐξέβησαν, ἀποστεῖλεν αὐτόν· προστάξας *αὐτῷ* ἵνα ὅταν ἀπέλθῃ εἰς τὸν Ἀρισταγόραν, εἴπῃ αὐτῷ πάλιν	ὁ γὰρ Ἱστιαῖος βουλόμενος τῷ Ἀρισταγόρῃ σημῆναι *ἀποστῆναι* ... καὶ *ἀνέμεινε* ἀναφῦναι τὰς τρίχας, ὡς δὲ ἀνέφυσαν τάχιστα, ἀπέπεμπε ἐς Μίλητον ἐντειλάμενος *αὐτῷ* ἄλλο μὲν οὐδέν, ἐπεὰν δὲ ἀπίκηται ἐς	Ἱστιαῖος δὲ βουλόμενος τῷ Ἀρισταγόρᾳ σημῆναι ... καὶ *ἐπέσχεν* ἕως ἀνέφῦσαν αἱ τρίχες, ⟨Ὡς δὲ ἀνέφυσαν⟩ τάχιστα, ἔπεμπεν εἰς Μίλητον, ἐπιστείλας *τῷ ἐπεστιγμένῳ* ἄλλο μὲν οὐδέν, ἐπειδὰν δ' ἀφίκηται εἰς Μίλητον

173 Compare e.g. Aineias 31.4–5 and 10–13 with respectively Philon D.79 (102.39–40) and D.80 (102.40–47). On Philon's style and language see Arnim (1912); Whitehead (2016) 25–33.

174 Dain (1937) 65 n. 5 (116 misnumbered 117); Dain and Bon (1967) 72 n. 1. Subsequently, de Foucault (1967) esp. 183–185 devoted a short study to these three texts and other versions of Histiaios' ruse, but he failed to notice the critical correspondences between *Taktika* 116 and Herodotos 5.35, absent from Aineias 31.28–29, and merely repeated Dain's view that *Taktika* 116 derived from Aineias via a lost monograph on cryptography. De Foucault (1967) 183–184 published an edition of *Taktika* 116 on the basis of an unpublished and incomplete text prepared by the late Dain (also cited in Foucault [1973] 282), then in de Foucault's possession thanks to the kindness of Mme Dain. The absence of an apparatus criticus leaves the manuscript authority uncertain. I plan to publish a critical text of *Taktika* 116 elsewhere.

(cont.)

Nikephoros, *Taktika* 116	Herodotos 5.35.3	Aineias 31.28
ξυρῆσαι τὰς τρίχας καὶ ἰδεῖν τὴν κεφαλὴν αὐτοῦ.	Μίλητον, κελεύειν Ἀρισταγόρην *ξυρήσαντά μιν τὰς τρίχας* κατιδέσθαι ἐς τὴν κεφαλήν.	πρὸς Ἀρισταγόραν, κελεύειν *ξυρήσαντα* κατιδεῖν εἰς τὴν κεφαλήν.

The source of the four unprovenanced cryptographic methods (*Taktika* 117–118.3 = *Syll. Tact.* 76.9, 11–13) is of no immediate concern here and will be discussed in a separate study; it suffices to observe that again there is nothing in their form, content or language to suggest an origin in a lost or putative work by Philon. Of more immediate importance, this is also demonstrably the case with the four methods (*Taktika* 112–115) that originate in Aineias' *Poliorketika*. It must be stressed that Vieillefond's sole rationale for identifying this suite of cryptographic chapters as a residue of Philon's lost book was the presence of material originating in Aineias' chapter 31, a known source for Philon's interest in cryptography. Vieillefond himself acknowledged, however, that some of this Aineias-derived content had, without doubt, entered Nikephoros' *Taktika* by an alternative route that did not involve the intermediacy of Philon. Vieillefond recognised that *Taktika* 112 was in fact derived (via the *Corpus Perditum*) from *Apparatus Bellicus* 51 (= Aineias 31.4):[175]

Nikephoros, *Taktika* 112	*Apparatus Bellicus* 51	Aineias 31.4
Πῶς ἵνα πέμψῃς κρυφὰ γράμματα	Περὶ κρυφίας ἐπιστολῶν εἰσπομπῆς	Περὶ ἐπιστολῶν κρυφαίων
Τοὺς ἔχοντας προδότας τινὰς εἰς πολέμια ἔθνη, ἀναγκαῖόν ἐστι γίνωσκειν πῶς δεῖ πέμπειν αὐτοῖς κρύφα γράμματα. ὀφείλει	Τοῖς κεχρημένοις προδόταις ἀναγκαῖον εἰδέναι πῶς ἐπιστολὰς δεῖ αὐτοὺς εἰσπέμπειν. Ἀπόστελλε γοῦν οὕτως.	

175 Vieillefond (1932a) 31–34. Nikephoros, *Takt.* 112 reproduces the contextualising passage inserted at the beginning of *App. Bell.* 51, with which the excerptor introduces the series of extracts on cryptography (51–53). This reference to the necessity of sending secret messages to traitors within the city, reflecting the perspective of a besieger, has no parallel at Aineias 31.4.

Nikephoros, *Taktika* 112	*Apparatus Bellicus* 51	Aineias 31.4
πέμπεσθαι πρὸς αὐτοὺς ἀνὴρ βαστάζων φανερὰ γράμματα περὶ ἄλλων ὑποθέσεων· …	Πεμπέσθω ἀνὴρ ἐν τῷ φανερῷ φέρων ἐπιστολήν τινα περὶ ἄλλων πραγμάτων …	⟨Ἢ⟩ πεμπέσθω ἀνὴρ ἀγγελίαν φέρων τινὰ ἢ καὶ ἐπιστολὴν περὶ ἄλλων φανερῶν· …

Vieillefond seemed content merely to admit this anomaly in his hypothetical reconstruction and did not consider its wider implications. The inconsistency was also disregarded or passed unrecognised in the work of Dain and subsequent commentators. Furthermore, Vieillefond knew the text of Nikephoros' *Taktika* only in *Monacensis gr.* 452, from which chapters 114 and 115 are missing. Had he been able to consult *Oxoniensis Baroccianus* 131, unknown to him at the time of writing, he would have discovered that *Taktika* 114 and 115 also derive from excerpts of Aineias in the *Apparatus Bellicus* (ch. 53.2–3), as clearly reflected in their language and consecutive sequence (i.e. *Takt.* 114–115 = *App. Bell.* 53.2–3, whereas in the original work these two passages occur widely separated at Aineias 31.23 and 31.14):

Nikephoros, *Taktika*	*Apparatus Bellicus*	Aineias
114 Τινες εἰς μολχάμιον ἢ χάσδιον γράψαντες γράμματα ψιλὰ εἰς τὸ ἐπώμιον τοῦ ὑποκαμίσου ὑποθέντες ἠπλωμένον. καὶ περιρράψαντες αὐτό, *ἀνύποπτον οὕτως ἐποίησαν τὴν* ἀποστολὴν τοῦ γράμματος.	**53.2** Ἄλλοι τινὲς εἰς βιβλίον γράψαντες ὡς λεπτότατον μακροὺς στίχους καὶ λεπτὰ γράμματα, ἵνα εὐογκότατα γένηται, εἶτα ἐπὶ τοῦ ὤμου τοῦ χιτωνίσκου ὑποθέντες καὶ ἀποπτύξαντες, *ἀνύποπτον ἐποίουν τὴν* κομιδὴν τῆς ἐπιστολῆς.	**31.23** Ἤδη δέ τινες εἴς τι βιβλίον γράψαντες ὡς λεπτότατον μακροὺς στίχους καὶ λεπτὰ γράμματα, ἵν' ὡς εὐογκοτάτη γένηται ἡ ἐπιστολή, εἶτα ἐπὶ τῷ ὤμῳ τοῦ χιτῶνος ὑποθέντες καὶ ἀποπτύξαντες τοῦ χιτωνίσκου ἐπὶ τῷ ὤμῳ, *ἀνύποπτος δὴ δοκεῖ εἶναι ἡ κομιδὴ τῆς ἐπιστολῆς καὶ ἐνδεδυκότος τινὸς τὸν χιτωνίσκον καὶ οὕτω φερομένου.*
115 *Ἄλλοι πάλιν* εἰς πινακίδιον ἀκήρωτον *γράψαντες* εἰς τὸ ξύλον ἅπερ ἤθελον. καὶ κηρὸν ἀναλύσαντες, ἐπείσχυσαν εἰς	**53.3** *Ἄλλοι πάλιν* ἐν τῆς δέλτου ζύλῳ *γράψαντες* κηρὸν ἐπέτηξαν καὶ ἄλλα εἰς τὸν κηρὸν *ἔγραψαν*. Εἶτα ὅταν ἔλθῃ παρ' ὃν δεῖ,	**31.14** Ἤδη δέ τις ἐν δέλτου ξύλῳ *γράψας* κηρὸν ἐπέτηξεν καὶ ἄλλα εἰς τὸν κηρὸν *ἐνέγραψεν*. Εἶτα ὅταν ἔλθῃ παρ' ὃν ἔδει, ἐκκνήσας

(cont.)

Nikephoros, *Taktika*	*Apparatus Bellicus*	Aineias
αὐτὸ καὶ ἄλλα γράμματα φανερὰ εἰς τὸν κηρὸν *ἔγραψαν*. εἶτα ὁ δεξάμενος αὐτὰ ἀνασπάσας τὸν κηρὸν καὶ ἀναγνοὺς τὰ γράφεντα εἰς τὸ ξύλον, ἀντιγράψας καὶ ἐκεῖνος *κατὰ τὸν ὅμοιον τρόπον* πάλιν *ἀπέστειλεν*.	τὸν κηρὸν ἐκκνήσας καὶ ἀναγνοὺς *ὁμοιοτρόπως ἀνταπέστειλεν*.	τὸν κηρὸν καὶ ἀναγνοὺς γράψας πάλιν *ὡσαύτως ἐπέστειλεν*.

With regard to Nikephoros' *Taktika* 112, 114 and 115, therefore, the alleged intermediacy of Philon can be discounted entirely. As Vieillefond himself had already recognised in the case of *Taktika* 112, these three chapters share peculiarities of lexis, syntax and arrangement with the excerpts of Aineias' chapter 31 found in the *Apparatus Bellicus*, to such a degree as to leave no doubt that it was the conduit by which this material was transmitted, via the *Corpus Perditum*, into the *Taktika* and, necessarily, also into the *Sylloge Tacticorum*. Indeed, it is striking that the four chapters of the *Taktika* which originate in Aineias' treatment of cryptography, 112–115, directly follow the other *Apparatus Bellicus*-derived chapters in the *Taktika* (89–111) and clearly form a contiguous sequence with them. That is to say, *Taktika* 112–115 never, in fact, belonged to a self-contained treatise on military cryptography, but are simply the concluding chapters of a block of material (*Taktika* 89–115) taken from the *Apparatus Bellicus*. Consequently, *Taktika* 116–118, partly Herodotean extracts, partly of unknown but not inevitably ancient origin, seem to constitute an appendix of miscellaneous material on cryptography, apparently assembled and inserted by the compiler of the *Corpus Perditum*, but detailed discussion of this question must be left to another occasion. In any case, insofar as the Aineias-derived pedigree of *Taktika* 112–115 supplied the original and only justification for thinking of Philon's lost book on cryptography, Vieillefond and Dain were chasing a chimera.[176]

176 Vieillefond's line of reasoning ([1932a] 29–30) as to why all this cryptographic material must descend from a single 'traité antérieur', even after he had identified contradictions, is especially uncompelling. It is also remarkable that Dain was clearly aware of, and even drew attention to, the excerpts on cryptography from Aineias 31 found in *App. Bell.* 51, 52 and 53, but he made no connection between these three chapters and the corresponding

Finally, a word of explanation is required regarding the remaining Aineias-derived chapter in this series, *Taktika* 113 (= Aineias 31.10–13), which now acquires a new potential significance. Unlike 112, 114 and 115, *Taktika* 113 finds no counterpart in the text of the *Apparatus Bellicus* that has come down to us in the direct tradition witnessed by codices V, P-B and N-E. The corresponding passage in the *Sylloge Tacticorum* (76.2), however, guarantees that it must have been present in the *Corpus Perditum*.

Nikephoros, *Taktika* 113	Aineias 31.10–13	*Sylloge Tacticorum* 76.2
Διακομίζονται πάλιν γράμματα κρύφα ὄυτως· φούσκαν ἔχουσαν ἴσον μέγεθος ῥογίου ἐπιτηδείου πρὸς ἔλαιον εἴτε μικροῦ εἴτε μεγάλου πρὸς τὸ πλῆθος τῶν μελλόντων γραφῆναι, ταύτην τὴν φούσκαν φουσκώσας καὶ ἀποδήσας, ξήρανον· εἶτα γράψον εἰς αὐτὴν εἴ τι θέλεις διὰ μέλανος ἐγκαυστοῦ· ἀφ' ἧς δὲ στεγνωθῆναι τὰ γράμματα, ἀποφούσκωσον αὐτήν, καὶ συσφίγξας αὐτήν, βάλε ἔσωθεν τοῦ ῥογίου· τὸ δὲ στόμα αὐτῆς ἔξω ἔχε· εἶτα φούσκωσον αὐτὴν ὡς ἔστιν ἔσωθεν τοῦ ῥογίου· καὶ βάλε ἔλαιον εἰς αὐτήν· εἶτα σκεπάσας, ἀπόστελλε ὅπου θέλεις· κενώσας οὖν ἐκεῖνος τὸ ἔλαιον καὶ φουσκώσας τὴν φούσκαν, ἀναγινώσκει τὰ γράμματα· εἶτα σπογγίσας αὐτὴν καὶ ἀντιγράψας, πάλιν ἀποστέλλει.	**10** Κομίζεται δὲ καὶ ὧδε. *Κύστιν* ἰσομεγέθη ληκύθῳ ὁπόσῃ ἂν βούλῃ πρὸς τὸ πλῆθος τῶν γραφησομένων *φυσήσαντα καὶ ἀποδήσαντα* σφόδρα *ξηρᾶναι*, ἔπειτα ἐπ' αὐτῆς *γράψαι* ὅ τι ἂν βούλῃ *μέλανι* κατακόλλῳ. **11** Ξηρανθέντων δὲ τῶν γραμμάτων ἐξελεῖν τὴν πνοὴν τῆς κύστιδος καὶ συμπιέσαντα εἰς ⟨τὴν⟩ λήκυθον ἐνθεῖναι· τὸ δὲ στόμα τῆς κύστιδος ὑπερεχέτω τοῦ στόματος τῆς ληκύθου. **12** Ἔπειτα *φυσήσαντα* τὴν κύστιν ἐν τῇ ληκύθῳ ἐνοῦσαν, ἵνα διευρυνθῇ ὡς μάλιστα, ⟨*καὶ*⟩ *ἐλαίου ἐμπλήσαντα* περιτεμεῖν τῆς κύστιδος τὸ ὑπερέχον τῆς ληκύθου καὶ προσαρμόσαι τῷ στόματι ὡς ἀδηλότατα, καὶ βύσαντα τὴν λήκυθον κομίζειν φανερῶς. Διαφανές τε οὖν τὸ ἔλαιον ἔσται ἐν τῇ ληκύθῳ καὶ οὐδὲν ἄλλο φανεῖται ἐνόν. **13**	Ὁ Σύλλας οὐροδόχον *κύστιν* χοίρου λαβὼν *φυσήσας* τε αὐτὴν ἰσχυρῶς *καὶ ἀποδήσας* ἄχρις ἂν *ξηρανθῇ*, γράμματά τε ἐν αὐτῇ περὶ ὧν ⟨ἂν⟩ ᾖ *γράψας μέλανι* ἐνκαυστῷ, εἶτα λύσας καὶ συμπτύξας αὐτὴν καὶ ἐλαιοδόχῳ ἐμβαλὼν ἀγγείῳ, εἶτα *φυσήσας* αὖθις *αὐτὴν καὶ ἐλαίου πλήσας* τῶν πιστοτάτων τινὶ τὸ ἀγγεῖον δέδωκε καὶ πρὸς τὸν (vacat) ἐξέπεμψεν, ἐπιτάξας εἰπεῖν ἐκείνῳ κατ' ἰδίαν τὸ ἀγγεῖον συντρίψαι. Καὶ ὃς τοῦτο δράσας καὶ τὰ γράμματα *ἀναγνοὺς* τόν τε Σύλλαν θαυμάσας τοῦ μηχανήματος κατὰ τὸν ἴσον καὶ αὐτὸς *ἀντέγραψε* τρόπον.

material in *Taktika* 112, 114 and 115, even though Vieillefond's original study had explicitly identified the dependence of *Taktika* 112 on *App. Bell.* 51. See Dain (1937) 64 ('Sauf dans un cas …', apparently aware that *App. Bell.* 51 > *Taktika* 112); (1939) 53 n. 4; Dain and Bon (1967) xliv; Dain and de Foucault (1967) 339.

(cont.)

Nikephoros, *Taktika* 113	Aineias 31.10–13	*Sylloge Tacticorum* 76.2
	Ὅταν δὲ ἔλθῃ παρ' ὃν δεῖ, ἐξεράσας τὸ ἔλαιον ἀναγνώσεται φυσήσας τὴν κύστιν· καὶ ἐκσπογγίσας καὶ κατὰ ταὐτὰ εἰς τὴν αὐτὴν γράψας ἀποστελλέτω.	

Taktika 113 is embedded in material that likewise concerns cryptography and originates from Aineias' chapter 31, while the text of 113 has also been modified using the same editorial methodology as all the other Aineias-derived excerpts found in the *Apparatus Bellicus*. There is no reason to suppose that for 113 alone, out of the twenty-seven chapters in this section (*Taktika* 89–115), the compiler of the *Corpus Perditum* uniquely departed from his model, the *Apparatus Bellicus*, and resorted to an ill-defined alternative source. If this analysis is correct, it supplies potential support for the thesis of Vieillefond that the compiler of the *Corpus Perditum* had at his disposal a somewhat different edition of the *Apparatus Bellicus* to that transmitted in the direct tradition, an edition that contained at least one additional excerpt of Aineias (and possibly more). Alternatively, and more plausibly, this chapter was simply lost or omitted in error from the text of the *Apparatus Bellicus* in a common ancestor of codices V, P-B and N-E. Such an accidental loss is rendered all the more likely by the recurrence of similar headings (Ἕτερον …, Ἕτερα …) and introductory phrases (Ἄλλοι τινές, Ἄλλοι πάλιν) in these chapters of the *Apparatus Bellicus* (51–53), which increased the chances of the copyist's error of *saut du même au même*; indeed precisely this phenomenon later occurred in the transcription of these chapters in the manuscripts of Nikephoros' *Taktika*.[177]

177 *Oxoniensis Baroccianus* 131 preserves all the chapters of this section (*Takt.* 112–118), whereas *Monacensis gr.* 452 lacks 114 and 115, presumably owing to *saut du même au même*, as all but the first chapter are identically entitled Ἄλλο.

Conclusion

Aineias was not a well-known or popular tactical author in Byzantium, indeed perhaps barely a name. The most significant factor in the reception of his writings is their tenuous textual transmission. The codicological evidence suggests that the *Poliorketika* circulated as an independent treatise by c. 400, while the testimonia and wider tradition of Greek tactical literature give no reason to suspect that other volumes of Aineias' oeuvre were known to late antique and Byzantine readers. The direct tradition of the *Poliorketika* through the Byzantine 'Dark Age' depends on a single majuscule exemplar, though in this regard the fate of Aineias' sole surviving book was no worse than that of some other classical *taktika*, including works by Arrian, Asklepiodotos and Polyainos. When the *Poliorketika* re-emerges from obscurity around the mid-tenth century, it appears that in the intervening period the text had attracted no editorial interest or fresh transcriptions that might signal a contemporary or recent readership. Tasked with assembling the remaining specimens of classical military literature, the staff of the imperial scriptorium managed to locate only a single long-neglected majuscule papyrus codex, probably dating to the early fifth century. This 'rediscovery' did not bring Aineias or his work to greater prominence. The text of the gravely corrupt and physically damaged exemplar, retaining all of its many errors and lacunae, was mechanically replicated in minuscule script into *Mediceo-Laurentianus gr.* LV–4, an officially sponsored codex compiled in the Great Palace c. 950 for deposition in the emperor's library. There it seems to have remained for several centuries before passing into private hands, never mentioned nor cited, and apparently unknown to the world beyond. By the time the first copy of this codex was made, in Florence c. 1510–1520, Byzantium had already passed into memory. The reputation of Aineias' *Poliorketika* fares little better in its indirect tradition. The treatise had almost no discernible impact upon Byzantine poliorcetic literature, except insofar as a residue of Aineias' thought lingered in the writings of his chief Hellenistic heir, Philon of Byzantion (c. 200). In this sphere it is perhaps true to say that Philon's reprise and expansion of Aineias' work had rendered his source to a large extent redundant. Only in an excerpted tradition, most probably originating in late antiquity (or at least drawing on a majuscule tradition different to that represented by *Mediceo-Laurentianus gr.* LV–4) does Aineias' *Poliorketika* enjoy an appreciable *Nachleben*. The selectivity exercised by the excerptor retained the overarching context of a city under siege but reconfigured Aineias' text into a suite of generally applicable prescriptions for stratagems and technological devices. Thereafter Aineias' work was transmitted as a selection of anonymous and repeatedly modified extracts via a succession of Byzantine military compendia, which inevitably had the effect of diluting the authentic wording and content. This excerpted tradition concludes with the *Taktika* of Nikephoros Ouranos (c. 1000) nearly fourteen centuries after the original date of composition, itself no mean legacy. In this respect the reception of Aineias' *Poliorketika* in Byzantium broadly corresponds to the transmission of the Greco-Roman military genre as a whole, whereby Byzantine writers and editors tended to favour ancient works and adaptations that were already popular or current in late antiquity. It may be no coincidence that the two most fashionable classical treatises on tactics and siegecraft, respectively Aelian's *Taktike Theoria* and Apollodoros' *Poliorketika*, are those works whose received texts most clearly bear the marks of late antique editorial interference. The excerption of Aineias' *Poliorketika* necessarily implies that some parts of the text were deemed more useful than others, and poses complex questions concerning late antique and Byzantine tastes, priorities and perceptions of utility. Although the evidence permits only impressionistic arguments, it is difficult to escape the inference that Aineias' besieged polis, with its purely localised governance, loyalties and perspectives, its omnipresent plots and revolutions, ubiquitous mercenaries and shifting, ill-defined opponents, belonged to a distant world of which Byzantine readers had limited knowledge, understanding or perhaps interest. This intuition is consistent with the Judeo-Roman focus of Byzantine historical and chronographical literature, in which classical Greek history had little or no resonance. To scholarly *Rhomaioi* in tenth-century Constantinople the Hellas of Aineias was, in every sense, another country.

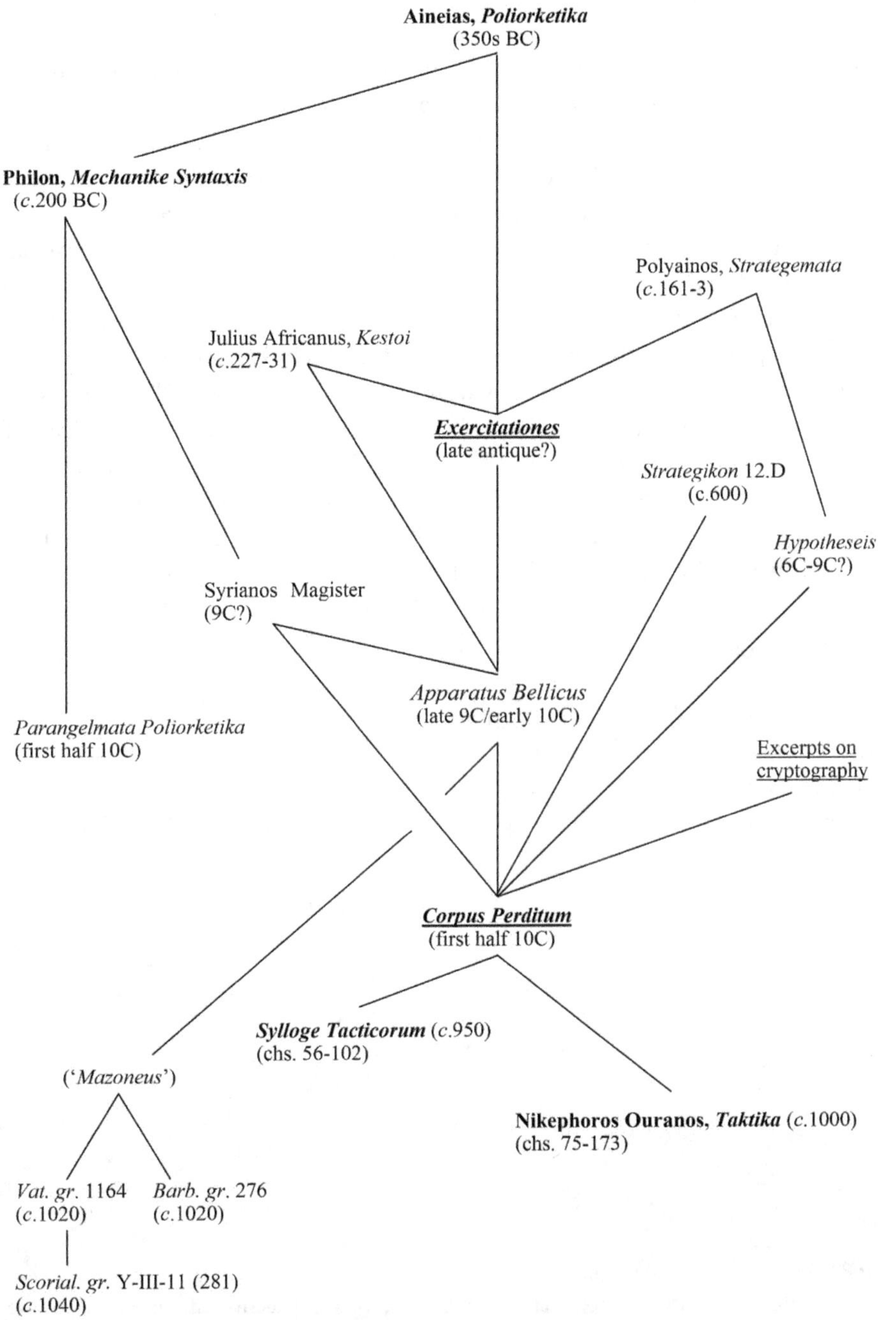

FIGURE 15.1 *The Transmission of Aineias'* Poliorketika *in Late Antique and Byzantine Military Literature.*
Key: works shown in bold contain material derived from Aineias' Poliorketika*; works underlined are hypothetical or no longer extant.*

Appendix

Edited below are the seven chapters of the *Taktika* of Nikephoros Ouranos that ultimately derive from Aineias' *Poliorketika* (104, 112–115, 172–173). They are assembled here for the first time; most have been edited before, but separately and from incomplete and/or defective manuscript evidence. In 1932 Vieillefond reproduced chapters 112 and 113 from *Monacensis gr.* 452, in which the text is in several respects corrupt, including a lacuna in 112 owing to *saut du même au même*.[178] These two chapters are re-edited here with the assistance of the superior text in *Oxoniensis Baroccianus* 131. Dain edited chapters 104 and 172–173 in an appendix to his critical edition of Aineias' work, drawing readings from both manuscripts but without providing an apparatus criticus.[179] A provisional edition of chapter 115 was published by Dain in 1937 without codicological reference; the text can only derive from *Oxoniensis Baroccianus* 131, but apparently with Dain's emendations and/or misreadings.[180] Chapter 114 has not previously been published. In accordance with editorial principles established by Dain and confirmed by McGeer, I have preferred the authentic recension of the text in O and resorted to alternative readings in N only where O is intrinsically erroneous.[181] The following sigla are used:

O *Oxoniensis Baroccianus* 131 (268^{v}, 269^{r}–269^{v}, 280^{v})
N *Monacensis gr.* 452 (139^{r}, 140^{r}–140^{v}, 156^{r})
Dain^{1} A. Dain, *La "Tactique" de Nicéphore Ouranos* (Paris 1937)
Dain^{2} A. Dain and A.-M. Bon (eds), *Énée le Tacticien, Poliorcétique* (Paris 1967)

178 Vieillefond (1932a) 30–31, 33–34.

179 Dain and Bon (1967) 102–104. An earlier and partly defective edition of ch. 104 was published in Dain (1937) 63.

180 Dain (1937) 65.

181 Dain (1937) 102–107, 128–130; McGeer (1995a) 83–86.

104 Πῶς δυνάμεθα σβέσθαι πῦρ

Ἂν ἐμπυρίσωσιν οἱ πολέμιοι διὰ πυρὸς σκευαστοῦ ἤγουν διὰ τοῦ λαμπροῦ τεῖχος ἢ ἄλλο τί ποτε, σβέσομεν αὐτὸ καταχύνοντες ὄξος. Ἂν δὲ προνοήσῃς τὸ ποῦ μέλλει γενέσθαι ὁ ἐμπυρισμός, χρῖσον ἔξωθεν ὄξος· καὶ οὐ μὴ ἅψηται ἐκεῖ τὸ πῦρ.

112 Πῶς ἵνα πέμψῃς κρυφὰ γράμματα

Τοὺς ἔχοντας προδότας τινὰς εἰς πολέμια ἔθνη, ἀναγκαῖόν ἐστι γίνωσκειν πῶς δεῖ πέμπειν αὐτοῖς κρύφα γράμματα. ὀφείλει πέμπεσθαι πρὸς αὐτοὺς ἀνὴρ βαστάζων φανερὰ γράμματα περὶ ἄλλων ὑποθέσεων· ὑποκάτω δὲ τῶν ὑποδημάτων τοῦ πεμπομένου, ἔσωθεν τοῦ κασσύματος, ἐμβαλλέσθω πιττάκιον καὶ καταρραπτέσθω· πρὸς τὸ μὴ ἐρημοῦσθαι καὶ μολύνεσθαι αὐτὸ μήτε ὑπὸ πηλοῦ μήτε ὑπὸ ὕδατος ὀφείλουσι δὲ γράφεσθαι εἰς πέταλον κασσιτέρινον τὰ γράμματα. οὕτως τίθεσθαι ὑποκάτω τοῦ ὑποδήματος· ὅταν δὲ φθάσῃ ὁ ἀποκομιστὴς μετὰ τούτων γράμματων πρὸς τὸν προδότην, ἵνα ἐκβάλῃ τὴν νύκτα ἐκ τοῦ ὑποδήματος τὰ τοιαῦτα γράμματα. ἐκεῖνος δέ, ὅπερ ἂν θέλῃ πάλιν ἀντιγράφων, οὕτως ὑποτιθέτω τῷ ὑποδήματῳ καὶ ῥαπτέτω. τὰ δὲ ἄλλα γράμματα τὰ φανερῶς ἀποσταλέντα ἀποδοὺς ὁ ἀποκομιστής, πρὸς ὃν ἀπεστάλη, ὑποστρεφέτω.

113 Ἄλλο

Διακομίζονται πάλιν γράμματα κρύφα ὄυτως· φούσκαν ἔχουσαν ἴσον μέγεθος ῥογίου ἐπιτηδείου πρὸς ἔλαιον εἴτε μικροῦ εἴτε μεγάλου πρὸς τὸ πλῆθος τῶν μελλόντων γραφῆναι, ταύτην τὴν φούσκαν φουσκώσας καὶ ἀποδήσας, ξήρανον· εἶτα γράψον εἰς αὐτὴν εἴ τι θέλεις διὰ μέλανος ἐγκαυστοῦ· ἀφ' ἧς δὲ στεγνωθῆναι τὰ γράμματα, ἀποφούσκωσον αὐτήν, καὶ συσφίγξας αὐτήν, βάλε ἔσωθεν τοῦ ῥογίου· τὸ δὲ στόμα αὐτῆς ἔξω ἔχε· εἶτα φούσκωσον αὐτὴν ὡς ἔστιν ἔσωθεν τοῦ ῥογίου· καὶ βάλε ἔλαιον

---

104 *App. Bell.* 38 = Aineias 34 **112** *App. Bell.* 51 = Aineias 31.4 **113** Aineias 31.10–13

2 ἤγουν ... λαμπροῦ O : om. N 3 ποτε O : om. N ‖ καταχύνοντες O : καταχέοντες N ‖ τὸ O : ὅ N 4 τὸ O : om. N 9 κασσύματος O : κασσήματος N 11 δὲ N : om. O ‖ κασσιτέρινον O : κασσιτήρινον N ‖ οὕτως ... γράμματα O : om. N ‖ ὑποκάτω ed : ὑπὸ κάτω O 14 ὅπερ ἂν θέλῃ πάλιν O : πάλιν ὅπερ βούλεται N ‖ τῷ ὑποδήματῳ O : τὰ ὑποδήματα N 15 δὲ ἄλλα O : δ' ἄλλα N 18 πάλιν N : om. O 20 φούσκαν O : φύσκαν N 20–21 εἰς αὐτὴν εἴ τι θέλεις O : εἴ τι θέλεις εἰς αὐτὴν N 21 ἀφ' ἧς δὲ O : μετὰ τὸ N 22 αὐτήν O : om. N ‖ βάλε O : βάλον N 23 αὐτῆς O : ταύτης N ‖ βάλε O : βάλον N

εἰς αὐτήν· εἶτα σκεπάσας, ἀπόστελλε ὅπου θέλεις· κενώσας οὖν ἐκεῖνος τὸ ἔλαιον καὶ φουσκώσας τὴν φούσκαν, ἀναγινώσκει τὰ γράμματα· εἶτα σπογγίσας αὐτὴν καὶ ἀντιγράψας, πάλιν ἀποστέλλει.

114 Ἄλλο

Τινες εἰς μολχάμιον ἢ χάσδιον γράψαντες γράμματα ψιλὰ εἰς τὸ ἐπώμιον τοῦ ὑποκαμίσου ὑποθέντες ἠπλωμένον. καὶ περιρράψαντες αὐτό, ἀνύποπτον οὕτως ἐποίησαν τὴν ἀποστολὴν τοῦ γράμματος.

115 Ἄλλο

Ἄλλοι πάλιν εἰς πινακίδιον ἀκήρωτον γράψαντες εἰς τὸ ξύλον ἅπερ ἤθελον. καὶ κηρὸν ἀναλύσαντες, ἐπείσχυσαν εἰς αὐτὸ καὶ ἄλλα γράμματα φανερὰ εἰς τὸν κηρὸν ἔγραψαν. εἶτα ὁ δεξάμενος αὐτά, ἀνασπάσας τὸν κηρὸν καὶ ἀναγνοὺς τὰ γράφεντα εἰς τὸ ξύλον, ἀντιγράψας καὶ ἐκεῖνος κατὰ τὸν ὅμοιον τρόπον πάλιν ἀπέστειλεν.

172 Πῶς ὀφείλουσιν νοεῖν οἱ πολεμούμενοι ἔσωθεν κάστρου τοὺς ἔξωθεν ὀρύσσοντας καὶ πῶς δεῖ κωλύειν αὐτούς

1 Ἁρμόζει τὴν ἔξω τείχους σοῦδαν ὀρύσσεσθαι εἰς βάθος πολὺ πρὸς τὸ καταντῆσαι εἰς αὐτὴν τὸ γινόμενον ὄρυγμα παρὰ τῶν ἐχθρῶν καὶ πρὸς τὸ φανῆναι τοὺς ὀρύσσοντας ὅταν φθάσωσιν εἰς αὐτήν. 2 Ἂν δὲ δύνασαι, ποίησον ἐκεῖ καὶ τεῖχος ἰσχυρότατον· ἂν δὲ οὐκ ἔχεις λιθάριον, ποίησον ξύλινον τὸ τοιοῦτον τεῖχος. 3 Ὅταν δὲ φθάσωσι τὰ ὀρύγματα τῶν πολεμίων εἰς τὴν τοιαύτην σοῦδαν, βάλε πάλιν καὶ ἄλλα ξύλα εἰς τὰ ὀρύγματα καὶ εἰς τὸ ξύλινον ἐκεῖνο τεῖχος, καὶ ἐμπύρισον αὐτὰ καὶ ἐπισκέπασον μετὰ τοῦ χώματος, ἵνα ὑπάγῃ ὁ καπνὸς εἰς τὸ ὄρυγμα καὶ βλάπτῃ τοὺς ὄντας εἰς

---

114 *App. Bell.* 53.2 = Aineias 31.23 **115** *App. Bell.* 53.3 = Aineias 31.14 **172** *App. Bell.* 54 = Aineias 37.1–5

---

1 ὅπου θέλεις O : ἔνθα βούλει N 9 πινακίδιον O : πινακύδιον Dain[1] || ἤθελον O : ἔθελον Dain[1] 10 ἐπείσχυσαν O : ἐπέσχυσαν Dain[1] 11 ἀνασπάσας Dain[1] : ἀνασπάσαι O 12 ἀπέστειλεν O : ἀπέστειλε Dain[1] 17 Ἂν O : Εἰ N || ποίησον … τεῖχος O : καὶ τεῖχος ἐκεῖ ποίησον N : ποίησον καὶ ἐκεῖ τεῖχος Dain || ἰσχυρότατον O : ὀχυρώτατον N 18 ἂν δὲ O : καὶ εἰ N 19 βάλε O : βάλον N 21 βλάπτῃ O : βλάψῃ N Dain[2] 21–363.1 εἰς αὐτό O : ἐκεῖ N

αὐτό. πολλοὶ γὰρ ἀποθνήσκουσιν ἐξ αὐτοῦ. 4 Βλάψεις δὲ τοὺς ὀρύσσοντας καὶ ἂν ἀπολύσῃς σφῆκας καὶ μελίσσας εἰς τὰ ὀρύματα. 5 Ἁρμόζει δὲ ἵνα εἰς ὃν ἂν ὀρύσσωσι τόπον ἀντορύσσῃς καὶ ὑπαντᾷς.

173 Δόλος ὑπὸ τῶν πολεμουμένων εἰς κάστρον γινόμενος

Οἱ πολεμούμενοι ἔσωθεν κάστρου ὀφείλουσιν εἰς τὰς πύλας εἰς τὸ ἔσω μέρος ὀρύξαι σοῦδαν καὶ καταλιπεῖν ἔνθεν καὶ ἐκεῖθεν διαβασίδια· εἶτα ἐκεῖνοι μὲν φεύγοντες ἵνα διαβῶσιν εἰς τὰ τοιαῦτα διαβασίδια· ἐκδεχόμενον δέ ἐστι τοὺς πολεμίους ἐπιδιώκοντας καὶ μὴ προϊδόντας τὴν σοῦδαν, ἐμπεσεῖν εἰς αὐτήν.

---

173 *App. Bell.* 56 = Aineias 39.1–2

---

2 σφῆκας ON : καὶ σφῆκας Dain^{2} || ἂν O : om. N 5 ἔσω O : ἔσωθεν N Dain^{2} 6–7 εἶτα ... διαβασίδια O : om. N 6 φεύγοντες O : διαφεύγοντες Dain^{2} 8 τὴν ... αὐτήν O : ἐμπεσεῖν εἰς αὐτὴν τὴν σοῦδαν N Dain^{2}

Bibliography

Arnim, M. (1912) *De Philonis Byzantii dicendi genere*. Emil Hartmann, Greifswald.

Aubreton, R. (1968) 'La tradition manuscrite des epigrammes de l'Anthologie grecque', *Revue des études anciennes* 70: 32–82.

Aubreton, R. (1969) 'Michel Psellos et l'Anthologie Palatine', *L'Antiquité classique* 38: 459–462.

Bagnall, R.S. (2009) *Early Christian Books in Egypt*. Princeton University Press, Princeton; Oxford.

Baldwin, B. (1988) 'On the Date of the Anonymous ΠΕΡΙ ΣΤΡΑΤΗΓΙΚΗΣ', *Byzantinische Zeitschrift* 81: 290–293.

Bandini, A.M. (1766) *Epistola de celeberrimo codice Tacticorum Bibliothecae Laurentianae*. Typis Regiae Celsitudinis, Florence.

Bandini, A.M. (1764–1770) *Catalogus codicum graecorum Bibliothecae Laurentianae*. 3 vols. Typis Caesareis, Florence.

Becker, J. (1915) *De Suidae Excerptis Historicis* (Diss. Bonn). Carl Georg, Bonn.

Bettalli, M. (1990) *Enea Tattico. La difesa di una città assediata (Poliorketika). Introduzione, traduzione e commento*. Edizioni ETS, Pisa.

Bevilacqua, L. (2013), 'Basilio *parakoimomenos* e i manoscritti miniati: impronte di colore nell'Ambrosiano B 119 sup.', in Rigo, A., Babuin, A., Trizio, M. (eds), *Vie per Bisanzio. Atti del VII Congresso nazionale dell'Associazione Italiana di Studi Bizantini, Venezia 25–28 novembre 2009*. Edizioni di Pagina, Bari. 1013–1030.

Blass, F. (1888) 'Die griechischen und lateinischen Handschriften im alten Serail zu Konstantinopel', *Hermes* 23: 219–233.

Blyth, P.H. (1992) 'Apollodorus of Damascus and the *Poliorcetica*', *Greek, Roman, and Byzantine Studies* 33.2: 127–158.

de Boor, C. (1912, 1914–1919) 'Suidas und die Konstantinische Excerptsammlung', *Byzantinische Zeitschrift* [pt. I] 21 (1912) 381–424; [pt. II] 23 (1914–1919 [1920]) 1–127.

Bosworth, A.B. (1983) 'Arrian at the Caspian Gates', *Classical Quarterly* 33: 265–276.

Brokkaar, W.G. (1972) 'Basil Lacapanus. Byzantium in the Tenth Century', in Bakker, W.F., van Gemert, A.F. and Aerts, W.J. (eds) *Studia Byzantina et Neohellenica Neerlandica* (Byzantina Neerlandica 3). Brill, Leiden. 199–234.

Browning, R. (1978) 'Literacy in the Byzantine World', *Byzantine and Modern Greek Studies* 4: 39–54.

Boura, L. (1989) ''Ο Βασίλειος Λεκαπηνός παραγγελιοδότης ἔργων τέχνης', in 'Ο *Κωνσταντίνος Ζ' ὁ Πορφυρογέννητος καὶ ἡ ἐποχή τοῦ* (Β' Διεθνής Βυζαντινολογική Συναντήση, Δελφοί, 22–26 'Ιουλίου 1987) Athens. 397–434.

Cameron, A. (1993) *The Greek Anthology: from Meleager to Planudes*. Clarendon Press, Oxford.

Cameron, A. (1997) 'New Themes and Styles in Greek Literature: Seventh-Eighth Centuries', in Cameron, A. and Conrad, L.I. (eds) *The Byzantine and Early Islamic Near East* 1: *Problems in the Literary Source Material*. The Darwin Press, Princeton. 81–105.

Casaubon, I. (1609) *Polybii ... Historiarum libri qui supersunt ... Aeneae, vetustissimi Tactici, Commentarius De toleranda obsidione*. Jérôme Drouart, Paris.

Chatzelis, G. and Harris, J. (2017) *A Tenth-Century Byzantine Military Manual: The Sylloge Tacticorum*. Routledge, London.

Cohn, L. (1900) 'Bemerkungen zu den Konstantinschen Sammelwerken', *Byzantinische Zeitschrift* 9: 154–160.

Cosentino, S. (2000) 'The Syrianos' Strategikon—a 9th-Century Source?', *Bizantinistica: Rivista di studi bizantini e slavi* 2: 243–280.

Cosentino, S. (2001) 'Per una nuova edizione dei *Naumachica* ambrosiani. Il *De fluminibus traiciendis* (*Strat.* XII.B.21)', *Bizantinistica: Rivista di studi bizantini e slavi* 3: 63–107.

Dagron, G. (1983) 'Byzance et le Modèle Islamique au xe siècle à propos des *Constitutions Tactiques* de l'Emperor Léon VI', *Comptes rendus des séances de l'Académie des inscriptions et belles-lettres* 127.2: 219–243.

Dagron G. and Mihăescu, H. (†) (1986) *Le Traité sur la guérilla* (*De velitatione*) *de l'empereur Nicéphore Phocas (963–969)*. Éditions du Centre National de la Recherche Scientifique, Paris

Dain, A. (1930) *Les manuscrits d'Onésandros*. Les Belles Lettres, Paris.

Dain, A. (1931) 'Les cinq adaptations byzantines des "Stratagèmes" de Polyen', *Revue des études anciennes* 33: 321–345.

Dain, A. (1933) *La tradition du texte d'Héron de Byzance*. Les Belles Lettres, Paris

Dain, A. (1934) 'Les Manuscrits des Traités tactiques d'Arrien', *Annuaire de l'Institut de Philologie et d'Histoire Orientales et Slaves* 2: 157–184.

Dain, A. (1934–1935) 'Les manuscrits d'Asclépiodote le Philosophe', *Revue de philologie, de littérature et d'histoire anciennes* [I] 8 = 60 (1934) 341–360; [II] 9 = 61 (1935) 5–21.

Dain, A. (1935) 'Les manuscrits d'Énée le Tacticien', *Revue des études grecques* 48: 1–32.

Dain, A. (1937) *La "Tactique" de Nicéphore Ouranos*. Les Belles Lettres, Paris.

Dain, A. (1938) *Sylloge Tacticorum quae olim Inedita Leonis Tactica dicebatur*. Les Belles Lettres, Paris.

Dain, A. (1939) *Le Corpus Perditum*. Les Belles Lettres, Paris.

Dain, A. (1940a) *La Collection florentine des Tacticiens grecs*. Les Belles Lettres, Paris.

Dain, A. (1940b) 'Mémorandum inédit sur la défense des places', *Revue des études grecques* 53: 123–136.

Dain, A. (1941) 'Le Parisinus 2522', *Revue de philologie, de littérature et d'histoire anciennes* (sér. 3) 15 = 67: 21–28.

Dain, A. (1946) *L'Histoire du texte d'Élien le Tacticien des origines à la Fin du Moyen Âge*. Les Belles Lettres, Paris.

Dain, A. (1946–1947) 'Inventaire raisonné des cents manuscrits des "Constitutions tactiques" de Léon VI le Sage', *Scriptorium* 1: 33–49.

Dain, A. (1953) 'L'encyclopédisme de Constantin Porphyrogénète', *Lettres d'Humanité* 12 (= *Bulletin de l'Association Guillaume-Budé* [sér. 3] 4): 64–81.

Dain, A. (1954) 'La transmission des textes littéraires classiques de Photius à Constantin Porphyrogénète', *Dumbarton Oaks Papers* 8: 33–47 (repr. in D. Harlfinger [ed.] [1980] *Griechische Kodikologie und Textüberlieferung*. Wissenschaftliche Buchgesellschaft, Darmstadt. 206–224).

Dain, A. and Bon, A.-M. (eds) (1967) *Énée le Tacticien, Poliorcétique*. Les Belles Lettres, Paris.

Dain, A. (1967) (complété par J.-A. de Foucault) 'Les Stratégistes byzantins', *Travaux et Mémoires*. 2: 317–392.

Dain, A. (complété par J.-A. de Foucault) (1968) 'Urbicius ou Mauricius?', *Revue des études byzantines* 26: 123–136.

Debidour, M. (2006) 'Le secret et les messages secrets dans la Poliorcétique d'Énée le Tacticien', in Olivier, H., Giovannelli-Jouanna, P. and Bérard, F. (eds), *Ruses, secrets et mensonges chez les historiens grecs et latins*. De Boccard, Lyon. 213–241.

Deissmann, A. (1933) *Forschungen und Funde im Serai; mit einem Verzeichnis der nichtislamischen Handschriften im Topkapu Serai zu Istanbul*. Walter de Gruyter, Berlin; Leipzig.

Dennis, G.T. (ed.) (1981) *Das Strategikon des Maurikios*, with Ger. trans. E. Gamillscheg. Verlag der Österreichischen Akademie der Wissenschaften, Vienna.

Dennis, G.T. (ed.) (1986) *Three Byzantine Military Treatises*. Dumbarton Oaks, Washington.

Dennis, G.T. (2010) *The Taktika of Leo VI. Text, Translation and Commentary*. Dumbarton Oaks, Washington.

Devine, A.M. (1989) 'Aelian's Manual of Hellenistic Military Tactics', *Ancient World* 19: 31–64.

Diller, A. (1974) 'The Age of Some Early Greek Classical Manuscripts', in Heller, J.L. and Newman, J.K. (eds) *Serta Turyniana. Studies in Greek Literature and Palaeography in honor of Alexander Turyn*. University of Illinois Press, Chicago; London. 514–524.

Eramo, I. (2007) 'Ὦ ἄνδρες στρατιῶται. Demegorie protrettiche nell'*Ambrosianus* B 119 sup.', *Annali della Facoltà di Lettere e Filosofia dell'Università di Bari* 50: 127–165.

Eramo, I. (2008) 'Omero e i Maccabei: nella biblioteca di Siriano Μάγιστρος', *Annali della Facoltà di Lettere e Filosofia dell'Università di Bari* 51: 123–147.

Eramo, I. (2010) 'Retorica militare fra tradizione protrettica e pensiero strategico', *Talia Dixit* 5: 25–44.

Eramo, I. (ed.) (2010 [2011]) *Siriano. Discorsi di guerra. Testo, traduzione e commento*. Dedalo, Bari.

Eramo, I. (2012) 'Composition and Structure of Syrianus Magister's Military Compendium', *Classica et Christiana* 7: 97–116.
Featherstone, M. (2002) 'Preliminary Remarks on the Leipzig Codex of *De Cerimoniis*', *Byzantinische Zeitschrift* 95: 457–479.
Featherstone, M. (2004) 'Further Remarks on the *De Cerimoniis*', *Byzantinische Zeitschrift* 97: 113–121.
Förster, R. (1877) 'Studien zu den griechischen Taktikern', *Hermes* 12: 426–471.
de Foucault, J.-A. (1949) *Strategemata*. Les Belles Lettres, Paris.
de Foucault, J.-A. (1967) 'Histiée de Milet et l'esclave tatoué', *Revue des études grecques* 80: 182–186.
de Foucault, J.-A. (1973) 'Douze chapitres inédits de la *Tactique* de Nicéphore Ouranos', *Travaux et Mémoires* 5: 281–312.
Fryde, E.B. (1983) 'The Library of Lorenzo de' Medici', in Fryde, E.B. *Humanism and Renaissance Historiography*. Hambledon, London.
Fryde, E.B. (1996) *Greek Manuscripts in the Private Library of the Medici 1469–1510*. National Library of Wales, Aberystwyth.
Garlan, Y. (1973) 'Cités, armées et stratégie à l'époque hellénistique d'après l'oeuvre de Philon de Byzance', *Historia* 22: 16–33.
Garlan, Y. (1974) *Recherches de poliorcétique grecque*. Bibliothèques de l'Ecole française d'Athènes et de Rome—Série Athènes 223, Paris.
Gatto, M. (2010) *Il Περὶ μηχανημάτων di Ateneo Meccanico. Edizione critica, traduzione, commento e note*. Aracne editrice, Rome.
Gentile, S. (1994) 'Lorenzo e Giano Lascaris', in Garfagnini, G.C. (ed.) *Lorenzo il Magnifico e il suo mondo (Convegno internazionale di Studi, Firenze, 9–13 giugno 1992)* Leo S. Olschki, Florence. 177–194.
Haldon, J.F. (1990) *Three Treatises on Imperial Military Expeditions*. Verlag der Österreichischen Akademie der Wissenschaften, Vienna.
Haldon, J.F. (1997) *Byzantium in the Seventh Century. The Transformation of a Culture*, rev. ed. Cambridge University Press, Cambridge.
Haldon, J.F. (2000) 'Theory and Practice in Tenth-Century Military Administration Chapters II, 44 and 45 of the *Book of Ceremonies*, Text and Translation', *Travaux et Mémoires* 13: 201–352.
Haldon, J.F. (2008) 'Military Technology and Warfare', in Jeffreys, E., Haldon, J. and Cormack, R. (eds) *The Oxford Handbook of Byzantine Studies*. Oxford University Press, Oxford. 473–481.
Haldon, J.F. (2014) *A Critical Commentary on the Taktika of Leo VI*. Dumbarton Oaks Studies 44, Washington D.C.
Hammerstaedt, J. (2009) 'Julius Africanus und seine Tätigkeiten im 18. Kestos (P.Oxy. 412 col. II)', in Wallraff, M. and Mecella, L. (eds) *Die Kestoi des Julius Africanus und ihre Überlieferung*. Walter de Gruyter, Berlin; New York. 53–69

Hercher, R. (1870) *Aeneae Commentarius poliorceticus*. Weidmann, Berlin.

Holste, L. (Holstenius) (1640 [1715–1719]) 'De libris optimis, & maximam partem ineditis Bibliothecae Mediceae ... cum esset Florentiae anno 1640', in Lilienthal, M. *Selecta Historica et Literaria*. Heinrich Boyen, Könisberg; Leipzig 1715–1719. I 85–113.

Hug, A. (ed.) (1874) *Aeneae Commentarius poliorceticus*. Teubner, Leipzig.

Hunger, H. (1978) *Die hochsprachliche profane Literatur der Byzantiner*. Beck, Munich.

Hunter, L.W. and Handford, S.A. (1927) *Αἰνείου Πολιορκητικά. Aineias on Siegecraft*. Clarendon Press, Oxford.

Huuri, K. (1941) *Zur Geschichte des mittelalterlichen Geschützwesen aus orientalischen Quellen*. Societas Orientalis Fennica, Studia Orientalia, Helsinki.

Irigoin, J. (1958–1959) 'Pour une étude des centres de copie byzantins', *Scriptorium* [I] 12: 208–227; [II] 13: 177–209.

Irigoin, J. (1962) 'Survie et renoveau de la littérature antique à Constantinople (IXe siècle)', *Cahiers de Civilisation médiévale* 5.3: 287–302.

Jackson, D.F. (1998) 'A New Look at an Old Book List', *Studi Italiani di Filologica Classica* (ser. 3) 16.1: 83–108.

Jeffreys, E., Haldon, J. and Cormack R. (eds) (2008) *The Oxford Handbook of Byzantine Studies*. Oxford University Press, Oxford.

Jenkins, T.E. (2006) *Intercepted Letters: Epistolarity and Narrative in Greek and Roman Literature*. Lexington Books, Lanham, MD.

Knös, B. (1945) *Un ambassadeur de l'Hellénisme—Janos Laskaris et la tradition greco-byzantine dans l'humanisme français*. Les Belles Lettres, Paris.

Köchly, H. (ed.) (1855–1856) *Rhetorica Militaris* (Δημηγορίαι προτρεπτικαὶ πρὸς ἀνδρείαν ἐκ διαφόρων ἀφορμῶν λαμβάνουσαι τὰς ὑποθέσεις) in *Index Lectionum in Literarum Universitate Turicensi ... habendarum*. Zürcher u. Furrer, Zürich.

Köchly, H. and Rüstow, W. (1853–1855) *Griechische Kriegsschriftsteller*. Engelmann, Leipzig.

Kollautz, A. (1985) 'Das militärwissenschaftliche Werk des sog. Maurikios', *BYZANTIAKA* 5: 87–136.

Korzenszky, E. and Vári, R. (eds) (1935) *Onasandri Strategicus*. Societas Frankliniana, Budapest.

Krentz, P. and Wheeler, E.L. (1994) *Polyaenus, Stratagems of War*. Ares, Chicago.

Kresten, O. (2000a) *"Staatsempfänge" im Kaiserpalast von Konstantinopel um die Mitte des 10. Jahrhunderts. Beobachtungen zu Kapitel II 15 des sogenannten "Zeremonienbuches"*. VÖAW, Vienna.

Kresten, O. (2000b) 'Sprachliche und inhaltliche Beobachtungen zu Kapiteln I 96 des sogenannten "Zeremonienbuches"', *Byzantinische Zeitschrift* 93: 474–489.

Kresten, O. (2005) 'Nochmals zu *De Cerimoniis* I.96.', *Jahrbuch der Österreichischen Byzantinistik* 55: 87–98.

Kroll W. (1917) in Kroll, W. and Sickenberger, J. 'Iulius[47] (Africanus)', *Paulys Realencyclopädie der classischen Altertumswissenschaft* 10: 116–125.

Kuchma, V.V. (2001) '"Византийский Аноним VI в.": Основные проблемы источников и содержания', *Vizantiyskiy Vremennik* 41: 68–91 (repr. *idem, Военная Организация Византийской Империи* [St. Petersburg]. 208–242).

Kuchma, V.V. (1982–1986) '"Стратегикос" Онасандра и "Стратегикон Маврикия": Опыт сравнительной характеристики', *Vizantiyskiy Vremennik* [I] 43 (1982) 35–53; [II] 45 (1984) 20–34; [III] 46 (1986) 109–123 (repr. *idem, Военная Организация Византийской Империи* [St. Petersburg 2001]. 139–207).

Kuchma, V.V. (1998) 'Феномен Элиана. К проблеме зффекта асинхроностн в развитии византийской военной науки', in *История Византии и византийская археология. Тезисы докладов X научных Сюзюмовских чтений, 25–27 марта 1998 г.*, Ekaterinburg. 16–19.

Lami, I.(G.) (1741–1763) *Ioannis Meursii opera omnia*. Regiis Magni Etruriae Ducis Typis, Florence

Lammert, F. (1940) 'Die älteste erhaltene Schrift über Seetaktik und ihre Beziehung zum Anonymus Byzantinus des 6. Jahrhunderts, zu Vegetius und zu Aineias' *Strategika*', *Klio* 33: 271–288.

Lammert, F. (1951) 'Julius Afrikanus und die byzantinische Taktik', *Byzantinische Zeitschrift* 44: 362–369.

Lammert, F. (1952) 'Polyainos', *Paulys Realencyclopädie der classischen Altertumswissenschaft* 21.2: 1432–1436.

Lauxtermann, M.D. (2003) *Byzantine Poetry from Pisides to Geometres* I: *Texts and Contexts*. Verlag der Österreichischen Akademie der Wissenschaften, Vienna.

Lawrence, A.W. (1979) *Greek Aims in Fortification*. Oxford University Press, Oxford.

Lee, A.D. and Shepard, J. (1991) 'A Double Life: Placing the *Peri Presbeon*', *Byzantinoslavica. Revue internationale des études byzantines* 52: 25–30.

Lemerle, P. (1971) *Le Premier Humanisme Byzantin. Notes et remarques sur enseignement et culture à Byzance des origines au Xe siècle*. Presses Universitaires de France, Paris.

Leoni, B. (2003) *La Parafrasi Ambrosiana dello Strategicon di Maurizio: L'arte della guerra a Bisanzio*. Vita e Pensiero, Milan.

Loreto, L. (1995) 'Il generale e la biblioteca. La trattatistica militare greca da Democrito di Abdera ad Alessio I Comneno', in Cambiano, G., Canfora, L. and Lanza, D. (eds) *Lo spazio letterario della Grecia antica* II. Salerno, Rome. 563–589

Lowe, C. (1927) *A Byzantine Paraphrase of Onasander*. Washington University Studies, St. Louis.

Mango, J.C. (1975) 'The Availability of Books in the Byzantine Empire, AD 750–850', in Mango, J.C. and Ševčenko, I. (eds) *Byzantine Books and Bookmen: A Dumbarton Oaks Colloquium 1971*. Dumbarton Oaks, Washington. 29–45.

Martin, Th.-H. (1854) *Recherches sur la vie et les ouvrages d'Héron d'Alexandrie*. Mémoires présentés par divers savants à l'Académie des Inscriptions et Belles-Lettres (sér. 1) 4, Paris.

Martini, A. and Bassi, D. (1906) *Catalogus codicum graecorum Bibliothecae Ambrosianae*. Hoepli, Milan.

Mazzucchi, C.M. (1978) 'Dagli anni di Basilio Parakimomenos (cod. Ambr. B 119 sup.)', *Aevum* 52.2: 267–316.

McCabe, A. (2007) *A Byzantine Encyclopaedia of Horse Medicine: the Sources, Compilation and Transmission of the Hippiatrica*. Oxford University Press, Oxford.

McGeer, E. (1991) 'Tradition and Reality in the *Taktika* of Nikephoros Ouranos', *Dumbarton Oaks Papers* 45: 129–140.

McGeer, E. (1995a) *Sowing the Dragon's Teeth: Byzantine Warfare in the Tenth Century*. Dumbarton Oaks Studies 33, Washington.

McGeer, E. (1995b) 'Byzantine Siege Warfare in Theory and Practice', in Corfis, I.A. and Wolfe, M. (eds) *The Medieval City under Siege*. Woodbridge, Suffolk. 123–130.

McGeer, E. (2003) 'Two Military Orations of Constantine VII', in Nesbitt, J.W. (ed.) *Byzantine Authors. Literary Activities and Preoccupations: Texts and Translations dedicated to the Memory of Nicolas Oikonomides*. Brill, Leiden. 111–135.

McGeer, E. (2008) 'Military Texts', in Jeffreys, E., Haldon, J. and Cormack R. (eds) *The Oxford Handbook of Byzantine Studies*. Oxford University Press, Oxford. 907–914.

Mecella, L. (2009) 'Die Überlieferung der Kestoi des Julius Africanus in den byzantinischen Textsammlungen zur Militärtechnik', in Wallraff, M. and Mecella, L. (eds) *Die Kestoi des Julius Africanus und ihre Überlieferung*. Walter de Gruyter, Berlin. 85–144.

Müller, K.K. (1880) 'Ein griechisches Fragment über Kriegswesen', *Festschrift für Ludwig Urlichs*. Stahel, Würzburg. 106–138.

Müller, K.K. (1882) 'Ein griechische Schrift über Seekrieg', in Gramich, V., Haupt, H. and Müller, K.K. *Festgabe zur dritten Säcularfeier der Julius-Maximilians-Universität zu Würzburg*. Stuber, Würzburg.

Müller, K.K. (1884) 'Neue Mittheilungen über Janos Laskaris und die Mediceische Bibliothek', *Zentralblatt für Bibliothekswesen* 1.9/10: 333–412.

Naudé, G. (1637) *Syntagma de studio militari*. Giacomo Facciotti, Rome.

Németh, A. (2011) 'The Mynas codex and the Bibliotheca Corviniana', in Gastgeber, C. et al (eds) *Matthias Corvinus und seine Zeit. Europa am Übergang vom Mittelalter zur Neuzeit zwischen Wien und Konstantinopel*. Verlag der Österreichischen Akademie der Wissenschaften, Vienna. 155–178.

Nesbitt, J.W. (2003) (ed.) *Byzantine Authors. Literary Activities and Preoccupations: Texts and Translations dedicated to the Memory of Nicolas Oikonomides*. Brill, Leiden.

Oldfather, W.A. (1923) *Aeneas Tacticus—Asclepiodotus—Onasander: with an English translation by members of the Illinois Greek Club* (Loeb Classical Library). William Heinemann and G.P. Putnam's Sons, London and New York.

Pasini, C. (2007) *Bibliografia dei manoscritti greci dell'Ambrosiana (1857–2006)*. Vita e Pensiero, Milan.

Pattenden, P. (1983) 'The Byzantine Early Warning System', *Byzantion* 53: 258–299.

Petrocelli, C. (2008) *Onasandro, Il generale. Manuale per l'esercizio del comando*. Dedalo, Bari.

Pryor, J.H. and Jeffreys, E.M. (2006) *The Age of the ΔΡΟΜΩΝ. The Byzantine Navy ca 500–1204*. Brill, Leiden.

Rance, P. (2000) '*Simulacra pugnae*: the Literary and Historical Tradition of Mock Battles in the Roman and Early Byzantine army', *Greek, Roman, and Byzantine Studies* 41.3: 223–275.

Rance, P. (2007a) 'The Date of the Military Compendium of Syrianus Magister (formerly the Sixth-Century Anonymus Byzantinus)', *Byzantinische Zeitschrift* 100.2: 701–737.

Rance, P. (2007b) 'The *Etymologicum Magnum* and the "Fragment of Urbicius"', *Greek, Roman, and Byzantine Studies* 47.2: 193–224.

Rance, P. (forthcoming a) 'Late Byzantine Elites and Military Literature: Authors, Readers and Manuscripts (11th–15th Centuries)', in Theotokis, G. and Yıldız, A. (eds) *A Military History of the Mediterranean Sea—Aspects of War, Diplomacy and Military Elites*. Brill, Leiden.

Rance, P. (forthcoming b) *The Roman Art of War in Late Antiquity: The Strategikon of the Emperor Maurice. A Translation with Commentary and Textual Studies* (Birmingham Byzantine and Ottoman Studies), Routledge, London.

Ravegnani, G. (1983) *Castelli e città fortificate nel VI secolo*. Edizioni del Girasole, Ravenna.

Roberto, U. (2009) 'Byzantine Collections of Late Antique Authors: Some Remarks on the *Excerpta historica Constantiniana*', in Wallraff, M. and Mecella, L. 71–84.

de Rochas, A. and Graux, C. (1879) 'Philon de Byzance', *Revue de Philologie* N.S. 3: 91–151.

Schettino, M.T. (1988) *Introduzione a Polieno* (Studi e testi di storia antica 10). Edizioni ETS, Pisa.

Schindler, F. (1973) *Die Überlieferung der Strategemata des Polyainos*. Österreichische Akademie der Wissenschaften phil.-hist. Klasse Sitzungsberichte 284.1, Vienna.

Schöne, R. (ed.) (1893) *Philonis Mechanicae Syntaxis, libri quartus et quintus*. Reimer, Berlin.

Schöne, R. (ed.) (1911) *Aeneae Tactici de obsidione toleranda commentarius*. Teubner, Leipzig.

Schreiner, P. (1975–1979) *Die byzantinischen Kleinchroniken*. 3 vols. Verlag der Österreichischen Akademie der Wissenschaften, Vienna.

Sekunda, N. (2001) 'The *Taktika* of Poseidonius of Apameia', in Sekunda, N. *Hellenistic Infantry Reform in the 160s BC*. Oficyna Naukowa MS, Łodz. 125–134.

Speck, P. (1981) 'Versuch einer Charakterisierung der sogenannten Makedonischen Renaissance', in Zeitler, R. (ed.) *Les Pays du Nord et Byzance (Scandinavie et Byzance). Actes du colloque nordique et international de byzantinologie tenu à Upsal 20–22 avril 1979*. Acta Universitatis Upsaliensis Figura N.S. 19, Uppsala. 237–242.

Speck, P. (1991) 'Über Dossiers in byzantinischer antiquarischer Arbeit, über Schulbücher für Prinzen, sowie zu einer Seite frisch edierten Porphyrogennetos', in *ΠΟΙΚΙΛΙΑ ΒΥΖΑΝΤΙΝΑ* 11: *Varia* III. Bonn. 269–292.

Sullivan, D.F. (1997) 'Tenth-Century Byzantine Offensive Siege Warfare: Instructional Prescriptions and Historical Practice', in Oikonomidès, N. (ed.), *Byzantium at War (9th to 12th c)*, Goulandri-Horn Foundation, Athens. 179–200.

Sullivan, D.F. (2000) *Siegecraft: Two Tenth-Century Instructional Manuals by 'Heron of Byzantium'*. Dumbarton Oaks Studies 36, Washington.

Sullivan, D.F. (2003a) 'A Byzantine Instructional Manual on Siege Defense: The *De Obsidione Toleranda*. Introduction, English Translation and Annotations', in Nesbitt, J.W. (ed.) *Byzantine Authors. Literary Activities and Preoccupations: Texts and Translations dedicated to the Memory of Nicolas Oikonomides*. Brill, Leiden. 139–266.

Sullivan, D.F. (2003b) 'Byzantium Besieged: Prescription and Practice', in Avramea, A., Laiou, A. and Chrysos, E. (edd) *Byzantium, State and Society, in Memory of Nikos Oikonomides*. Centre for Byzantine Studies, Athens. 509–521.

Sullivan, D.F. (2010) 'Byzantine Military Manuals. Perceptions, Practice and Pedagogy', in Stephenson, P. (ed.) *The Byzantine World*. Routledge, London. 149–161.

Taragna, A.M. (2004) 'Λόγος e Πόλεμος: eloquenza e persuasione nei trattati bizantini di arte militare', in Creazzo, T. and Strano, G. (eds) *Atti del VI Congresso nazionale dell'Associazione Italiana di Studi Bizantini, Catania-Messina, 2–5 ottobre 2000*. Numero speciale di *Siculorum Gymnasium* n.s. 57, Catania. 797–810.

Thévenot, M. (1693) *Veterum mathematicorum Athenaei, Bitonis, Apollodori, Heronis, Philonis et aliorum opera*. Typographia Regia, Paris.

Tougher, S. (1997) *The Reign of Leo VI (886–912): Politics and People*. Brill, Leiden.

Treadgold, W. (1979) 'The Revival of Byzantine Learning and the Revival of the Byzantine State', *American Historical Review* 84: 1245–1266.

Treadgold, W. (1984) 'The Macedonian Renaissance', in Treagold, W. (ed.) *Renaissances before the Renaissance: Cultural Revivals of Late Antiquity and the Middle Ages*. Stanford University Press, Stanford. 75–84.

Trombley, F. (1997) 'The Taktika of Nikephoros Ouranos and Military Encyclopaedism', in Binkley, P. (ed.) *Pre-modern Encyclopaedic Texts*. Brill, Leiden. 261–274.

van den Berg, H. (ed.) (1947) *Anonymous De Obsidione Toleranda*. Brill, Leiden.

Vári, R. (1898) *Bölcs Leó császárnak 'A hadi taktikáról' szóló munkája*. Értekezések a történeti tudományok köréből XVII.10, Budapest.

Vári, R. (1901) *Incerti scriptoris byzantini saec. x. tractatus de re militari*. Teubner, Leipzig 1901.

Vári, R. (1906) 'Zur Überlieferung mittelgriechischer Taktiker', *Byzantinische Zeitschrift* 15: 47–87.

Vári, R. (1917–1922) *Leonis Imperatoris Tactica*. Typis Regiae universitatis scientiarum budapestinensis, Budapest. 2 vols.

Vieillefond, J.-R. (1932a) 'Adaptations et paraphrases du Commentaire d'Enée le Tacticien', *Revue de Philologie* 6: 24–36.

Vieillefond, J.-R. (1932b) *Jules Africain, Fragments des Cestes provenant de la Collection des Tacticiens grecs*. Les Belles Lettres, Paris.

Vieillefond, J.-R. (1970) *Les Cestes de Julius Africanus*. Sansoni, Florence; Didier, Paris.

Wallraff, M. and Mecella, L. (eds) (2009) *Die Kestoi des Julius Africanus und ihre Überlieferung*. Walter de Gruyter, Berlin.

Weitzmann, K. (1935) *Die byzantinische Buchmalerei 9. und 10. Jahrhunderts*. Gebrüder Mann, Berlin. Reprinted (1996) with vol. 2 Addenda und Appendix. Verlag der Österreichischen Akademie der Wissenschaften, Vienna.

Wescher, C. (1867) *Poliorcétique des Grecs*. Imprimerie impériale, Paris.

Wheeler, E.L. (1981) 'The Origins of Military Theory in Ancient Greece and China', *International Commission of Military History, Acta 5, Bucarest 10–17 VIII 1980*: 74–79.

Wheeler, E.L. (1988) *Stratagem and the Vocabulary of Military Trickery* (*Mnemosyne* Supplement 108). Brill, Leiden.

Whitehead, D. (1990) *Aineias the Tactician, How to Survive under Siege. Translated with Introduction and Commentary*. Clarendon Press, Oxford. [Updated edition Bristol Classical Press (2001) = Duckworth (2002)]

Whitehead, D. (2008a) 'Fact and Fantasy in Greek Military Writers', *Acta Antiqua Academiae Scientiarum Hungaricae* 48.1–2: 139–155.

Whitehead, D. (2008b) 'Apollodorus' *Poliorketika*: Author, Date, Dedicatee', in Krieckhaus, A., Schellenberg, H.M. and Hirschmann V.E. (eds) *A Roman Miscellany: Essays in Honour of Anthony R. Birley on his Seventieth Birthday*. Akanthina, Gdańsk. 204–211.

Whitehead, D. (2010) *Apollodorus Mechanicus: Siege-matters* (*Poliorkētika*) (*Historia* Einzelschriften 216). Franz Steiner Verlag, Stuttgart

Whitehead, D. (2016) *Philo Mechanicus: On Sieges* (*Historia* Einzelschriften 243). Franz Steiner Verlag, Stuttgart.

Wilson, N.G. (1978) 'A Byzantine Miscellany: MS. Barocci 131 described', *Jahrbuch der Österreichischen Byzantinistik* 27: 157–179.

Wilson, N.G. (1983) *Scholars of Byzantium*. Duckworth, London.

von Wölfflin, E. and Melber, I.(J.) (1887) *Polyaeni strategematon libri octo* (second edition; reprinted 1970). Teubner, Leipzig.

Zuckerman, C. (1990) 'The Compendium of Syrianus Magister', *Jahrbuch der Österreichischen Byzantinistik* 40: 209–224.

Zuckerman, C. (1994) 'Chapitres peu connus de l'Apparatus Bellicus', *Travaux et Mémoires* 12: 359–389.

Index locorum

Passages which are discussed in the main text are listed in bold

Index

Names mentioned in Aineias Tacticus' *Poliorketika* are in bold